"The most comprehe [barcode: D0985552]
written account of Ruby
Mr. Walter writes with fluency and grace, and he manages
to make even minor characters spring to life. . . . While in
no way absolving the government of its errors, Mr. Walter
reveals that Mr. Weaver's paranoia and stubbornness . . .
helped to escalate the situation to its fatal end."

—*The Washington Times*

"A brilliant, cautionary tale of the dangers of conspiracy
thinking—by people and by governments."

—Dennis Prager
Talk show host and author of *Think a Second Time*

"Just to take a case as complicated and controversial as
U.S. v. Weaver and explain it would have been enough.
But Walter does more. [His] opening is so strong and his
writing is so good. . . . *Every Knee Shall Bow* is an effort
that leaves us to draw our own conclusion about who was
right and wrong on Ruby Ridge."

—*Ft. Lauderdale Sun-Sentinel*

Every Knee Shall Bow

THE TRUTH AND TRAGEDY OF RUBY RIDGE AND THE RANDY WEAVER FAMILY

Jess Walter

HarperPaperbacks
A Division of HarperCollinsPublishers

HarperPaperbacks   *A Division of* HarperCollins*Publishers*
10 East 53rd Street, New York, N.Y. 10022

A hardcover edition of this book was published in 1995 by ReganBooks, an imprint of HarperCollins*Publishers.*

Cover photograph © 1996 CBS Inc. CR: Tony Esparza

Unless otherwise indicated, all photos are courtesy of David and Jeane Jordison and are used by permission.

Map by Vince Grippi

First HarperPaperbacks printing: May 1996

Printed in the United States of America

HarperPaperbacks and colophon are trademarks of HarperCollins*Publishers*

❖ 10 9 8 7 6 5 4 3 2 1

acknowledgments

*T*his book could not have been written without the help and encouragement of a great many people. First, I must thank my researchers: chiefly, Dean Miller for his reporting and his invaluable insights and for helping to write much of Chapter 22; and Bill Morlin, for his remarkable knowledge of Idaho's white separatists and for his law enforcement contacts. I am also grateful to the reporters, editors, and photographers at the *Spokesman-Review*, primarily Kevin Keating, J. Todd Foster, and Richard Wagoner; and to Dan Rubin of the *Philadelphia Inquirer*. Several people helped by reading early chapters, including Phil Gruis, Kevin Gilmore, Jim Lynch, Jim DeFede, and Deb Rose. My thanks go to Judith Regan, Nancy Peske, and everyone at ReganBooks and HarperCollins. I'd like to thank my daughter Brooklyn for her patience, and most of all, my wife Anne Windishar for her editing and her unwavering support.

This book was reported over a three-year period while I covered the Weaver case for the *Spokesman-Review*— from the standoff to the trial to the aftermath. It is based on trial transcripts, government documents, wiretap transcripts, personal letters and hundreds of interviews, from Sara Weaver to Lorenz Caduff to Dave Hunt. I would like to thank them all, especially Julie and Keith Brown and the rest of Vicki Weaver's family—which I hope finds peace.

Every Knee Shall Bow

IDAHO

Boise

IDAHO

Canada
40 miles

Bonners
Ferry

WASHINGTON

Priest
Lake

95

RUBY
RIDGE
AREA

Sandpoint

Coeur d'Alene
60 miles

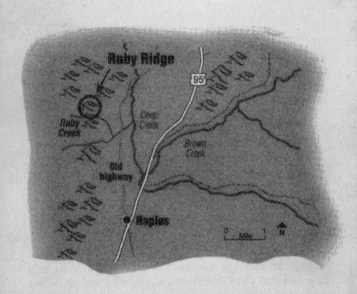

Ruby Ridge

95

Ruby
Creek

Deep
Creek

Brown
Creek

Old
highway

Naples

0   Mile   1

N

introduction

Ruby Creek is a stone-bedded scratch that traces like a finger the base of Ruby Ridge, in that part of North Idaho that aims like a barrel at Canada, the part of Idaho known as the Panhandle. The ridge isn't much different from the mountains around it. "No Trespassing" signs are nailed to the trees, in whose gaps flicker the suggestion of cabins and trailers, connected by driveways that disappear into the woods like smoke. Beneath Ruby Ridge, one unsure lane leaves the old highway. A road at first, it crosses the dark creek and becomes a narrow swipe of dirt, switching back every few hundred feet along imposing, root-veined banks. The path climbs easily for two miles, until it becomes nothing more than two tracks, which cut through a mile of woods so dense they choke off the midday sun. Take the right fork, and the path breaks through the forest into a brief meadow and opens to a steep, wooded field that climbs impossibly and spills out finally on a rock-strewn knob with a predawn view of everything: a 120-degree, 40-mile window on Idaho, Montana, and Canada. On this point, 3,100 feet above sea level, the sky is close and cloudless. The sun bakes the forest floor and the crowns of ponderosa pines until nightfall, when the tired heat slips away and the deep chill of granite bedrock refills the forest. Cold gusts run like liquid off these wooded peaks, merging into wind-whipped rivers of air that can tip the plume from a cabin woodstove and defy common sense by dragging smoke downhill.

An empty cabin rests on this point, a ramshackle construction of weathered plywood, sawmill waste, and two-by-fours, wedged into the hillside among an outcrop of boulders. Hinged windows are set into the cabin walls seemingly at random, like afterthoughts. A stovepipe chimney rises from the peak of a corrugated steel roof, which is rusted in places and dips at the corners. No electricity up here. No phone. You could survey forever and not find a piece of ground flat enough for a home on this ridge and so the cabin is built on stilts that are like the legs of a sitting dog—longer in front to reach out over the shoulder of this cliff and level the house. A few years ago, there was a scattering of outbuildings on this point, including "a birthing shed"—a guest house built like a tiny barn, used by the religious mother of this family as a retreat when she was menstruating, when she believed she was unclean.

That woman is gone now. Her family has moved away and the cabin is vacant, the yard grown over, the point empty. Standing alone on this craggy bluff in thick Idaho forest, it is difficult to imagine how important this rickety place has become, to conjure the events that took place here: the frightened prophecies, gunfire, and death; the Green Berets, FBI snipers, and angry skinheads. But images do come finally, twisted and tragic—as perplexing as your first view of smoke racing downhill.

Randy Weaver was a thin, hollow-eyed woodcutter who decided one summer to drive down from this cabin to a summer meeting of the Aryan Nations—an organization whose followers believe Jews are the children of Satan and that white America should have its own homeland.

The seventy-mile drive forever changed his life. Weaver and his family, in turn, unwittingly changed the way many Americans view their freedoms, their values, and their government. The Weavers' road led to an Old West gunfight, three deaths, and the invasion of a remote county in North Idaho by an army of state and federal agents. It channeled America's deepest insecurities and drew thousands of people into a whorl of fear and stubbornness, mistakes and misjudgments, lies and a cover-up that shook the top levels of the FBI.

The road detoured in a wild court case featuring the brilliance and bombast of Gerry Spence, the fringed Wyoming lawyer who argued passionately that America needs people like Randy and Vicki Weaver to show us the line where our freedoms begin and end. We need to constantly be aware of that line. In the Weaver case, it fades in and out and switches back dizzyingly, like the old logging road that leads away from the top of Ruby Ridge.

*T*he same road passes through Waco, Texas, where ninety people were killed in the spring of 1993, in a gunfight and standoff between federal agents and a religious sect. The road leads, too, through Oklahoma City, where on April 19, 1995, a bomb tore in half the federal building and killed 168 people. That act of terrorism warped the line Spence talked about—the one that connects our freedoms to our collective values. In the days that followed the bombing, many Americans heard for the first time about militias, freemen, and the patriot movement—a vein of disgruntled gun owners and conspiracy theorists, people who believe, to varying degrees, that government is their enemy. And they heard about what most angers and frightens those people—the bloody standoff in Waco and the standoff with Randy Weaver's family on Ruby Ridge. For those uninitiated in conspiracy thinking, the revelation that angry Americans were behind the Oklahoma City bombing was more chilling than if international terrorists had been to blame. It was as if we were at war with ourselves.

If that is the case, the standoff at Ruby Ridge has become a battle cry. Its details hum in Web sites on the Internet, scream from right-wing newsletters, fill government reports, and resonate in the minds of people who have no trouble imagining *themselves* in that besieged cabin. The case doesn't just touch the fringe, though. When elections are waged on the premise that government is out of control, the Weaver case is a chilling example. More even than the deaths at Waco, the Weaver standoff brings paranoia into the mainstream. For how can you convince people that their government isn't trying to kill its own citizens, when, on Ruby Ridge, the FBI gave itself permission to do just that? How can you tell people to trust their government when it continues to cover up details of the case? There is little wonder it has become a symbol for government tyranny.

But symbols are just half-truths, and they fall well short of explaining a place as hard and remote as Ruby Ridge. From this jagged point, the Weaver case is not proof of broad government oppression and tyranny, but of human fallibility and inhuman bureaucracy, of competitive law enforcement agencies and blind stubbornness. The Randy Weaver case is a stop sign, a warning—not of the danger of right-wing conspiracies or of government conspiracies—but of the danger of conspiracy thinking itself, by people and by governments.

Ruby Ridge is also a good place to view the path of paranoia. For the Weavers, that trail cut right through our backyards, through patriotism, idealism, the military, and fundamentalist Christianity. Randy and Vicki's story is a map of disenfranchisement. More than crime, more than moral decay, they were driven by the loss of self, a refraction of what is normal and right. They were seduced by conspiracy and by a religion called Identity, by beliefs steeped in racism and fear of government oppression, beliefs that helped bring about the very thing they feared.

You come to the Weaver story along the same trail Randy and Vicki took, from the heart of Christian Iowa to the deep woods of North Idaho. There is much to ponder along the way—the accountability of government and the danger of paranoia, the villainy of coincidence and the desperate need to decide, every day all over again, where our society's lines will be drawn. Up a twisting, rutted dirt road, past gnarled pine trees and scrub grass, you come finally to a sign at the edge of the old Weaver property. Two sets of unbending law clashed on the mountain, two incompatible views of the world that are still at odds, outlined by defiant red letters painted on a plywood sign: "Every knee shall bow to Yashua Messiah."

one

Sara Weaver snapped awake, felt for her rifle and hoped she'd at least get the chance to shoot one of the bastards before they killed her. She had no idea whether it was day or night. The Bible was open on the floor where she'd left it and she quickly found her place and resumed her prayers to Yahweh, the stern and unbending God of the Old Testament. If she'd slept at all, it was only for a few minutes; that's all she allowed herself. Had it been three days, now? Four? A noise brought her eyes up to the windows, covered with the denim curtains that she and her mother had hung to keep the enemy from seeing them. Still, a few shards of unnatural light cut through the room and lit the cabin like constant dusk. Sara looked across the beamed living room at her friend Kevin Harris, who'd been like a brother to her for half of her sixteen years. He was still coughing blood. Sara'd given him herbs, tea, and cayenne pepper. She'd dressed and cleaned his gunshot chest and arm, but he was still too pale and had lost too much blood. He would probably die.

Her father was in better shape, awake, but staring off toward the kitchen. His gunshot wound was healing, but he seemed distant and tired, and Sara was afraid that he blamed himself for what had happened. It wasn't his fault. She knew he just wanted to protect the family. But there was no way she was going to let him feel so bad that he would surrender to the Beast. Her ten-year-old sister Rachel was asleep at last, curled up on the floor next to her. Sara was glad for that. The baby was asleep too and had finally stopped crying "Mama."

The voice startled Sara as it blew through the cabin like a December gust. There had been so many noises: tanks and trucks and helicopters echoing through the canyon. But it was the steady voice of the negotiator that was making her crazy—so calm on that PA system somewhere outside the cabin. "Pick up the phone," he kept saying, as if they were insane. "We've thrown a phone onto the porch. Pick it up." He sounded Mexican to her. Wouldn't that just figure; the Babylonian One World Government sends a Mexican to talk a white separatist out of his cabin. They will do anything to break us, she thought. Yesterday, he had called himself Fred. A Mexican negotiator named Fred talking on a PA system every fifteen minutes, trying to get them to step outside. The FBI had made it perfectly clear what happened when the family stepped outside. Agents blasted away at them. It was ridiculous and horrifying at the same time. Rachel stirred as the one-sided conversation began again, and she cried as the cruel, taunting words settled on the cabin.

"Good morning, Mrs. Weaver," the voice called. "We had pancakes this morning. And what did you have for breakfast? Why don't you send the children out for some pancakes, Mrs. Weaver."

Why were they doing this? Sara stared at her mother's body covered with an old army blanket and pushed underneath the kitchen table. Soon, Sara would have to crawl through her mother's blood to the cupboards to get canned apricots and tuna fish to feed their family. As the voice tormented them, Sara's anger fled, and she prayed for her mother's strength. Her mom had practically built this cabin, pieced together the walls from mill scraps, made the quilts they were huddled upon on the floor, canned the food that was keeping them alive, and shaped the cupboards where Sara had to get the food.

She couldn't let her dad go in there. If he moved in front of a window, they'd kill him, finish what they'd started outside the cabin. Sara sat up, her long, black hair in a ponytail, her eyes tender and puffy from crying, her lips drawn tight. It had been so long since she'd spoken in more than a whisper, and now she wanted to scream. She knew she should go to the kitchen, but she didn't want to get off the floor.

Beneath her—where an open basement was framed with thick timbers—Sara listened for the agents of Babylon, who had crawled under the cabin with their goddamn listening devices, trying to get any edge. She thought she heard their muffled whispers and wondered for a moment if they were really there. She wished she could yell at them or pound on the floor or something. She was just too tired. Too tired to crawl through the blood into the kitchen. Too tired to shield her dad when he stood in front of the windows. Too tired to rock the baby to sleep, to tend to Kevin's wound. For the first time, her fatigue seemed stronger than her anger, and she wished Yashua the Messiah would just come and end this suffering.

And so she prayed to Yahweh as her parents had taught, thanking Him for His blessings and asking for deliverance. Lying on the floor with what was left of her family, Sara Weaver looked across the long room at the bullet hole in the kitchen window and she prayed that they not be picked off one at a time any more, that they be taken together to Paradise. She prayed that the evil agents of ZOG just get it over with. She prayed that they firebomb the house.

A long the denim curtains, across a narrow gully on an adjacent hillside, the barrel of a bolt-action, high-powered rifle traced the breadth of the cabin, looking for any movement. Behind the gun, a compact, muscular sniper watched the windows through a magnified, ten-power scope. Nothing. Lon Horiuchi knelt camouflaged and still in the low underbrush and rocky ground, separated from the cabin by two hundred yards. He ran his scope along the house again, from the covered back deck, which leaned out over a steep hillside, along the plywood walls of the house. It had been two days since he'd fired any shots, two days since he'd seen the target flinch and he'd called into the radio that he thought he had hit one of them.

There were ten other snipers on the hillside across from the cabin, another twenty agents crawling over the knob where the house itself sat. First light settled evenly on the grayish brown cabin and glinted off its small offset windows as the sniper watched for any movement.

Behind him, the hill broke at a severe pitch, covered by clinging mountain grass and leaning timber, cut occasionally by a logging road or a plunging stream, down the slope a mile, to a meadow where deputy U.S. Marshal Dave Hunt paced and smoked, killing another Marlboro 100 with a few grave steps. He paused in the middle of a meadow packed with sagging army tents as though a dull green circus had come to town. A couple hundred camouflaged federal agents and state cops filed in and out of the tents, catching some sleep before going back to the line or to the sniper positions. Any minute, Hunt expected more white separatists to break into the meadow from the woods and begin firing. It was like a war zone. Slope-shouldered and frowning, Hunt watched a handful of busy men across the meadow, FBI brass and investigators who climbed the steps to the trailer command post. None of them was interested in Dave Hunt's opinion.

It wasn't right that he was on the outside now. He knew this case. He knew Randy Weaver and his family like no one else in law enforcement. For eighteen frustrating months, he'd butted up against their religious fervor, their government paranoia and their unbridled stubbornness. He knew their beliefs and the language they used. He knew the weapons their children carried. He knew Randy was a coward and a straw man and that if they wanted to end this thing, they had to negotiate with his wife, Vicki. He knew that unless they convinced Vicki to give up the kids, this thing might only get worse.

He watched trucks of all kinds—moving, army, pickups and motor homes—beat the mountain field into dust. They broke through the forest one at a time on that narrow dirt road and began looking for parking in the perpendicular rows, which by now contained more than 100 vehicles: cars, trucks, Humvees, armored personnel carriers, and bulldozers lined the edges of the tent city. More agents were showing up all the time to secure the mountain and they reported here, to an encampment surrounded so completely by pine-covered ridges that it seemed entirely possible there *was* an enemy out there.

Doubt broke constantly into Hunt's thoughts. He'd done everything to bring Weaver in, hadn't he? The second-guessing carved away at him until he slid another tan-filtered cigarette into his mouth, lit it, and began pacing again.

He just wanted to get as far away from here as possible, to grab his wife and hold her. Soon, he and the other guys would be leaving for the funeral. The shoot-out flashed in his mind like someone flipping through snapshots: the Weaver men stroll down the hill with their rifles. The dog barks, cold at first, as if he'd just caught a whiff of something. The radio squawks, "I think the dog's onto us." And then nothing. For five awful minutes. Nothing. Then a gunshot. Two more. And then bursts of gunfire like a loud shuffling of cards. And Hunt runs panting through the woods. Near the bottom of the hill, another barrage of gunfire drops him to the ground and lands him back in Vietnam, the shots cracking over his head like a round of suppressive fire.

No, he'd done everything possible. That was true enough. This case had gotten out of control because Randy and Vicki Weaver wanted it this way. But Dave Hunt couldn't scare off the self-doubt as he churned up his own dust pacing and smoking in a mountain field that had been nicknamed years earlier Homicide Meadow.

*T*hrough the meadow, the dirt road dipped and twisted for two more miles, widened and came to a halt at a bridge that crossed Ruby Creek and T-boned into the old highway, where two armed sides faced off and the threat of violence hung in the air like fall.

The two sides—Randy Weaver's supporters and the government agents—were separated only by the yellow police tape and the rippling water. On the bridge, cordoned off by the plastic tape, police cars, a military truck, and a motor home blocked the only road up Ruby Ridge.

A dozen state and federal agents with machine guns and bulletproof vests watched a crowd that grew larger and angrier by the minute. The agents lowered the tape so another green Humvee—a short, fat, military truck—could rumble past, and immediately the yelling started again.

A ponytailed man with a day's blond growth stepped forward from a pack of bitter, bearded men in jeans and mesh baseball caps. He jammed his index finger at the windows of the military vehicles and yelled at the drivers. "Baby killer! Baby killer! Which one of you is going to shoot the baby?" On the other side of the tape, the federal agents shifted their rifles like bachelors holding infants. The protesters—perhaps fifty now—closed in around the roadblock and held up signs—"You Could Be Next!" and "It's Time For War!"—some with misspellings that made the agents laugh nervously: "Our mountians were peacefull untill now." At one end of the police tape, a camouflaged ATF agent videotaped a bowlegged man in a black T-shirt, who, in turn, videotaped the ATF agent videotaping him.

Behind them, a nineteen-year-old boy from Las Vegas argued with his buddies over whether they should be wearing brown shirts or the black ones that all the other skinheads were wearing. Hitler's original force had been the Brownshirts, the boy explained, cocking his shaved head. "Maybe black *is* scarier, but I wanted to wear brown, to let the Jews and niggers know who's coming after them."

On their own shoulder of the road, television satellite trucks hummed and photographers circled around the yelling protesters, who fanned out against the banks of the old highway. Reporters and photographers milled about, wishing they could get closer to the cabin, stuck here interviewing anyone who didn't look too violent. They half expected shooting to break out here at any time.

One of Vicki Weaver's friends held court with three reporters, explaining that Vicki was nothing short of a prophetess, touched by her creator and convinced that whites should separate themselves from other races. Yes, Vicki Weaver was a white separatist. But hadn't Malcolm X preached separation of races? Didn't Indian leaders fight for their own sovereign nations? Why were they the subject of flattering movies while the Weavers were the subject of a government setup and attack? "Randy and Vicki and the kids are a family," she said, pointing across the police tape. "Those are the criminals."

In the road's other grassy shoulder, children in denim jackets and cowboy boots penned protest signs—"Go Home" and "Zionast pig—Why don't you shoot me too?"— and held them up for their parents, who stood in clusters, talking about what was happening on the mountain. "They're going to gas the whole family!" Tax protesters, constitutionalists, Vietnam veterans, mountain folk and skinheads raged and shook their fists, wondering what they could possibly do to stop this injustice, wondering which of them would be next.

A car trolled along the old highway, past the tense balance of protesters and law enforcement. The car slowed, but there was nowhere to park because other vehicles were strung along either side of the roadblock for several hundred yards—cars from fourteen states and one with a license plate that simply said HEAVEN. The driver had to take a wide corner before finding a place to park, a quarter mile from the roadblock, just down the narrow road, near the dark-wooded Deep Creek Inn, from whose doorway a confused Swiss chef named Lorenz Caduff watched protesters and law enforcement officers parade past.

Caduff had only been in North Idaho six weeks, and it had seemed like paradise until now. Now it was crazy, FBI agents running in and out of his business, neo-Nazi skinheads living in tents in his yard, strangers sleeping on his floor. He'd sent his wife and kids away and worked twenty hours a day, serving up plates of biscuits and country gravy to people who told him the government was trying to kill them. An earnest, friendly man, Lorenz liked everyone, trusted everyone and struggled to understand who was right.

How could people in the greatest country in the world fear their own government? And yet, why did the government use tanks and armies to arrest one man? That's when Lorenz began to wonder: Will they come for me too?

Another car cruised past the Deep Creek, a Jeep Cherokee filled with five skinheads from Oregon, all of them between the ages of nineteen and twenty-four. They were looking for a gun shop to buy some rifles and ammunition and to ask for directions up to Randy Weaver's cabin.

*T*he old highway ran parallel to the new one, an asphalt ribbon that actually had lines painted on it and that bisected the glacial valley for another twenty miles before finding civilization in the resort town of Sandpoint. In a motel at the edge of town, a solid Iowa farmer named David Jordison peered at the television with his wife, Jeane, waiting for some news. Why hadn't the FBI agents come by? The Jordisons had driven fifteen hundred miles straight through as soon as they heard and now they were left hanging, nervously awaiting any word about their daughter Vicki Weaver and their grandkids.

Vicki's brother, Lanny, and her sister, Julie, talked about the children and about why in the world federal agents would treat the family like criminals. The people in that motel were fourth-generation Iowans, good Americans who didn't condone Vicki's beliefs, but also didn't think it took an army to deal with a husband, a wife, and their four kids. "They're little girls," Julie said, maybe trying to convince herself. "These are normal children. These aren't trained guerrillas."

Maybe they should drive down to the roadblock, Lanny suggested. Maybe they could get more information there. But the television pictures of those protesters, skinheads, and hill people scared them. So they waited, hoping the FBI agents would call or come back or something. Julie and Lanny just wanted to see their sister and her kids again. They hadn't seen them in nine years, since Vicki and Randy stopped by on the way to the mountains, their entire life packed on the back of a moving truck, wild fear in their eyes.

This was usually the time of year that Vicki's parents drove from Iowa to Idaho to help Randy and Vicki work on the cabin and to see how the grandchildren were growing. David Jordison knew the mountain, and he knew his daughter, and he wondered when the FBI was going to come get him, drive him up the scarred driveway and let him settle this thing. Sitting on the bed in a sterile motel room in Sandpoint, Idaho, David Jordison would have given anything to be on top of that mountain with his oldest daughter. If anyone could talk Vicki out of the cabin, her family knew, it was her dad. And Vicki was clearly the one they needed to talk out. The family could stay on that mountain forever unless Vicki received word from God that they should come down.

Perched on the jagged rim of Ruby Ridge, ringed by zinnias, marigolds, and FBI snipers, Randy Weaver's cabin was quiet as morning swung into afternoon. An armored personnel carrier—a squat military vehicle like a tank without guns—sat in the driveway, twenty yards from the house.

Near the APC, FBI agents monitored the listening device in the telephone that had been placed on the porch near the cabin. They heard muffled voices, footsteps, a whining dog. At least someone inside was alive. The hostage negotiator, Fred Lanceley, had never been on a case quite like this one. A former street agent, he had worked as a negotiator primarily on overseas kidnapping cases, about a dozen of them, most often in cities. During one of the last cases he did, Lanceley worked as backup negotiator in a situation similar to this one, where children were caught in the middle. They negotiated with a man on the other side, but when it was over, they found that a child inside the building had died of thirst.

Lanceley asked for Vicki again, offering to do anything for the baby—food, water, medicine. Still no answer.

Lanceley knew he had to get some sort of dialogue going, even if it was just yelling back and forth. The first day, they'd driven up to the porch of the cabin and thrown the telephone up there. Every fifteen minutes the phone rang, but they couldn't get Randy to pick it up. The intelligence reports had identified Vicki as the strong one, had shown she talked her husband out of surrendering, and indicated that she might be so fervent in her religion that she would kill her own kids. Randy did most of the talking, according to intelligence, but they were Vicki's words.

That was probably the most important part of Lanceley's job, knowing whom he was negotiating with and having some way of sparking trust in that person. That's why this case was so frustrating.

He could yell all he wanted and there was no way to get any response from the cabin. He'd tried pleading, making them mad, and asking about their beliefs: nothing worked. He knew that if they didn't talk soon, the FBI agents would have no choice but to raid the cabin.

Lanceley needed to get inside their heads, but right now, he simply lacked the information. Sure, they had Randy Weaver's army file—some Special Forces training in explosives and engineering. They knew that a family friend, Kevin Harris, was in the cabin and that Vicki and the children were in there as well. But until a couple of days ago, Frederick Lanceley had never heard of the people inside the cabin or their religion, Christian Identity. Now he struggled to find the right words to defuse the most difficult case he'd ever seen. And he knew this mysterious woman was a key to ending the standoff. It was Randy Weaver's wife who defined their intense beliefs— that they were God's chosen Israelites, and that Jews were impostors, the spawn of the Devil. If only Lanceley knew where such beliefs came from, if he could see how Vicki Weaver had gone cartwheeling over the edge, how a pretty wife and secretary from Iowa suddenly came to believe that an evil, shadow government was chasing her toward Armageddon.

Fred Lanceley held the microphone to his mouth and gave it another try, his words echoing off the rocky knob, into a canyon that bent sound and distorted perspectives, settling finally on the simple, plywood cabin. Inside, the bloodied and frightened family winced every time they heard her name.

"Vicki?"

two

She was born in a white, two-story farmhouse set behind a row of wind-break trees, between the Central Iowa towns of Fort Dodge and Coalville, on land gently pitched like the quilt on a made bed. Her father, David Jordison, was the third generation of his family to live there, but when he moved his pretty young wife, Jeane, into the house, in 1945, there was still no electricity. Seemed neither the Coalville nor Fort Dodge power companies wanted to string a line to one lonesome house in the middle of a lake of corn and soybean. So they got by until 1949, when the crews buried the last pole in the ground, connected the wire to the house, and hooked the Jordison farmhouse up to the twentieth century.

Vicki arrived June 20, 1949, about the same time as the electricity.

She was farm pretty, with long, straight black hair, bubble cheeks, and narrow but searing dark eyes that seemed to know everything. She spoke early and with a midwestern slur, like someone with a piece of straw in her mouth. From the beginning, she was Dad's girl, short, stoic and serious, like her father. Julie, the cute one, was born four years after Vicki, and Lanny, the athletic one, came along eighteen months after that. There were cousins everywhere, but even in a far-ranging family of handsome and talented kids, Vicki stood out. She could do anything: sew, cook, knit, finish furniture; she picked up things without ever seeming to study them. Her mom, Jeane, could barely sew a button on, and yet Vicki took to it like she was born with her foot on the pedal. It was said she could do anything with her hands. She was distinctly feminine—domesticated, you might say—and yet there she was, crouched on the roof of the hog shed, driving nails into the new shingles with her grinning father. And smart? From the beginning, her parents felt they were having conversations with a little adult. She read everything.

She was eerily earnest: she never talked back and rarely got in trouble. Of course, that meant her younger sister had to do both. By the time Julie came along, Vicki had already mastered most of the responsible lines, so the only role open was Rebel, the one who *didn't* do everything right, antagonist to Vicki's sappy protagonist. Julie worshiped her sister, but she also felt jealous of Vicki's looks and her accomplishments and the two argued and fought the way sisters can. Vicki would measure her thighs, then make Julie sit down before measuring hers, pronouncing them bigger than her own. Julie thought she was the only one who could see through her sister's practiced perfection.

Still, the girls got along more often than they did not and shared the most intense and deep feelings between them, like the time, in the early 1960s, when the same electrical company that so slowly brought power to the Jordisons informed them that the family's favorite tree—an ancient oak with limbs that reached down and practically demanded that children swing—would have to be taken out to make way for more, bigger power lines. The girls cried and tried to convince their dad to protect the tree, but there was little he could do, and it was cut down. It was Vicki's first lesson in the glacial power of government and progress, powers which—like electricity—some people feel better off without.

*L*ife revolved around the farmhouse. It was built by Vicki Jordison's great grandfather, an English coal miner who settled eighty miles north and west of Des Moines, in Coalville, the best place in Iowa to practice his trade. Eventually, he saved up enough money to leave the mines and buy 160 acres, and in the late 1800s he built the farmhouse on the edge of his property. By the time Vicki's dad, David, came along, the coal mines were shutting down, and the Jordisons had switched completely to farming. Toward the end of World War II, David bought the original Jordison farmhouse from an uncle and set to work on the land. He was a tireless and fair man who preached hard work, responsibility, and the faith of the Reorganized Church of Latter Day Saints.

He gave Julie and Lanny outdoor chores—feeding the chickens and cows, working in the fields—but Vicki did more work inside, helping Jeane with some of the cooking and sewing and cleaning. Like most farm families, the noon meal was called dinner and was eaten around the table—roast beef, potatoes and gravy, corn, slaw, biscuits, and tea. Having lived through the Depression, Jeane and David Jordison stockpiled food in cupboards and freezers, and they lived off the top layer of the deep freezer and the top row of canned foods.

They were a religious family, but there was a quiet discomfort around the house when it came to church. David was a devoted member of the reformed Mormon church. Jeane was a Congregationalist. And so, on Sundays, the family would split up, David and the kids trudging off to the RLDS church while Jeane stayed home. There were a few arguments, but Jeane would not change her mind.

Julie would have liked to have stayed home as well, but she had to go to church. She assumed Vicki felt the same way, although Vicki never would have said anything to her dad about it. After church, the kids went to Sunday School and the adults to their own Bible classes, where they talked about current events and church prophesy. "We are God's chosen people," the elders would say, spreading the Mormon doctrine that one of the lost tribes of Israel came to America and that Jesus had appeared in the New World. It was the thing that always stuck in Julie's mind about those sermons, the strange notion that Americans alone were God's chosen.

After church on Sunday, David would tinker around the house—as close as he ever got to resting—before sitting down in front of the television, turning on the football game, and grabbing the book of Mormon. Then, he'd talk about the things he'd heard in his Bible study, detailed predictions of Judgment Day and the worlds reserved for the true believers. He talked about a universe inside the earth, something he'd heard about at church, and he looked for proof of it in the Bible. His favorite topic was Revelation, and he read from it in his cracked, leaning-over-the-fence, midwestern voice. Julie ignored it and figured her sister was doing the same, but later she wondered if Vicki could dismiss anything their dad said. Clearly, some of the church teachings had settled with Vicki. "You can't tell me that Joseph Smith advocated race mixing," Vicki wrote to a cousin years later. "There were no black RLDS elders until our generation."

David didn't read a lot, but he got his daughters hooked on a science fiction series by Edgar Rice Burroughs. And, every summer, he took the family to the mountains in the camper he'd built from scratch, driving to the Rockies, or the Grand Tetons, camping under the stars, fishing, and exploring. "I love the mountains," he'd say. On the camping trips, like other times around the house, Julie would be struck by how close her dad was with Vicki. It was evident when she worked outside with him and when he stood behind her, watching her cook or sew. She sewed like a musician playing by ear, and David would shake his head as he watched her tailor a suit.

"Look at that," he'd say. "Amazing." And she'd turn

back and smile at him, and Julie would have the distinct impression she was sister to the queen.

Vicki Jordison's Iowa couldn't be any more different than the wooded mountains of North Idaho. Straight asphalt roads cut a near-perfect grid across the farmland, interrupted occasionally by groves of hardwood trees and shallow river channels; there is no place in Iowa to hide. Small towns like Coalville remain fundamentally unchanged from a century before, when they sprouted up a half-day's horse ride from the most outlying farms. White, gabled houses top the town's hills and American cars troll straight highways at just below the speed limit. The homecoming parade from the one high school winds through brick, downtown blocks, and each small town seems to have a hall of fame enshrining a wall of white, solid, short-haired athletes, politicians, businessmen, and farmers. The hardware store goes through six times as much white paint as any other color; the little newspaper gives better play to the photos of local boys off to Marine boot camp than it does crime stories; and next year, when that big farm kid is a senior, the football team is perpetually going to state.

"*H*ow can they do this?" Julie Jordison asked her sister. Again, just like they had with the power company and the oak tree, Julie and Vicki felt like progress and government were trampling the family.

"I don't know," Vicki said. The teenage girls paced around the living room, waiting for their parents to come home from a town meeting where they were discussing the new interstate highway that was supposed to run alongside Coalville. The Department of Transportation had notified families whose farms were in the way of the new highway, and the Jordisons had been shocked to find out they would be displaced. In fact, the Fort Dodge off-ramp was going to slice right through the farmhouse, the very symbol of the Jordison family in Iowa. Like other families whose property was in the way, David and Jeane would be paid a "fair-market value" for their farm, but it wasn't enough. It certainly didn't pay for the four generations of Jordisons who'd already lived there and the ones that David—looking over his pasture in the late afternoon—imagined would work the land when he was gone.

Jeane and David were upset, but taciturn. "I guess you can't stop progress," said Vicki's dad, a lifelong pragmatist.

Vicki and Julie weren't taking it so well. They wondered how this could happen in America, how it could happen to their dad's farm.

"Does our Constitution allow this?" Julie asked.

"I don't know," Vicki answered.

But Vicki wasn't going to stop fighting until they bulldozed the house. She suggested her father look for gypsum on the land and then make a claim of more value than the Department of Transportation was offering. Perhaps if the house was worth too much, they wouldn't tear it down. Then Vicki suggested that her dad hire an attorney, but he was a small farmer and didn't think he could afford to fight the government. They'd just have to move somewhere else.

But a neighbor came to their rescue. A widow who owned ten different farms in the area hired an attorney and, in the end, the government allowed one highway in Iowa to be laid a little crooked. Yet Julie and her sister never forgot how close they'd come to losing their farm.

In the 1960s, Coalville's prime was well behind it, and it was little more than a cluster of trees and grain silos, hiding 250 people in a few blocks of houses. Fort Dodge, six miles north of the Jordison farm, was the regional center for farming and banking, home to 25,000 people. Coalville kids like Vicki went to their own tiny grade school until ninth grade, when they were suddenly thrust into a 3-A high school—the biggest classification in the state—where everyone seemed to know everyone else and being from Coalville was not conducive to being homecoming queen.

But Vicki's place was always in the family anyway, and she really didn't seem to mind that she wasn't as popular in school. She got *A*'s and found things to be involved in, becoming vice president of the Future Business Leaders of America and the unquestioned star of the Pleasant Valley Pixies 4-H. Yet by high school, she had run headlong into the first problem she had no solution for: boys.

Julie couldn't understand it. Vicki was the pretty one, the talented one, the one with the smaller thighs, and yet, when the boys came over, they wanted to see Julie. Four years younger, she was growing into a cheerleader and a ringer for Marlo Thomas in *That Girl*. From kindergarten on, boys sought out Julie, and by eighth grade or so, the younger sister began wondering why Vicki never had any boyfriends.

In 1967, Vicki graduated from Fort Dodge High School and decided to go to the little community college they'd tacked on to the end of it. Iowa Central Community College had just opened the year before, in a cluster of temporary buildings on the north end of Fort Dodge. In a student body of 1,181 farm kids, the boys—studying agricultural sciences, auto mechanics, and liberal arts—outnumbered girls—future teachers, nurses, and secretaries—almost five-to-one. Even with those odds, Vicki had trouble with boys. She was eighteen, not engaged and from a farm in Iowa, a combination that meant she could easily be on her way to being an old maid. And her personality, charming to the family, could seem controlling to the boys who drove up the Jordison driveway to take Vicki to the Playmoor ballroom for a dance or to Dodger Lanes for a Coke. Some of them ran from her domesticity, her self-reliance and her doting, their arms full of baked cookies and knitted sweaters. Her first boyfriend, Dave, was charming, but Vicki fell much harder for him than he did her, and when he broke up with her, she was devastated. Always so steady and even-tempered, she threw herself down on her bed and cried. Julie was surprised by that side of her sister: self-confident and strong around the family, all she really wanted was to find a man to devote herself to. In 1968, Vicki graduated from ICCC with a two-year degree in business and took a job as a secretary at Sears. That's where she met her second boyfriend, a mechanic named Bob. But there was no spark for Vicki.

Then came the one she thought was right. He was a few years older than Vicki and as charming as Dave had been, but with a darker side. As before, Vicki turned herself over to him completely, even though the rest of the family didn't trust him. He seemed slick. He hung around for months until, one day, a county sheriff's deputy pulled up to the farmhouse, arrested him, and charged him with raping his sister-in-law.

"That woman's been throwing herself at him," Vicki explained patiently. "And when he rejected her, she accused him of this." Vicki was so upset, she talked her dad into putting the farm up as collateral for his bail money. The charge was dismissed when the woman didn't show up for court. Eventually, Vicki left him.

She was miserable, turning twenty-one, feeling old and alone.

And then she bumped into Pete.

He was charming, funny, as good-looking as her first boyfriend, with dark, close-cropped brown hair that curled at the edges when it started to get long, and bushy brown brows that covered devilish eyes. His real name was Randall, but he hated it and so everyone called him Pete. He was a year older than Vicki and had gone to Iowa Central for a year before he dropped out and took a job driving a school bus. After that, he enlisted in the army, and now he was home on leave. He came back to Fort Dodge in 1970, muscled and serious, tooling through town in his red Mustang, a fast-talking planner who was ready to do something with his life.

Vicki was in love.

"How could you be in love?" Julie asked. After the fiasco with her last boyfriend, she figured Vicki was throwing herself at the first guy she saw. "You don't even know this guy."

Sure she did. She'd even gone out with him once or

twice, between Bob and her last boyfriend. "He was really wild and I didn't think it was the right thing for me," Vicki said. And he hadn't been interested in a long-term relationship anyway.

"He told me that when he went out with me before, he knew I was the kind of girl you married," Vicki said. "And he wasn't ready to get married yet. Now he is." The kind of girl you marry? After all she'd been through the last three years, Vicki couldn't have heard anything that sounded better. To her, Pete Weaver was perfect.

Well, Julie thought, for my big sister, nothing else would do.

Ask anybody: Randall "Pete" Weaver was just a regular guy. He didn't really have the build to be an athlete, but he was wiry and strong from long summers of farm work. "I was always the little kid in school and I hated a bully," he said. Randy started working in the fields when he was ten, and his dad was never so proud as he was the day Randy stood up to a farmer who tried to pay him less than the bigger boys. The Weaver family lived in Grant, a small town in southwestern Iowa, where Randy hid from his sisters, played Little League baseball, and goofed off with the local farm boys.

Clarence Weaver was an agricultural supply salesman who waited patiently through three girls before Randall was born, when the old man was forty. He passed on his compact toughness and his sharp features, and he invested much patriotism and Christianity in his kids. Randy, especially, strived to please him, and he was treated with the deference of the only boy and the youngest child. Clarence and Wilma Weaver were fervent and practiced Christians, with Bible Belt intensity, especially Clarence. He bounced the family among Evangelical, Baptist, and Presbyterian churches, trying to find a denomination that mirrored his own rigid faith. When Randy was eleven, he made his dad proud again, walking down the aisle at church and accepting Jesus as his savior. His dad cried. "He was a good boy," Clarence said. "He always believed in God, always did right by Him. We didn't stand for anything else."

In 1962, Clarence moved the family north, to a gray, two-story house with a square-post porch on Vine Street in Jefferson, Iowa, about fifty miles from Fort Dodge. It was the perfect town for Randall, who was popular enough in school—one of those decent guys who would joke around with anyone, even though he came from one of the most religious families in town. He fit in with his new classmates and went to the Presbyterian Church youth group and Sunday School, tinkered with his car, worked summers in the fields, and tried beer with his buddies. "It was like growing up on *Happy Days,*" one of his friends remembered. "It was Friday night football, get the crops out of the field, and wait for the next parade."

Jefferson had been founded in 1854 by farmers who picked a high spot between the Raccoon River and Hardin Creek, bragging that it might be the only place in Iowa safe from both flood (because of its elevation) and tornado (because it was surrounded by water). There was only one tree on the whole 3,000-acre town site, and so they nailed a plaque to it. But being on high ground on the great plains meant there was no break from the wind, and though they were relatively safe from flood and tornado, winter storms raked the town, until, more than a century after Jefferson was founded, a midwestern windstorm called a derecho finally blew down the founding tree.

But by 1962 there were plenty of shade trees in Jefferson. It was one of those perfect, self-enclosed farming towns, with a bank holding $15 million in deposits from the 4,500 mostly white people spread among 1,400 tidy, mostly white houses. There were eleven churches—ten Protestant, one Catholic—and no taverns. A few years later, buoyed by the completion of a thirteen-story bell tower, the chamber of commerce changed the town's slogan from "Home of the Horn of Plenty" to "City on the Rise." The population dropped by 250.

• • •

Randy Weaver graduated from Jefferson High School in 1966 and enrolled in Iowa Central Community College, driving the fifty miles to class every morning. Although most students were farm kids, Fort Dodge and ICCC—unlike Jefferson—also had a small black population. Randy got along with everyone, no matter their color. At night, he and a high school buddy, Dave Luther, shagged the drag—cruising back and forth on the assigned street in Randy's Mustang—and scooped the loop—turning the car painfully slowly around the square at the end of the drag, trying to get the attention of every girl in Fort Dodge. They went to beer parties and dances at the Playmoor, the town ballroom where the Fort Dodge custom was for a group of guys to surround a couple of dancers, at which point the boy would step back into the chain of guys and the girl would pick a new dance partner from the spinning, laughing circle around her. It was called the Trap. Girls did the same thing to boys, and in such ways, everyone knew everyone else.

Randy Weaver took a job driving a school bus and was given the Otho route, right next to Fort Dodge. Denise was a junior at Fort Dodge High and when the bus driver asked her out, he was so funny and handsome, she barely hesitated. They only went out a couple of times; once, he picked her up in the Mustang, and they drove to a friend's mobile home in the Sunset Trailer Court, near campus, where three or four other couples sat around in the low light, drinking Schlitz beer and talking. She was uncomfortable around these college kids, and Randy seemed to understand that without her even having to say anything, and he suggested they leave. Even though he was older than she was and hung around with these drinkers, she was impressed that he was such a gentleman.

Denise was short and pretty, with dark hair and eyes. And so was another girl Randy started dating. Although he and Vicki went out only a few times, friends said they belonged together, this small, attractive pair of Iowa kids. Randy's buddies called them the all-American couple without a bit of sarcasm.

But Randy wasn't ready to settle down, in part, he said, because he felt his country calling. By the time he dropped out of Iowa Central for good, in 1968, he'd already enlisted in the army. One high school friend, John Milligan, was drafted about the same time Randy enlisted. When they talked about the war later, Randy told him that anyone who wouldn't serve their country wasn't doing his share. "I was ready to fight, ready to go over to Vietnam," he told Milligan.

Randy was trained as a combat engineer, volunteered for airborne training and, later, passed the rigorous training for Special Forces. He told friends he was a Green Beret, part of the most elite fighting unit in the country. Back home, Clarence was more proud than he'd ever been before.

In his training, Randy learned to survive on almost nothing, to make explosives and prepare fortifications. But mainly, he was a construction equipment operator. He was a good soldier and rose to the rank of sergeant, qualifying as an expert with the M-14 rifle and as a sharpshooter with the M-16 and the .45-caliber handgun. His military record was spotless, and he was given a National Defense Service Medal and a parachute badge.

But, by 1969—after the bloody Tet Offensive—American sentiment was turning against the war, and troop reductions were already in the works. Strangely, Randy Weaver never went to Vietnam. At Fort Bragg, in North Carolina, he watched a trickle of other men go over to Vietnam and a flow of body bags come home. It seemed as if the government and the people protesting the war were in some way trying to lose it. An idealist, Randy was disillusioned by the grayness and corruption of military life. Once, he told friends, he was part of an army intelligence drug bust on the base. He noticed that all the confiscated drugs weren't turned over to authorities. When he told a superior, Randy was instructed to mind his own business. They were all in on it, he decided.

He came home on leave in 1970, already planning to get out of the army. One afternoon, he showed up at the house of Denise, the high school girl he'd asked out when she rode his bus. They talked outside by his car—Randy, Denise and one of Denise's friends. He was a man now, hardened and world-weary, seated low in the bucket front seat of his Mustang, his hair military close and a green beret perched on his head. After a while, Denise went into the house to get something and her friend stayed outside, talking to Randy in the middle of the driveway. When the friend came inside, she said Randy had come home to find a wife. Denise peeked out the window at him and refused to go out again. She was afraid he meant her.

But Randy had his eye on a different compact brunette. During his short leave from Fort Bragg, he and Vicki went out almost every night. He met her parents, and they talked about their lives, and she made plans to go see him at Fort Bragg.

On that trip, he gave her a ring and they were engaged to be married.

"Are you crazy?" Julie Jordison asked her sister. "You just met this guy."

She told Julie how they'd dated before, but the timing hadn't been right. Now it was. After his stint in the army, Randy moved home, grew his curly hair, and managed some sideburns and a decent mustache.

They were married in November 1971. Still uncomfortable with the division in her parents' religious lives, Vicki had the wedding at the First Congregationalist Church in Fort Dodge but had two ministers conduct the ceremony, a pastor from her mom's church and one from the Reorganized Church of Latter Day Saints. It was a small wedding, mostly family. The bridesmaids wore purple, Vicki's favorite color.

As Julie watched them say their vows, she felt as if her sister had restored herself to perfection again, and after three tough years of sour relationships, she'd gotten everything she wanted from life. After the wedding, Randy and Vicki moved up to Cedar Falls, on the other side of the state, where Randy was going to use his G.I. Bill loan to go to college, at Northern Iowa University. Vicki was planning to work as a secretary for a while, but, deep down, she said, she only wanted to be a housewife and mother. Randy wanted to right the wrongs he'd seen in the army, and so he decided to go into federal law enforcement, to work for the Secret Service or the FBI.

Randy and Vicki came home from Cedar Falls for Thanksgiving that first year of marriage, and Julie was never so excited to see her sister. As usual, they'd argued a little in the past few years—this time, because Julie had begun smoking and drinking beer. The straight-laced Vicki always chided her about it. But lately everything seemed to be going well for both girls. Vicki was a happy newlywed, and Julie was going out with a boy named Jeff, a handsome wrestler who everyone agreed was perfect for her.

It was a regular Jordison family Thanksgiving, Jeane racing to get all ten courses on the table, and everyone talking at once about everything. After dinner, they moved to the living room to digest, and Randy could barely hide his excitement.

"Now I have something I want to show you," he said. He ran out to the car and came back with a film projector, which he set up in the living room. "You can't believe this stuff. It's great. Let me show you this."

For the next two hours, with Randy and Vicki smiling at the incredible opportunities available to them, the family watched a film strip about Amway detergents, cleaning supplies, and other products. Vicki and Randy explained they were going to make a living selling it.

My God, Julie thought, they want us to buy that stuff. It wasn't until an hour into the film that she understood she was mistaken. No, they want us to *sell* that stuff. She'd never seen Randy so animated and Vicki seemed to feed his zeal with her mastery of Amway products. They were a perfect team, Vicki's studious knowledge and Randy's tireless energy. Julie wondered if they were going to have to listen to the pitch every time they got together from then on.

"It's good for the environment," Vicki offered. But Randy did most of the talking, in his fast sales voice, squaring out how much they could make in a year, in five, in ten. "It's really a wonderful opportunity."

They presented Amway to all their friends and talked some, like Randy's high school and army buddy, John Milligan, into selling it. But the Weavers' interest faded quickly when they started making money elsewhere.

Randy went to school for only two quarters in 1972 before dropping out and applying for a high-paying job at a John Deere tractor factory in Waterloo, the industrial town that butts up against Cedar Falls in eastern Iowa. He got the job. Vicki was still working as a secretary, and they seemed happy and well-off. It was the kind of life Julie had begun imagining for herself.

She and the wrestler, Jeff, were engaged to be married. But in 1973, while they were camping at a nearby lake, Jeff drowned. Julie was in shock when her parents showed up and drove her back to Fort Dodge, where Randy and Vicki were waiting. Julie slumped down on the couch and stared off, catatonic. Then, the television news came on and announced what had happened and she broke down, sobbing and thrashing around.

Vicki sat down on the couch and held her sister, rocking her back and forth. "It's okay. It's okay." After years of watching Vicki's devastation over her awful love life, the family now turned its attention to Julie, and it was Vicki who was able to get through to her. Despite all the competition, all the petty squabbling and posturing, Julie realized that her sister was the only one who could have comforted her then.

*E*ven though they were friends, Julie Jordison hadn't voted for Keith Brown when he ran for school president of Fort Dodge High in 1968. Keith was a radical. In fact, in Fort Dodge, he was *the* radical. He was from a bigger city—Omaha, Nebraska—and played in a rock band. He wasn't too happy when his family moved to

Fort Dodge, smack dab in the middle of boring farm country, smack dab in the middle of Iowa.

On National Moratorium Day in 1969, when people across the country demonstrated against the Vietnam War, Keith was the only student in Fort Dodge High School to wear a black armband. It lasted through class pictures, until some football players held him down and tore it off his arm.

He and Julie Jordison were as different as any kids in the school. Her whole family was straight and square. She was a good-looking cheerleader and wasn't about to date the Abbie Hoffman of Fort Dodge.

After high school and after Jeff died, Julie's friend talked her into going to hear Keith's latest group, a four-piece cover band called Locust. She didn't recognize him when she saw him: wavy brown hair almost as long as hers and a bushy walrus mustache. Julie felt strange being there, and they tried to leave before he saw her. But on the way out, she bumped into him, and Keith leaned close in the noisy bar and asked if she'd go out with him. They began dating, fell in love, and were married the next year.

Keith's band played Beatles and Cream songs in a circuit of taverns that took them through Cedar Falls occasionally. Once, while the band's roadies were setting up at a college bar called The Circle, Keith looked up to see this beautiful woman walk through the door. She was wearing a tight, fitted leather coat with stitching on the wide lapels, skin-tight bell-bottom jeans, and high-heeled boots. Her hair was curly, black and long, and the guys in the band stopped tuning and watched her move through the bar.

She set a large bag down on the counter, and when Keith finally saw her face, he realized it was his sister-in-law, Vicki. Randy came in then, his arms full of paper sacks, too. Inside were hamburgers for the band, the roadies, anyone who wanted them. Keith was impressed. He always thought of Randy and Vicki as too rigid and conservative to think much of his band. They didn't even know any of the guys. In fact, they barely knew Keith, and yet here they were, springing for burgers for everyone. They drove off in Randy's muscular Corvette, and Keith could tell that the rest of the band members were impressed. Keith had a weird sort of pride about his sister- and brother-in-law, weird because he was surprised to be so fond of people that serious, patriotic, and conservative.

O n Sundays, the Jordison kids and their spouses met at the farmhouse for dinner and conversation. They all arrived in the afternoon, Vicki and Randy, Julie and Keith, and Lanny with his wife, Melanie. David would come home from church, sit in front of the football game, and talk about what he'd heard at the RLDS Sunday School. Sometimes, Keith and Randy would go downstairs to smoke, and Lanny would join them for a game of cutthroat on the pool table. The conversation quickly turned to politics. Lanny and Randy were both conservative, and they tried to convince the liberal Keith of something or other, but he was unbendable, especially when talk swung around to the war in Vietnam. With the war winding down and Americans cynical and tired, Keith and Randy could find almost no common ground.

"It's the people back here with the signs and everything that are losing that war," Randy said. He talked often about Vietnam, so passionately that Keith assumed he'd served over there, although he never came out and said so one way or the other.

By that time, Randy was hauling in money from John Deere. And he was spending it. He was always driving new sports cars: the low-riding, burnt orange Corvette for a while, then a Triumph Roadster and a 240Z. He bought trucks and motorcycles and snowmobiles and fishing gear. Keith watched his brother-in-law with envy, wondering how a guy got a good-looking wife, a good job, and all those toys at the same time. Randy and Vicki had recently bought a house, too, a well-kept rancher in Cedar Falls, the kind of house Keith had begun imagining himself and Julie in. But Keith saw a side of the Weavers that worried him, too. They were idealistic and, in a way, naive, throwing their substantial energies blindly into whatever they became interested in at the time, whether it be sports cars or Amway. Randy's latest obsession was silver. It was the best investment around, he lectured Keith in the mid- and late-1970s. Soon the currency would be devalued, and precious metals would be the only salvation for people who wanted to survive the economic shock. As the price of silver began to climb, he gave silver medallions and coins as gifts and talked Lanny into investing heavily. The price of silver collapsed in the late 1970s, but by then the Weavers' obsessive personalities had moved on to something else.

After those Sunday meals, the women sat around the table drinking coffee, or they went shopping. Sara was born in the middle of March 1976, and that took up most of their conversation. Vicki was crazy about babies. She didn't just talk about them, she gave lessons, instructing Julie on morning sickness, diaper rash and breast feeding, in her careful, scholarly way, the same way she instructed Randy on Amway before he spread the word to everyone else.

In families, years pass like that, on Sunday afternoons that blend together in a haze of barbecues, good-natured arguments, and baby diapers. They were the happiest times, from 1974 to 1978, when everyone was young and got along, and the only talk of Armageddon was from David, who lectured peacefully away in an empty living room, while his family went about the business of growing up.

"You've got to read this book," Vicki said. She and Julie were sitting around the farmhouse sometime in 1978 or so, watching little Sara play.

Julie had heard that many times before from her sister—that she should read this book or that one. The last time, it had been some novel by Taylor Caldwell. But this one sounded different because among the things Vicki and Julie had in common was that they were both seekers, people who looked a little harder for truth than did most others.

"It's called *The Late Great Planet Earth*," Vicki said. She told her sister that it answered a lot of questions that perhaps she hadn't thought about.

Written and published in 1970 by Hal Lindsey with C. C. Carlson, *The Late Great Planet Earth* introduced thousands of people to the idea of Old Testament prophecy. In the middle of America's born-again movement, Lindsey's book applied the words of biblical prophets to the world of the 1970s and came to the conclusion that people were living in the "end time." It was a huge bestseller.

Julie tracked down a copy of the book and, for the first time since they were teenagers, became worried about her sister. It wasn't an inflammatory book, but its message seemed aimed right at idealistic seekers like Vicki and Randy, and she wondered what such beliefs might do coupled with their intensity. They were becoming more involved in their Baptist church and believed the world was an evil place, decaying before their eyes. Lindsey's book didn't just impart information; it was written almost like a long letter, directly to readers like Randy and Vicki, stopping occasionally to ask if they were getting it, if it was starting to come together.

"In this book I am attempting to step aside and let the prophets speak," Lindsey wrote. "If my readers care to listen, they are given the freedom to accept or reject the conclusions."

The book acknowledged that the world was a mess and said the solution was in seeking out biblical prophecy. "Bible prophecy can become a sure foundation upon which your faith can grow—and there is no need to shelve your intellect while finding this faith." The book detailed the words of Old Testament prophets like Isaiah, Ezekiel and Micaiah, men who believed that God spoke directly to them and warned them of times to come. It ascribed modern definitions to Old and New Testament words: for instance, Gog, the evil empire spoken of in Ezekiel, was the Soviet Union and the ten horns of the beast from Revelation described the ten nations of the Common Market of Europe.

The book showed how biblical prophecy could be used to predict an Arab-Israeli war that would trigger a nuclear holocaust between the United States and the Soviet Union, bringing about Armageddon. It was pretty familiar stuff for kids who grew up in the RLDS church, Julie thought. Still, Lindsey showed how everything was in place in the late twentieth century for the return of Jesus and, most frighteningly, the rapture and the great tribulation, when true believers are snatched up by God, and the Earth is subjected to every manner of plague and violence. "What a way to live!" Lindsey wrote. "With optimism, with anticipation, with excitement. We should all be living like persons who don't expect to be around much longer."

Other friends noticed the change in Randy and Vicki. At a gathering of classmates from Jefferson High School in 1978, they sat in a circle of friends, talking about everything they'd gone through, the turbulent sixties and seventies. Their lives had turned in a lot of directions: drug use, family life, careers. No one thought what the Weavers had to say was particularly strange, given the range of their class. Their classmates had been fodder for a very difficult time in history, when the world seemed to be ripped apart and hastily glued back together again. True, they were from Iowa, the wholesome core of America, but they were also Vietnam veterans, counterculturalists, born-agains.

Still, no one from Jefferson High's class of 1966 had any news like Randy and Vicki Weaver's. They talked about living on a wooded mountaintop where there were no other people, but where they were in danger from the evil, false government, and the hordes of desperate people living below. They talked about the great tribulation, when Christians would be hunted down simply because of their beliefs, when those who stockpiled food would be the only safe ones. They talked about their children, each with a biblical name, who lived on the mountaintop with them.

Where did all this come from, someone asked.

"We've been having this vision," Vicki began.

three

The lawn would be shaggy, the curtains pulled, and the clothes falling off the line when the neighbors would start to joke that maybe the Weavers had died in that house. Finally, someone would sneak over and find the television on, tuned to Jerry Falwell or The PTL Club, and the family in the living room, curled up on that beige and orange shag, poring over a bunch of mail-order books and an open Bible. The Weavers would finish the Bible lesson and burst outside, doing the lawn and laundry in a rush of Iowan good will and Christian warmth: "God bless you" to everyone who passed.

Fourth in a row of five tidy houses, Randy and Vicki Weaver's white and brick rancher was set back from a tree-lined avenue in Cedar Falls, Iowa. University Avenue was a busy strip of tire stores and pizza parlors that drifted in and out of residential neighborhoods and edged neat lawns. In 1973, the Weavers paid $26,000 for the best home on the block, impressive for a couple still sneaking up on thirty. Like an old-fashioned family, they "came-a-calling" on their new neighbors—Randy with his bushy brown hair and mustache, Vicki with her raven hair, parted in the middle and falling straight down to the middle of her back—both of them impressing everyone with their friendliness and manners. They drove the elderly people in the neighborhood to the store and back and helped them buy groceries. When little Sara was born in March of 1976, most everyone in the neighborhood agreed, she was the cutest and smartest baby, with her mama's seriousness, her narrow, dark eyes, and jet-black hair.

But for Randy and Vicki, something was missing. Sports cars, toys, and Amway didn't give purpose to their lives. Society didn't offer anything better. The antiwar movement had been idiotic and all the hippies and Yippies made a mockery of Randy and Vicki's generation. And now the '70s? Neither Randy nor Vicki could condone such shallow, hedonistic lifestyles. Unhappy with what they saw around them, Randy and Vicki returned to their religious upbringings. Every Sunday, the Weavers drove their Oldsmobile east toward Waterloo and pulled into the gravel parking lot of the Cedarloo Baptist Church, on a hill between Waterloo and Cedar Falls, took their place in the pews, and listened to the minister. But there seemed to be no fire or passion, no sense of what was really happening in the world. They'd tried other churches and found congregations interested in what God had done 2,000 years ago, but no one paying attention to what God was doing right then.

Certainly, churches weren't addressing the crime in Cedar Falls, the drugs, or the sorry state of schools and government, not to mention the kind of danger that Hal Lindsey described. They would have to find the truth themselves. They began doing their own research, especially Vicki. She had quit work to raise Sara, and later Samuel, who was born in April 1978. When Sara started school, Randy and Vicki couldn't believe the pagan things she was being taught. They refused to allow her to dress up for Halloween—Satan's holiday—and decided they had to teach Sara at home. But that was illegal in Iowa.

A booster shot of religion came with cable television

and The PTL Club, the 700 Club, and Jerry Falwell. The small television in the kitchen was on all the time for a while, but most of Vicki's free time was spent reading. She'd lose herself in the Cedar Falls public library, reading the science fiction her dad had introduced her to as a kid, the novels and self-help books friends recommended, biblical histories, political tracts and obscure books that she discovered on her own. Like a painter, she pulled out colors and hues that fit with the philosophy she and Randy were discovering, and everywhere she looked there seemed to be something guiding them toward "the truth," and, at the same time, pulling them closer together.

She spent hours in the library, and when she found something that fit, she passed it along first to Randy, who might read the book himself and then spread it to everyone—the people at work, in the neighborhood, at the coffee shop where he hung out. They read books from fringe organizations and groups, picking through the philosophies, taking what they agreed with and discarding the rest. Yet some of the books that influenced them came from the mainstream, such as Ayn Rand's classic libertarian novel *Atlas Shrugged*. Vicki found its struggle between the individual and the state prophetic and its action inspiring. The book shows a government so overbearing and immoral that creative people, led by a self-reliant protagonist, go on strike and move to the mountains.

"'You will win,'" the book's protagonist cries from his mountain hideout, "'when you are ready to pronounce the oath I have taken at the start of my battle—and for those who wish to know the day of my return, I shall now repeat it to the hearing of the world:

"'I swear—by my life and my love of it—that I will never live my life for the sake of another man, nor ask another to live for mine.'"

Another time, she told friends to read some short stories by H. G. Wells. Most of Wells's popular stories were about time travel, space and hidden civilizations, the kinds of science fiction that had been passed on by her dad when she was young.

But Wells was the author of some lesser known tales, too, several religious and prophetic short stories, such as "A Vision of Judgment," in which a man is pulled from his grave and taken before God; "The Story of the Last Trump," in which a handful of characters fails to see Judgment Day approaching; and the bizarre story, "A Dream of Armageddon." In it, two men meet on a train and one begins to tell the other about his vivid dreams of the future. "'Your dreams don't mix with your memories?' he asked abruptly. 'You don't find yourself in doubt; did this happen or did it not?'"

In Wells's story, the man dreams he is a great leader living with a woman in the future on a 1,000-foot cliff with a view in several directions. On his cliff, men come to him and tell him they are at war with him. "'Why cannot you leave me alone,'" the dreamer asks. "'I have done with these things. I have ceased to be anything but a private man.'"

"'Yes,'" the other man answers. "'But have you thought?—this talk of war, these reckless challenges, these wild aggressions—'"

Later in the story, the man and woman flee, but they are followed—"'There is no refuge for us,'" he says in his dream. They escape down their hill into Naples, Italy, and are followed by airplanes. In the end the couple stays together in the face of horrible danger—"'Even now I do not repent. I will not repent; I made my choice, and I will hold on to the end.'" Finally, the man and woman are killed and the dream, like the short story, ends.

Vicki told friends that H. G. Wells's fifty-year-old stories had a lot of relevance in modern America, especially for someone like herself, someone who had begun getting messages from God while she took baths and who was having dreams of great violence and a cabin on a mountaintop.

• • •

Carolee Flynn was thirteen years older than Vicki but nowhere near as learned about Scripture and the Lord's plan and all that business. When she and her husband, Dewey, moved in next door to the Weavers in 1979, Carolee—a short, Pall Mall–voiced bank clerk and pizza waitress—took them for just a couple of nice kids going about the business of raising a family. Randy and Dewey went fishing a few times, but Carolee and Vicki became the real friends, leaning across their driveways for "How you beens" that turned into forty-minute soul-searching conversations and explorations of faith.

It was Vicki who taught her to coupon, clipping and saving and challenging herself to find the best deal. And garage saleing, too—that was Vicki. But mostly Vicki was her teacher, trying to bring her to the Lord and make her understand how Christianity fit in with everything happening in the world. Carolee was glad to be there for Vicki when she needed a friend, like the time—after Sam was born, right around the time the visions became stronger—when Vicki miscarried. She was so brave and strong and smart, she hardly talked about losing the baby, except at first. And then she just said it was God's will. They became best friends over gallons of iced tea, sitting in the backyard, watching Sammy and Sara play.

Randy was always talking about money, how it was going to be devalued, how the banks were going to collapse. If you had to let someone keep your money for you, Randy said, the credit union might be the best bet. Randy loved to talk. He could charm any room, and he had a new best friend every week—factory workers, cops, professionals, even a black guy he worked with at John Deere who came by the house sometimes. If Randy knew something, he found it impossible not to tell people about it. Everything with him was right or it was wrong, and he tossed off authoritative opinions like so much small talk.

At night, he'd sneak off when the kids were getting ready for bed, when Vicki was in the tub, and walk down University, a Bible in his hands, a cigarette dangling from his mouth, until he got to Sambo's, an all-hours restaurant and coffee shop that filled every evening with the guys from the Deere plant, grumpy retirees, off-duty cops, and neighborhood Christians. Randy was known well enough there that he could pour himself a cup of coffee and cruise along the booths, talking to the people eating a late dinner and to the other restless Christians who came down to witness. He'd warn anyone who listened about current politics, tell them to repent, and launch into biblical prophecy and the coming end time. It wasn't long before he found a group of about ten people who felt the way he did, born-again coffee-swilling Christians who met at the restaurant at night, debating and sharing Scripture. Sambo's became the center of the radical born-again movement in Cedar Falls, and Randy became a spirited recruiter.

For instance, it was Randy who brought Vaughn Trueman to Sambo's. Vaughn owned a gun store—The Bullet Hole—at the other end of University, and in 1980, Randy started coming in, shopping for guns to protect himself during the end time. The first time Randy introduced Vaughn to Vicki, she looked him in the eyes and said she'd seen him already—in a vision—coming to the Lord. They talked more and more about the Bible until something sparked inside Vaughn, and in 1981, he found himself holding hands with one of Randy's friends, accepting the Lord as his personal savior. Such miracles were happening up and down University, and for some people the Lord's work seemed to be centered around that little white and brick rancher.

Another member of the Sambo's Bible study was Mike Roethler, a tall, gentle Cedar Falls police officer who was undergoing his own religious transformation. Roethler didn't know Weaver very well until one day when he brought a homeless man into Sambo's and asked some of the regulars to help the man out. While the others just sat there, staring at their coffee, this short, quick guy, Randy Weaver, pulled out a twenty-dollar bill. Randy was that way with strangers, always bringing home some lost soul to save. He brought them into his house, gave them

clothes and food, even a place to stay. Even after one of
them stole from the family, Weaver brought other strays
home.

That's how Shannon Brasher started coming around.
Shannon was a former Marine who'd been in Vietnam and
was a security specialist at the Deere factory. Shannon was
in the middle of his third divorce and hanging out in
Sambo's when Randy struck up a conversation with him.
He recognized Randy from the plant, and the two became
friends. Later, Randy even helped him root out some
thieves at the plant. Randy said it didn't matter that
Shannon had been married and divorced three times. God
loved him anyway. Soon, Shannon accepted Jesus and
became a regular at the Sambo's late-night Bible studies,
which quickly became too animated—with healings and
people speaking in tongues—to hold in the middle of a
family restaurant.

*T*hey called themselves legalists, because they
believed that the Bible was the literal word of God
and that it all must be taken as the truth, even the
laws of the Old Testament, which many churches treated as
arcane and pointless dogma. One of the beautiful mysteries
of the Old Testament was how often God just up and spoke
to people, told them what to do, when to do it, and—most
fulfilling and frightening of all—what was going to happen
in the future. It made the religion far more dynamic when
God was telling people directly what to do. To the Cedar
Falls legalists, if God's word could come that way 10,000
years ago, there was no reason to believe it couldn't come
that way now. So when Vicki decided her family would
follow Old Testament law and stop eating unclean meat
like pork and oysters ("The Lord says, 'Don't eat it'—He
knows it's got trichonomas and isn't good for your body,"
Vicki wrote to a friend), no one in the group thought she'd
come about the decision from anywhere but Scripture and
His divine will.

There would be anywhere from four to ten people at the

Weavers' house, sometimes as often as four nights a week. Randy led the Bible study most of the time, but everyone read chapters and commented on what they might mean. Vicki was clearly the scripturalist and scholar of the group. It was as if she had memorized the whole thing, from Genesis to Revelation, Acts to Zechariah.

They read only the King James Version of the Bible, because Vicki said other translations weren't divinely inspired and were pagan-influenced. By 1981, the Old Testament books were opening up for Randy and Vicki, not as outdated stories, but as the never-ending law of the Maker. He was opening their eyes to what was happening now, in the United States, just as Hal Lindsey had foretold. The forces of evil (the Soviet Union, the U.S. government, Jewish bankers) were ready to strike at any time against American people. From Ezekiel, they read: "Son of man [Christian Americans], set thy face against Gog [the grand conspiracy]. . .

"Be thou prepared, and prepare for thyself, thou, and all thy company [their Bible study group] that are assembled unto thee, and be thou a guard unto them. After many days thou shalt be visited: in the latter years thou shalt come into the land that is brought back from the sword [somewhere in the American West], and is gathered out of many people, against the mountains [the Rockies] of Israel [the United States], which have been always waste [the desolate mountains of Montana? Colorado? Idaho?]: but it is brought forth out of the nations, and they shall dwell safely all of them."

In that way, the Scriptures spoke to the Bible group in the early 1980s, through Randy and especially Vicki. There were tears and laughter and languages no one understood. Sore backs were healed, and the Weaver house filled with a spirit they'd never felt before. But God was trying to warn them of something darker, too. The couple agreed that all the signs that Hal Lindsey warned of were there: Some force was moving them to action, trying to gather the believers for the coming end time! From Ezekiel:

Thou shalt ascend and come like a storm, thou shalt be like a cloud to cover the land, thou and all thy bands, and many people with thee.

And I will call for a sword against him throughout all my mountains, saith the Lord God; every man's sword shall be against his brother's.

Clearly, they would need weapons. First, to fight the Communists, who would likely come through Canada, and then, once the tribulation started, government agents and nonbelievers who would come for them, and scavengers with guns who would be roaming the countryside. Randy began sleeping with a loaded pistol under his pillow.

The book of Daniel, the Old Testament prophet, sounded an alarm even more shrill:

. . . and there shall be a time of trouble, such as never was since there was a nation even to that same time . . .

And Matthew!

. . . and then shall the end come . . . And ye shall hear of wars and rumours of wars: see that ye not be troubled: for all these things must come to pass . . . For nation shall rise against nation, and kingdom against kingdom: and there shall be famines and pestilences, and earthquakes in diverse places.

Yes! All those things were happening. Randy and Vicki shook news clippings and applied news events to the war, famine, and pestilence litmus test. It all fit! Praise be His glorious name!

*All these are the beginning of sorrows. Then shall
they deliver you up to be afflicted, and shall kill you:
and ye shall be hated of all nations for my name's
sake. And then shall many be offended, and shall
betray one another, and shall hate one another.*

Yes! Already, they had been mistreated at church and
Randy had gotten in trouble for preaching at work. The
persecution was beginning.

*When ye therefore shall see the abomination of
desolation, spoken of by Daniel the prophet, stand in
the holy place: (whoso readeth, let him understand:)
    Then let them which be in Judea flee into the
mountains.*

Matthew 24 stopped them cold. Dear God! There it was,
right in front of them. Vicki shook with the message. It was
so clear! And it wasn't just in their readings.

"We have dreams," Vicki confided in Carolee Flynn.
Randy's visions, especially, were vivid and profound. He
dreamed of a configuration of buildings on a hillside, a
cabin and outbuildings. For her part, Vicki took baths and
the spirit showed her an empty cabin that would need to
be fully supplied for the coming tribulation. There were
cynics in her family and in the neighborhood who
imagined what might happen to a couple who believed
that every loose thought, every inexplicable picture that
popped into the subconscious was a message from the
Creator.

But to Randy and Vicki, the inspiration came only in
close association with the Scriptures. There were dozens of
biblical chapters warning of the evils of images and so,
when He showed Vicki that she should purge herself and
her house of all images, they knew it was true. And so, the

TV—that most devilish purveyor of images—was sold. Many of the children's toys were simply images, too. It wasn't that Vicki wanted the children to give up their teddy bears, but God had spoken and such bears were images of real bears and therefore a disrespect unto Him. Vicki and Randy went through the house that way, getting rid of photographs and coffee tables with images of leaves on them. One day, Vicki gathered up her Bird of Paradise dishes—with tiny, beautiful bluebirds painted on each one, including the gravy boat and the decorative wall dishes she'd scoured garage sales for—and she knocked on Carolee's door. Carolee felt awful, but in the end Vicki pleaded, and so Carolee traded her plain, ugly plates for Vicki's beautiful set. Carolee got end tables and a glass lamp because of God's message.

But the most important message was about the time. Of course, the time! Vicki pored over the Bible and prayed endlessly, begging to know exactly how much time they had before the end of the world, how much time she had to get ready. From Daniel—"Blessed is he that waiteth and cometh to the thousand three hundred and five and thirty days."

And, of course, Revelation.

> And he said unto me, Thou must prophesy again before many peoples, and nations, and tongues, and kings . . . and the holy city shall they tread under foot forty and two months.

Forty-two months! Thirteen hundred days! Three and a half years! Dear Heavenly King!

They had to leave as soon as possible.

Carolee Flynn was amazed by the energy and commitment at Vicki and Randy's Bible studies the few times she attended, but she never quite felt like she fit in. It all flew over her head. Still, she and Vicki talked all the time about spiritual matters, and Carolee believed she was learning a lot from her neighbor and best friend.

One Christmas, Carolee watched the Weavers go about their lives without decorations or presents or anything.

"Why don't you have a tree?" she finally asked. "It's Christmas. Aren't you supposed to be celebrating the birth of the Lord?"

Vicki told her that Jesus was probably born in September, certainly not December. "It's considered a pagan holiday to put shimmery things on a tree," she said.

Even if she didn't quite agree with everything, Carolee learned a lot from the Weavers. At one Bible study, Randy talked about Revelation, chapter fifteen, verse thirteen.

The beast "causeth all, both small and great, rich and poor, free and bond, to receive a mark in their right hand or on their foreheads: And that no man might buy or sell, save he that had the mark, or the name of the beast, or the number of his name . . . and his number is Six hundred threescore and six." Six-six-six.

The beast, Randy said, was a metaphor for computers. Soon, everything would be catalogued on computer: births, schooling, purchases, homes. And every credit card, connected as they are by computers, would mark people with the number of the beast, Vicki said. Of course, once the currency was devalued and finally changed, no man could buy or sell without a credit card, without 666.

Later, Carolee admitted she was confused. "My credit card doesn't have six-six-six on it." Vicki, patient as always, explained that every card would have numbers that were derivatives of 666.

Randy and Vicki were trying to reach others as well, especially their families. Sunday dinners at the Jordisons had become theological debates over the form of the coming tribulation—David's moderate RLDS version vs. Randy and Vicki's survivalist Christian beliefs. The whole family would eat dinner first, David, Jeane, their kids, and a blossoming flock of grandchildren. Then the women would go shopping or form a circle in the kitchen and drink coffee, leaving the men alone to do what they did best together—argue and debate.

There was soon going to be a social breakdown, Randy would say. The government would use the opportunity to declare martial law, crushing democracy and killing the good Christian Americans. People will be rioting in the streets and the traitorous government would turn against its own people. The only protection would be clusters of good Christians with guns—

And that's where Julie's husband, Keith, wouldn't be able to hold his tongue any longer. Reeling over the death of John Lennon, he'd say all guns ought to be illegal, tossing out statistics that showed how many people were wounded by their own guns—

And, on that point, Lanny would have to agree with his brother-in-law Randy about the gun thing, although his beliefs—

His beliefs are all wrong, David would point out. You can't prepare physically for the return of Jesus, only spiritually—

And somehow, farm subsidies would come up, and everyone would switch sides—

And then Randy would throw out something that ended the argument because it was so ridiculous. "Someone ought to kill the Supreme Court justices." Or: "The Holocaust never happened."

That was too much for Vicki's dad, who had been a young man during World War II and who knew real evil when he saw it. "Are you crazy? I was alive then, and I will tell you, it happened." And that subject was closed.

But mostly they argued with Randy about his plan to leave Iowa and move up into the mountains of Oregon, Montana, or Idaho. David had seen that Randy wasn't the most mechanical guy in the world, and he figured the family would starve to death as soon as they got five miles out of Iowa. "You've got a family to think about, you bonehead," David said. But Randy wouldn't budge. And if his in-laws wanted to survive the great tribulation, if they didn't want their children turned into slaves of the New World Order, the Jordison family would be wise to follow them.

In the kitchen, the women rarely talked about such stuff. Julie knew when they stopped celebrating Christmas that her sister was becoming more radical, but she also knew there was no way to talk her out of something once she was so deeply into it. And in every theological breakthrough, Julie saw her sister's personality. So while everyone else in the family argued with Randy about the great tribulation and moving to Idaho, Julie could tell Randy's bluster was coming from Vicki's ideas. They'd make eye contact and she'd give him a small nod or correct some point, and off he'd go again on a wild tangent about holidays being the work of Satan. Once, when Randy was losing an argument, Vicki stormed into the living room and told her dad and brother to leave him alone.

Julie knew her sister, and she guessed there was another side to this transformation. She wondered if Vicki liked the way she and Randy revolved around each other while they were becoming the instruments of the Lord. It had seemed to her that in the mid seventies, Randy had all these other things going on in his life—his job, his toys, his friends— and she knew Vicki was threatened by all of that. While she didn't doubt their sincerity, Julie thought she saw the psychological tracks that led to where the Weavers were going: Vicki with her tight family—alone in the woods— and Randy with his toys—every kind of gun he could afford—and his macho lifestyle, the survivalism.

In 1982, Julie had begun dabbling in astrology, and Keith was still playing in a rock-and-roll band when, one afternoon, Randy knocked on the door. He made some small talk and then got to the point. "You're letting the devil into your lives," he told Julie. He pointed upstairs to where Julie's baby, Emily, was sleeping. "The devil might be up there right now, possessing Emily."

Julie stopped him right there.

"Get out and don't come back, Randy." She was furious.

"It was your sister that sent me down," Randy admitted on the way out. Vicki never said a word to Julie about it.

Julie, like everyone else, realized that the less they said the better. Vicki gave birth to Rachel in 1982, and once again babies were a safer subject of conversation. Despite their growing paranoia, Randy and Vicki were still great parents. And anyway, Julie rationalized, she wasn't going to persuade people who believed God was talking to them. If they just let the arguments wind down, they'd last only as long as the red faces, and then they could move on to less flammable subjects.

By the early 1980s, Julie began a debate with herself that would last for years. She never once sat down with her sister and said, "What are you doing with your life?" She didn't think Vicki wanted that conversation. And so she just let it pass, asking polite questions as the Weavers became more and more excited about heading for the mountains.

Forever, Julie and Keith would talk about whether anything could have been done or said. But at the time, even they weren't sure how far out there Randy and Vicki Weaver were swinging.

*T*he beliefs revealed themselves slyly, one book, one tape, one conspiracy at a time. The Weavers came across the staples of radical right-wing thought, ideas that have been around for decades, in some cases longer, lurking just off the edge of the mainstream, a pattern of marginalization that in itself made the material more powerful because people value what they have to work to find. Like thousands of others, Randy and Vicki Weaver were not stumbling into rehashed ideas and hoaxes, but discovering great truths hidden away from the mainstream, ideas that did a much better job crafting a cohesive and universal explanation of the world than the media's mainstream message. They slipped imperceptibly from Christians to conspiracists.

Few people who don't follow conspiracy theories comprehend their attraction: they create a framework for understanding everything by tying coincidence and accident together. If every event is part of the fabric of the conspiracy, then everything must have a reason, a meaning. And so, when the same unlikely details from one mimeographed pamphlet show up on a tape or in a mail-order book, it comes not only as confirmation but as revelation: "Here it is again! The Illuminati!" For true believers, the conspiracies seem no more unlikely or illogical than other things that are considered truth.

The Weavers uncovered a conspiracy that began in America with the Masons, the kind of secret society the prophets warn about in the Bible, that buried its tentacles in the farthest reaches of government. (Nearly every president had been a Mason.) The tool for this conspiracy, they read, was the Illuminati, founded May 1, 1776, a secret society of socialists that led to the Council on Foreign Relations, and later, the Trilateral Commission—shadowy super governments that pulled the strings of every level of government and law enforcement. All of these groups, in turn, were controlled by evil, money-grubbing Jews. The Weavers bent biblical prophecy to fit their new beliefs, until Vicki and Randy and a few others in the Bible study could imagine the Beast, the many-tentacled, Satanic government spoken of in Revelation. Who wanted to pay taxes to the Beast?

Mail often brought a book, pamphlet, or tape from some obscure mail-order house. One day it was "Satan's Angels Exposed," a sort of clearinghouse for radical right-wing conspiracies connecting everyone from George Washington to Gandhi, with crude illustrations of white Christians being strung up during the great tribulations. Another day, it was comic books for the kids to read on the carpeted floor: *Betrayed* and *Doublecross,* comics about how the Jews had killed Jesus. In the back of one book would be an address for another small publishing house, and that day Vicki would walk across the front yard and drop a letter in the mailbox, requesting more. They listened to a half dozen tapes made by a conspiracist named John Todd, and Randy even arranged for Todd to visit Cedar Falls and speak in a half-empty banquet room at the Holiday Inn. They worked so hard to get Todd into town that Carolee felt bad for Vicki and baked a batch of rolls. But when she brought them over, she saw John Todd pacing in the Weavers' living room, packing a gun, of all things. She told Vicki she didn't like him or the people he attracted, and Vicki agreed, pointing at one person and saying, "Watch out for him. He's a neo-Nazi."

Christianity could be so passive. It was always about someone else, about the disciples, the community, loving thy neighbor. The Weavers saw a vibrant, dangerous world, a judgmental, vengeful God, and churches that lay down in front of evil and refused to do battle. For the Weavers and the Bible study group they had formed, religious experience was active, a trip of self-discovery and a heightened sense of their own place.

Of course, it came to them as a revelation, the place where all this had been leading. Combined with their new sense of self and of divine prophesy, combined with the visions, combined with the inescapable pattern of conspiracy and coincidence in their own lives, Randy and Vicki Weaver began to find it difficult to believe there could be any other chosen people but themselves.

*R*andy slept peacefully, a flak jacket and helmet beside his bed, the loaded gun still under his pillow. But during the day, there was much less peace. Work was becoming intolerable. The guys he supervised were slovenly and immoral, sneaking off to read dirty magazines when they should have been working. His bosses told him to quit preaching and passing out literature to the other factory workers, a few of whom turned and ran when they saw Randy coming. It was no secret that some people wanted him fired. Just as God had shown them, the pressure and suspicions of the world were coming to bear on them because they chose to follow the truth.

For instance, there was Shannon's girlfriend, whose parents didn't like the talk they were hearing. They called the Cedar Falls police department, whose detectives had already heard of "The Group," as some locals called it, and police were quietly looking into whether the Weavers and their friends had formed a cult.

Randy knew someone was watching him, and he became convinced his telephone was tapped. The forces of evil were gathering. They had to hurry. One weekend, the Weavers visited an Amish community to learn how to live without modern conveniences and how to store food for long periods of time. Vicki began dehydrating fruits and vegetables and stacking them against the basement wall. They planned to have a root cellar in the mountains and to hunt wild game, because they knew that in three and a half years, there would be no grocery on the corner. In fact, there might not even be any corner left. Randy and Shannon had been collecting the weapons they would need to hunt and to defend themselves: 30.06 rifles, pump shotguns and Mini-14 semiautomatic assault rifles. Randy bought about ten guns and thousands of rounds of ammunition. At any given time, a half-dozen other people were preparing to go with the Weavers, including Vaughn and Shannon. And Randy seemed bent on warning everyone else in Iowa.

Pretty soon, the local newspaper heard about The Group. Dan Dundon, a reporter for the *Waterloo Courier*, sat down with Shannon Brasher and the Weavers in their living room in December 1982, turned on his tape recorder and, for the next hour or two, listened to stories of the Great Tribulation.

"The Bible teaches us," Randy said, "that somewhere near, during the reign of the One World Leader, God will free the hands of Satan to wreak havoc with the peoples of the Earth."

Dundon asked about the rumors that they were forming a cult, and the Weavers strongly denied it. There may be others going to the mountains, they said, but they weren't trying to form any sort of group. "The ones who seek the truth will accept us, the ones who do not want the truth will call us crazy," he said. They called themselves Christian survivalists, and Randy and Shannon talked about the strategic home they would need and a plan for defending it. They talked about a "kill zone," a 300-yard defensible space around the home.

When the story came out, the Weavers were upset. It didn't calm the rumors floating around but ignited them. Vicki told friends that the reporter quoted every wild thing they said and ignored the fact that, at heart, they were still a quiet, wholesome family, good Christians just trying to find their way in the world. They denied saying anything about a "kill zone." Since the Weavers wouldn't allow themselves to be photographed, next to the story was a drawing of a Bible and two bullets.

"We're servants," Vicki said in the story. "And what the Lord tells us to do, we will do. He has told us we have to pull up our roots and leave. I don't want to leave my home, but if we are obedient, then He will protect our children."

*T*heir friends were slowly falling away. Shannon, who had moved in with the family, had gotten into a disagreement because Vicki and Randy made his girlfriend stay in a different room. It became clear Shannon wasn't going with them. Another family that had considered going decided against it, too.

And Vaughn was starting to have doubts. His wife didn't like the idea, and besides, he didn't have the money to stockpile food and guns. Then, in February 1983, driving past a church he'd never seen before, the gunshop owner stopped his car, walked inside, sat down in a pew, and asked for God's guidance. Seated next to him, his wife started crying. He talked to the pastor of the church and realized God was telling him not to go.

The Weavers didn't push him. Randy said he would honor Vaughn's own vision and that it wasn't for him to judge. Vicki admitted that in her visions of the mountain, she didn't see Vaughn anyway. The gunshop owner began going to the new church and accepted that the birth of Jesus meant people didn't have to live the Old Testament law anymore. Like everyone else in the congregation, Vaughn came to believe that he was saved by grace, not by following the impossible law of the Old Testament.

Cedar Falls police officer, Mike Roethler, had come to the same conclusion. He still liked Randy, and they got together to talk, but Roethler could feel his friend moving in a different direction. Once, Randy leaned forward and told Roethler that Jews were the product of Satan. "What? Jesus was a Jew," Roethler said. "How can you believe that?" Roethler kept witnessing at the old Sambo's, but his message had changed. More and more, it seemed as if Randy and Vicki were alone.

In March of 1983, Carolee watched Randy and Vicki pound the "For Sale" sign into their front yard. Oh my, she thought, they're really doing this. They held garage sales every weekend, Randy standing in the front yard with a Bible, selling everything that wasn't coming to the mountains with them. A few weeks later, Carolee and Vicki sat out on the back swingset, motionless in the still summer air.

"You're really going?" Carolee asked. Vicki was exhausted. She had lost fifteen pounds getting everything ready for the trip, taking care of the children, finding exactly enough clothes to last the three years before Jesus came back.

"How am I going to learn any more about God without you?" Carolee wondered.

Vicki told her to relax. Everything she needed to know was in the Bible. "The King James Version, not the standard version," Vicki said. "God's word doesn't change."

But the King James Bible, with all its "thees" and "foresakens" was so hard for Carolee to understand. "It's easier the other way," she said.

"Not if you're a true believer," Vicki said. Before she left, she gave Carolee a King James Bible, with especially large type.

Carolee said she'd never pick it all up. Inside, she knew she just didn't believe the same way the Weavers did, and to be truthful, she didn't understand all this Illuminati and Freemason stuff.

Vicki swung slowly and looked over at her friend. "Write down every question you've got and I'll try to answer them."

They talked about the things the Weavers would need to survive, and Carolee admitted she was worried about them. How would they eat? Where would they work?

"I'll see you again." Carolee tried to sound hopeful, for herself as much as anything, but, as she often did these days, Vicki seemed distant.

"You know Carolee, for our beliefs we could be killed. For our beliefs."

*T*hey traded their car for an old moving truck, a long, blue grain truck with a tarp thrown over the top that they backed up against the front porch. Sammy—who was a sickly kid—fell off the truck as soon as they got it and broke his leg. So the children stayed with Carolee while Vicki and Randy loaded the truck. "We goin' to da mou'tains," little Sammy said, and it was all Carolee could do to keep from crying. Meanwhile, Randy and Vicki worked themselves ragged packing food, clothes, and a little bit of furniture. Mike Roethler came over and helped Randy push an old woodstove to the back of the truck. They loaded guns and ammunition, kerosene lamps, everything that a family heading west might have taken eighty years earlier.

They got $50,000 for their house and cleared $20,012. When Randy showed up to get his money and they tried to give him a check, he asked for cash. "This is just a piece of paper," he said, holding the check. "It isn't worth anything." As he left, he said he was going to transfer the money into gold and silver.

The Sunday before they left, the Weavers drove to Fort Dodge for one more Sunday dinner. Lanny stood outside and barbecued steaks, and this time, there were no arguments.

Always the pragmatist, Vicki's dad offered advice. "Gear down when you're driving through the mountains."

Randy promised he would.

In the kitchen, Vicki said they were going to find a place in Idaho or Montana where home schooling was legal. God would show them where.

Jeane was going over everything. Would they have enough fruit? Vegetables? Where would they stay on the way out? Did they have enough clothes?

"I've got enough clothes for three and a half years," Vicki said. "That's all we'll need."

Still, Julie didn't talk about it with her sister. Maybe they needed to get out in the woods to settle down and return to reality. Surely there was nothing wrong with becoming less materialistic and more spiritual. "If it doesn't work out, you can always come back," she said.

Sammy, who was four, was excited for the adventure, hobbling around on his cast. But Sara seemed sad; she was seven and old enough to know what they were leaving behind—security and loved ones. Julie watched Rachel, who was still a toddler. You won't know me, she thought, but I will always be your aunt.

David and Jeane were going on vacation, and so they said good-bye. David gave one more warning to Randy that he'd better take care of his family, and then Randy and Vicki headed back to Cedar Falls, to finish packing.

A few days later, they stopped by Fort Dodge again on their way out of town. Randy was driving the moving truck, and Vicki followed him in a pickup that pulled a trailer behind it. They eased off the interstate and parked in Lanny's front yard. Randy climbed out of the moving truck in jeans and T-shirt with a cigarette pocket. Vicki got out of the pickup truck in one of the long denim skirts she'd taken to wearing. They stood on the porch with Lanny and Melanie, Keith and Julie.

Julie looked at the moving truck, loaded down with every possession they hadn't sold or given away. It reminded her of the Okies moving west, of *The Grapes of Wrath*, and she couldn't believe her smart, perfect sister had been reduced to this.

Julie hugged Randy and the kids, and they made their

way to the caravan until only Julie, Lanny, and Vicki were left on the porch.

Vicki and Julie cried as they hugged, and Julie held her sister tight. "I'm just so afraid I'll never see you again."

"I'm never coming back," Vicki said. "You'll have to come see me."

Julie cried harder and Vicki tried to comfort her. "Don't worry. We'll see each other again."

Vicki hugged her brother, Lanny, too. There never was so solid an Iowa farmer as Lanny Jordison, and he wasn't one to listen to intuition, but as he and Julie watched their sister drive off, they both had the feeling they would never see her again.

four

Vicki Weaver was worried. Here it was, the first of September, and they seemed no closer to their mountain sanctuary. Soon winter would come, choking off the mountain roads and making it too late to begin building anything at all. Jesus kept telling her to be patient, but it was trying. There was just so little time. God had made it known to Randy that they would find a place by the Feast of Trumpets, which fell on September 7. That didn't seem likely now. All through Montana, Randy and Vicki had looked for undeveloped land, but it was $1,000 an acre and more, a price that didn't even take into account that you'd have to drill for water on most of the property. They continued west across the Idaho border—checking the rearview mirrors to make sure they weren't being followed—and found North Idaho more affordable. Land was promising around Wallace, an old mining town in the middle of the Panhandle. But every hillside in Idaho that looked like a good place to wait out the Great Tribulation was owned by the government and marked with National Forest Service signs. All this remote wilderness and none of it for the people. It was a chilling sign of what Vicki and Randy already believed. They rumbled further west in the old moving truck, staying at motels with kitchenettes, where Vicki cooked soup and turkey dogs for the kids.

Finally, on September 1, 1983, after being on the road more than a week, the Weavers' moving truck happened to roll up Idaho State Highway 95, past range land and thick forest, into the town of Bonners Ferry. They found a motel and began asking about land and the possibility of jobs. That first day, the Weavers met a nice family—three kids and a couple who seemed to share their apocalyptic beliefs. The children all played together, and the parents politely asked the family to dinner. Vicki wondered if perhaps God had delivered them after all.

"There are a lot of people here who say they're Christians and that the Lord sent them here," Vicki wrote home. "They just smile and don't think we're crazy at all! They say they don't really understand what the Lord is doing up here." A local saw mill was adding a shift around the first of October and so Randy applied, and he and Vicki set about looking for somewhere to live. They visited real estate agents and looked at land, but it was all too expensive.

"As for me, I got really rested up the 2 weeks plus it took for the Lord to guide us here," Vicki wrote home. "My Lord Jesus Christ knew how hard I worked before we left and knew how badly my body and mind needed rest. I was so tired and I was forced to rest on the way out here."

Finally, on September 6, the day before God had insisted they have a place to live, Vicki and Randy found it. Their friends drove them south of Bonners Ferry, about seven miles, to a dirt road called County Road Number 12, which jogged off the old highway and etched its way up a steep hillside. A couple miles up the road, at a mountain meadow, they turned off the road onto a primitive logging path that ended in a stand of trees. For several hundred yards they walked up the wooded hillside until they found the spot their friends wanted to show them. It was the view that hit them first, profound and familiar, as if they'd known it all their lives.

"When we drove up to it, Weaver couldn't believe it," Vicki wrote. "It's just what the Lord showed him it would look like. The only way buildings will fit is the way the Lord showed him last August."

The Indian summer sun beat down through gaps in the washboard clouds, and Randy and Vicki Weaver stood on the rocky bluff, looking out thirty miles to the south, to Sandpoint, and twenty miles to the north, toward the wooded Canadian border. In between was a glacial valley, laid out before them with green pastureland and stands of ponderosa pine, white birch, and buckskin tamarack. It was a view worthy of the Bible, Ayn Rand, and H. G. Wells. All around the bluff there were boulders every few feet, excellent places to defend the hilltop, should that become necessary. There was a spring on the property, with fresh cold water, and Randy and Vicki talked about tapping it to bring running water into the cabin. There were fifteen acres for sale, at $500 an acre—$7,500 total—a great price considering Californians were already running up land prices all over the Northwest. God really had delivered them, and Vicki and Randy felt better than they had in months.

On the way down the hill, the Weavers and their friends stopped at the meadow about a mile from the land they'd just seen and met another family trying to scratch together a home. Wayne and Ruth Rau were working on their roof when the Weavers pulled in. The Raus were from California and, like a great number of people who were coming to North Idaho then, lived in a trailer while they finished their log cabin.

Everyone who met the Weavers insisted they had to meet another couple who lived up on that ridge, Arthur Briggs and his wife. They lived in a trailer between the Raus' place and the land the Weavers were buying. Like the Weavers, Arthur Briggs was a legalist, a follower of Old Testament law. But the Briggses were out of town when the Weavers arrived and they made plans to meet them the following week.

Wayne and Ruth Rau didn't think much of the Weavers' religious beliefs. But the new family was planning to home school, which the Raus did, and right away they liked the Weaver children. Vicki seemed nice, but the Raus didn't think much of Randy, who talked melodramatically about the trouble that would come and the need to be prepared. Vicki agreed with her husband, and they made eye contact as Randy let forth with his view of the world. But she also tried to quiet him when he got too animated by saying gently, "Aw, Weaver." He called her "Mama."

At dinner with the Raus that week, Randy was adamant about the Great Tribulation, how the government would turn on its people, and bloodshed would be visited upon white Christians. Then he talked about the beautiful piece of land he was buying and said, "Armageddon's gonna end on that hill."

*R*uby Ridge juts out over the town of Naples, Idaho, eight miles southwest of Bonners Ferry, like a proud, tree-covered chin. Some locals insist the knob is really part of Caribou Ridge and produce old maps to confirm it, but the Forest Service lists Caribou as another rocky knob that rises just across the creek.

Either way, it was Ruby Ridge that stuck, the peak misnamed after Ruby Creek, which was misnamed when a prospector found a strange red gem and decided there must be several tons more where that came from. These creeks, which feed from the mountains into one another like the veins of a leaf, were the scene of some mad panning around the end of the last century, before locals figured out logs were the most substantial treasure this far north in Idaho. But the good names were gone by that time, and so Gold Creek runs parallel and just a few miles north of Ruby Creek. For such grizzled, hard-luck men, the prospectors were certainly optimists.

The woods around Naples are so rugged, and the roads and streams so crooked, loggers who named one strip of water Twenty-Two Mile Creek had to change the name to Twenty Mile after they'd straightened the road enough to actually measure it.

The land, of course, is older than any names, billion-year-old metamorphic rock giving way to rising granite masses that seventy million years ago formed a valley framed by the Selkirk and Cabinet mountain ranges.

There were already trenches between the mountains when 7,000-foot-thick slabs of ice—Pliocene glaciers—oozed in between the granite peaks and carved some of the deepest freshwater lakes in the world, like 1,300-foot-deep Lake Pend Oreille, which still cuts away pieces of mountain as if to remind where the real power lies.

The first white settlers here found a band of northern Kootenai Indians, a tribe with a very musical language and a rich diet. They lived in mat-covered houses and made occasional war with their Flathead cousins to the east, but they spent most of their time fishing the lakes and streams for salmon, trout, suckers, and sturgeon and hunting the ridges for sheep, goats, grizzly, deer, elk, and moose.

And caribou. The last American caribou once roamed here in Boundary County, until, like everything else, the endless flow of people and logging and modern life—more steady and eroding than any ice age glaciers—pushed almost all of them out, except in a few remote places too hardscrabble for most people to settle. Places like Caribou Ridge, or, as it became known as the memory of caribou began to fade, Ruby Ridge.

Randy and Vicki Weaver hunched over the table of a motel room in Bonners Ferry on September 8, 1983, mapping out the house they were going to build. It was going to be big—Randy figured forty feet wide by fifty-five feet long, a one-story log home with lots of storage space for everything they would need during the end time. Three bedrooms, a bathroom, a workshop, a sewing room, and a pantry. The kitchen and living room would be one open area. It seemed extravagant to Vicki, especially if they only needed it for two or three years at the most. But Randy explained that they would be building on free rocks on the mountaintop and that the logs on their land would be free. Like every other house around there, the roof would be metal sheeting, so the snow would slide off it.

The only real cost will be floorboards and roof boards, Randy said.

Luckily, God had shown Vicki an empty cabin, and so she had brought everything she could think of to Idaho. Still, there was so much to buy, and everything out here was expensive and in short supply. Sometimes, the pressure of preparing for the end was almost too much for her. She was still exhausted from preparing for their trip out here, and now, it became clear, the hard work was only beginning.

By October, Arthur Briggs let them move into an eight-foot-by-fifty-foot mobile home about a mile from their new land, and they prepared for their first mountain winter. Sam's broken leg had healed, and as soon as he got his cast off, he and Sara played with the Rau kids and with a nanny goat named Amanda who followed them everywhere.

After some wrangling over the land, the Weavers ended up with twenty acres and the owner agreed to part with the land for $5,000 and the moving truck they'd brought from Iowa.

That fall, the Weavers got up every morning, packed a lunch, and drove or hiked up the steep logging road and then through the wooded field that led to their land. They built a slash pile of old timber and brush, set it on fire, and worked at clearing off the rocky point where the cabin would go, taking some chunks for firewood, others for the slash piles and a few that they saved for the building itself. The guy selling the land had a Caterpillar bulldozer and he agreed to scratch out a sort of driveway from the logging road up to the bluff.

"Sara and Sam are in school," Vicki wrote. "My school: Readin, Writin, Arithmetic and the Bible." The kids pored over the old textbooks Vicki had brought with her, and the precocious Sara, especially, seemed to thrive, quickly reading books well past her seven-year-old reading level.

One night that fall, an earthquake jostled much of northern Idaho. The Weavers were far from the epicenter, and they slept through it. But the next night, Vicki went to town and called her mother in Iowa, who told her they'd heard about an earthquake on the news. It was a sign: "That's just one more of the birth pangs of Matthew 24," Vicki wrote in a letter to Carolee. Matthew 24 promised "famines and pestilences and earthquakes in diverse places," before the Great Tribulation. Matthew 24 also promised wars, and Vicki said the invasion of Grenada was another clear sign the end was coming. Vicki urged her parents and friends back in Iowa to keep a notepad handy and write down all the news events, so she could apply them to biblical prophesy and her visions from the Lord.

It was becoming clear to her how the end would come about. "I still think the Russians are going to invade the United States from Canada," she wrote.

> Probably all the way across our border. We've heard they built a highway down from Alaska that's 10 lanes wide. (To carry an army, perchance?) They're building a brand new 4-lane bridge in Bonners Ferry which connects Canada traffic to the U.S. through Idaho. The old bridge was O.K., but probably not big enough for an army (only two lanes).

On November 6, Randy and Vicki celebrated their twelfth wedding anniversary, and although she was always tired now, for the first time in years Vicki was beginning to feel safe. But she missed her family, missed friends like Carolee. A few days after their anniversary, Vicki sat down and wrote Carolee a twelve-page letter, telling her not to worry about her insecurities over her faith and to keep up in her battles trying to follow the path without Vicki.

"You are fighting the whole world," Vicki wrote.

The whole world lies in wickedness and we fight spiritual battles like the one I just explained. . . . I spent my whole adult life in that house next to you—but the *house isn't my God*. "Jesus Christ the Lamb" is and whithersoever He goeth and sends me I will follow Him. Even though sometimes it hurts. We're all gold that has to be tried and refined through fire. The Lord is really perfecting (or trying to perfect) patience in Randy and I right now—because we're anxious to get busy and can't really yet.

Please take care—We are fine and healthy and happy. May the Lord watch over you both.

Love from all of us.

Vicki

*F*ear thrives in Boundary County. It takes form in tall mountain grass, peers down from granite crags, and waits in shaded creek beds. It jostles and pops along dirt roads and stares unflinchingly through stands of pine. Fear is the last cash crop left in North Idaho, the last big predator, the last roadside attraction.

From all over the country, fear dragged people away from cities and into the mountains of North Idaho. In the early 1980s, when the reasonable folk of North Idaho noticed all these strange newcomers, they quickly saw the hatred: of racial minorities, of the government, of the decadent society. But what many failed to notice was the fear, the choking paranoia that made young, reasonable families seek out a place where they felt in control of their lives again.

There were no zoning codes in Boundary County. No sewer. No fast-food chains. No building codes. Not even a stoplight. No one flinched when a man walked into a store wearing a pistol on his hip. The state itself held just more than a million people—only 3,000 of them black—in an area as big as New England. Such places have always attracted recluses, but until the early 1980s, those people were coming from the other end of the political spectrum: hippies, draft dodgers, an entire back-to-the-land movement.

But Boundary County doesn't discriminate. Anyone can hide there. In fact, the county—like much of North Idaho, like much of the West—always attracted people whose only common trait was the overwhelming desire to just *get away*. Sometimes it was more than a desire. The convicted spy Christopher Boyce found support and a place to hide in Boundary County, and there were always others trying on new names and identities. Boundary County defies stereotyping. It is the home of survivalists, but also of pacifist Mennonites. Democrats usually win the elections, but most residents would probably tell you they're conservative. Left and right swing out as far as they'll go, and then connect in Boundary, where people take the opposite political tracks to the same conclusion—that they want to be left alone.

In the early 1980s, it was Randy and Vicki and people like them who were looking for Boundary County and places like it, looking for a ridge top on which to hide out and build a life. A blurring continuum of home schoolers, Christian survivalists, apocalyptics, John Birchers, Posse Comitatus members, constitutionalists, tax protesters, Identity Christians, and neo-Nazis found one another at the army/navy surplus store in Sandpoint or the barter fair in Northport or the bookstore at the Aryan Nations church at nearby Hayden Lake, Idaho. From California, Florida, Indiana, and Iowa, they talked of reading the same things, coming to the same understandings, and they picked up beliefs and ideas from each other. For many, it was confirmation of everything they had been thinking in the wilderness of civilization. "If they believe it, too, it must be true."

The Weavers had a close group of friends, including the Tanners, whom they'd met their first week in Bonners Ferry; Terry Kinnison, an Indiana transplant who, with his wife, shared similar beliefs to the Weavers; and the Kumnicks, Frank, a janitor and handyman from Florida, and his wife, Mary Lou.

There aren't a lot of rules in Boundary County, but there is this one: No house is ever finished. The Weavers watched people who spent three, four, five years on the prototypical log homes, and they realized they didn't have time for that. No one did anymore.

Happily, it was a mild winter, and the Weavers made good progress on their cabin, although it changed quickly from their original plans. Instead of being forty feet by fifty-five feet, it was twenty-five-by-thirty-two, with a sleeping loft above the main floor. Inside, the house was beamed with logs as knotty and bent as arthritic knees. The entire house was built up on top of vertical timbers, as if the family were awaiting a great tide to wash right up underneath the house. Randy didn't work during that time, except on his cabin, and the money from their house in Iowa was quickly running out. But it didn't matter much to the Weavers. Their land was paid for, their cabin was coming together, and the only money they needed was for kerosene for their lamps. Soon, they said as they worked on their cabin, money would be useless anyway. They nailed the two-by-four frame together and, instead of using logs for the walls, put up a rack of uninsulated plywood, set the windows in, and hammered the metal sheet on top for a roof. It was not a house built to last more than a few years. But Randy and Vicki didn't figure they needed much more time than that.

The snow melted earlier than usual, and by early February the ridge was clear. It was a beautiful spring, and by the middle of March 1984, they were ready to move in. It was a common sight in those days to see a line of trucks negotiating some muddy mountain road, delivering the necessities of life to a family that would live without phone and running water. With coaching and help from Terry Kinnison, Randy bought a horse to help with the logging, and then he built a corral. Vicki unpacked and got the house ready, setting her dishes out on the L-shaped kitchen counter that she and Randy and their friends had built. Finally, they were in their mountain retreat. Sammy and Sara loved it. They fished and played in the woods and said they never wanted to go back to Iowa. But two-year-old Rachel had a tougher time and refused to call the primitive cabin home.

Finishing the cabin was rewarding, but Randy and Vicki were worn down trying to get ready. It was worse knowing approximately when and how the world would end. "The past six months have really been a trial," Vicki Weaver wrote.

> None of it has been easy. But in every little area, I see the Lord gently pushing and opening paths to enable us to do what we must do.
>
> I can't help but think things are shortly going to come to pass. I feel like we were utterly put in a situation where we *had* to get this house built quickly—otherwise we would have taken our time. (But there may not be time.)

The Weavers resumed their Bible studies, this time on Friday nights with their new friends, who believed, as they did, that Jesus was the savior of Israel—whose people were, of course, really American Christians—and that they should obey the Old Testament, especially the Ten Commandments.

If the Weavers were uncertain about the role of racism in their beliefs in Iowa, they had no misgivings about it in Idaho. They believed God was telling them that Jews and gentiles, blacks and whites, Asians and Indians should all be separate, and that mixing was forbidden by God—as represented in a description of the Hittites in Deuteronomy, chapter seven, verse three: "Thou shalt smite them and utterly destroy them; thou shalt make no covenant with them, nor show mercy unto them. Neither shalt thou make marriages with them; thy daughter thou shalt not give unto his son, nor his daughter shall take unto thy son."

Other justifications of racism came from biblical apocrypha, the so-called lost books of the Bible. The number of books accepted as canon varies from church to church, but the Weavers began believing in Christian Identity, a racist religion that accepts obscure, ancient writings that few churches recognize as legitimate, like the book of Enoch, in which dark angels descended to earth, took human wives and "the women became pregnant and gave birth to great giants whose heights were three hundred cubits [about fifty feet]. These giants . . . turned against the people in order to eat them. And they began to sin against birds, wild beasts, reptiles and fish . . ." Many Identity Christians say that was the beginning of the African race, which they called "mud people."

Quickly, the Weavers fanned through the views of the radical people they met, accepted the ones that fit their ever-changing philosophy, discarded the ones that didn't, and worked to become the most studied, serious, and spiritual racists in all of Idaho.

V icki was getting worked up again by current affairs and the conversations she was having with other people in northern Idaho, many of whom were further along in the track of radical right-wing belief. The election was coming up, and Vicki wrote her friends in Iowa for more information about "this Gary Hart person," who she believed might have been foretold by one of the prophets. She found herself drawn to Daniel, chapter five, wherein the original Babylon is destroyed. The Hebrew phrase "Mene, Mene, Tekel" appears in Daniel, and it began popping into Vicki's mind "when I least expect it. I think Jesus is telling me the U.S.A. is going to fall soon."

The first sign, the Weavers and their friends agreed, would be the devaluing and changing of money right after the November 1984 election.

"I prayed in Iowa that He would let me know so I could finish getting supplies before our money was useless," Vicki wrote. "We've heard they'll give you a new dollar for every ten of the old. It'll hit people with savings and the retired the hardest. The factory workers with debts will be fine. They'll just make their new phony money every week."

Toward the end of every summer, Vicki's parents drove 1,500 miles west to visit the Weavers in the mountains. That first trip, David turned off the highway, crossed Ruby Creek and started along the dirt road, up the mountainside, until they reached a clearing, and he figured, this must be it. But according to the directions Vicki had sent, they were still a mile away. That last mile was nearly impassable, an old logging road that David thought couldn't get any worse until it would jog around a tight switchback and be even more rutted, even more overgrown. Finally, they came to the rocky point where Vicki and Randy had settled, and one thought crept into David's mind.

"Why in the world," he asked, "would anyone build a cabin up here?" The view was incredible, but it was no place to live. Why not build the cabin down below and walk up here to look at the view, he asked Randy. The house was built on logs leaning out over the bank, and the logs weren't even tied down. With all the nice, accessible land below, his daughter and son-in-law had chosen an awful, rocky point and built a house that looked like it was going to slide right down the cliff.

They unloaded the supplies and books they'd brought and helped work on the house. David even tied down the logs on which the house was built. While they still disapproved of Randy and Vicki's lifestyle and their ever-hardening beliefs, they reached a sort of truce and enjoyed their visits, watching the kids chase each other through the woods and fish for palm-sized trout in the mountain streams. Sara and Sammy would wait hungrily for their grandparents and then run off with whatever books they'd brought, often finishing them before David and Jeane had even left the mountain. On those warm summer nights, Vicki and her mom stayed up until 2 a.m., canning and sewing and just talking about things. Vicki was clearly the one who kept things going up there, Jeane thought. On those summer trips, she began seeing those same qualities in Sara, who was growing up like a piece cut from her mother.

Randy got jobs on a road crew, splitting wood and running errands. His longest stint was at Paradise Dairy in Bonners Ferry, where he spent a year driving farm equipment and doing chores. He was a good worker and didn't talk much about his beliefs, but there was so much to do on his own cabin, he barely had time for a job.

*I*n 1984, Randy brought home another stray. Kevin Harris's father had died when he was two years old. His mother was only twenty-two and expecting her fourth baby, and it was difficult for her to raise children when she was growing up herself. Kevin had trouble in school in Spokane, was placed in foster homes, and began running away from home at nine. He lived with some mutual friends in Idaho for a while, and that's where he met Randy and Vicki—when he was fifteen. Kevin told them he needed help getting off the streets and away from drugs. They said he could stay with the family whenever he wanted. On Ruby Ridge, he found a family as warm and structured as he'd always wanted, along with horses, dogs, chickens, and guns. As Randy began to share the family's beliefs with him, they made sense to Kevin and explained the world in a way he'd never thought of before. Even though he'd known black and Jewish kids in Spokane, Kevin began to agree with the Weavers that the races ought to be separate. He was as quiet and loyal as a German shepherd, an impressionable young guy looking for a father figure like Randy and for someone who saw him as a big brother, as Sara, Sammy, and Rachel did.

His mother, Barb, a manicurist in Spokane, and his stepfather, Brian, a paralegal for the county prosecutor's office, were glad Kevin was staying somewhere and was off the streets, but they couldn't believe he'd fallen for that ridiculous religion. The Pierces visited the cabin once and were impressed by how close the family was and glad to see Kevin gardening and baking. But they were sickened by Randy's beliefs, as he explained that black people were created as a slave race and that Jews controlled all the money in the world. Brian had to keep from laughing when Randy told him the Jewish hook nose was "the mark of Cain."

Over the next nine years, Kevin Harris would drift around the western states, working as a logger and a laborer. He'd come and go from Ruby Ridge, often walking eighty miles from Spokane on blistering feet. He'd be gone for a couple months and then he'd show up again, always seeming to know when the family was feeling persecuted and needed his strong, quiet friendship.

*B*ack in Cedar Falls, Iowa, Carolee Flynn read the letters from Vicki and worried the family wasn't getting enough to eat. Across town, Vaughn Trueman read his own letters from the Weavers explaining how they'd come to believe that only white Christians were chosen and how Jews were the spawn of Satan. He still thought the Weavers were the kindest, hardest-working people he knew, but he wished they would find the peace of mind that he'd found in the love of Jesus. Mike Roethler got a phone call from Randy, and even though his own spirituality seemed static since the end of the Bible study sessions, he felt bad that his old friends had become mired in hatred and fear. He wondered how he could have been on the same track at one time.

In Fort Dodge, David and Jeane reported to the others that Randy and Vicki really were living on a mountaintop. But it seemed they were doing okay and sometimes the apocalyptic talk subsided, although it was being replaced by more fervent racism and survivalism. Jeane hoped they would just settle down and eventually move down the mountain and into town. David hoped they didn't freeze to death once winter came.

Julie and Keith tried to comprehend that the Weavers' racial beliefs had gotten even more radical, but they couldn't. They cringed at the descriptions of Sara and Sammy learning to shoot guns. Julie just couldn't believe that Vicki would let her babies fire guns! They had seen

Randy and Vicki obsessed and overwrought about everything that caught their attention, but it was hard to imagine that they'd become the thing that liberals like Julie and Keith simply had no use for: outright racists.

Julie remembered how Vicki had come out of the bad times, and she just hoped that—once she and her husband realized the government wasn't coming after them—she would come to her senses, drive down the mountain, and come home to Iowa. Maybe, she hoped, the peace of country living would bring her sister back from the edge.

*C*ity people like to believe that once you get out to a place like Boundary County, where you might be a half-mile from your nearest neighbor, everyone gets along in a sort of Mayberry bliss. Most rural sheriffs will tell you that's not the case. There are always disagreements, land disputes, misunderstandings, and grudges. Oftentimes, the sheriff can't even figure out who started it or what the fight is about.

Almost from the time they arrived, the Weavers were at the center of those kinds of feuds. Seemed every six months or so, the sheriff was called up to Ruby Ridge to talk to the Weavers or one of their neighbors. Someone was shooting guns at night. Someone else ran pipe over that person's land. Those people dumped garbage on the road. That guy cut firewood off this guy's property. This guy shot that guy's goat.

Sam Wohali was a judge for the Kootenai Indian tribe who moved his wife and seven children onto land near the Weaver cabin in the summer of 1984. A big guy with waist-length braids, he was always being told he didn't look Indian. That summer, Randy helped Sam build a fence around his place, but their friendship lasted only a couple of weeks. Wohali—who said he was half Indian, half German Jew—didn't appreciate hearing about Randy's racist beliefs, and he got tired of all the gunfire coming from the Weaver cabin. Once, when a border patrol agent was visiting Wohali, the judge heard automatic weapon fire come from a stand of trees where Weaver and another man had just driven.

By August, Sam Wohali had had enough gunfire and racist talk. He saw Randy standing in the Naples general store with a bunch of Randy's friends, who also wore fatigues and sidearms.

"Can I talk to you in private?"

Randy said no.

Wohali told him to stop shooting at his house.

"What'll you do if I don't?" Randy asked.

"I'll kick your ass," Wohali said. He was furious. "Am I correct in understanding that you believe the end of the world will come sometime in the next two months, that this will start to take place—something between the blacks and the Hell's Angels—and that it will end up at your house, on your front door, and that it will be the start of the end of the world?"

"More or less," Randy said.

"And you will have to take your groceries by violence and shoot your neighbors?"

"More or less."

"Randy, whatever God you serve, when you come down that mountain, and He tells you it's time to shoot your neighbors, turn left at my driveway, don't turn right. Because if you come to my house . . . I'll tie you up by your big toes, I'll cut off your fingertips, and I'll drip you dry, you sucker. I'll torture your ass. Don't screw with my family. That goes for you and all of you G.I. Joe haters. I'm tired of it. You're not going to intimidate me."

Later, Sam Wohali and Weaver made up. But Terry Kinnison never got over his problems with the Weavers. He said Randy sold him half interest in the Weavers' twenty acres for $3,000, just so Terry would have a place to move a trailer and build a barn. And then, Kinnison said, Randy kicked him off the land. Randy said it didn't happen that way at all. Whatever happened, Kinnison and his family moved up to the mountaintop and then quickly got into a fight with the Weavers.

Finally, an argument over some firewood nearly started a fight between the two men when Randy came into the barn and told Kinnison he was a liar and a cheat. Terry held up the hammer in his hand but decided against it. "You're not even worth it."

After that, Kinnison moved off the mountain, filed a lien for the $3,000 he said he'd invested, drove into Bonners Ferry, and told the sheriff that Randy was nuts and was possibly even going to kill the president.

Kinnison wrote letters and spoke to the FBI, the Secret Service, and the local sheriff, claiming Randy had threatened President Ronald Reagan and Idaho Governor John Evans. He said Weaver had a cache of guns and ammunition, including automatic weapons, and had rigged his driveway with bombs. Wohali talked to the agents, too, and said he'd heard many of the same things, that Randy's wife was "a crack shot" and that anyone who went up there would face three armed people: Randy, Vicki, and six-year-old Sammy. The federal agents interviewed several other friends and neighbors but had to wait to interview the Weavers because a late winter storm snowed the family in. Finally, though, the Weavers drove down the hill to talk to federal agents.

So this was how the end would come, Randy and Vicki realized, through betrayal by their former friends, who brought an invasion of federal agents—toting guns and extending the rule of the Jewish-controlled government. They had chased Randy out of the John Deere factory, bugged the family's phones, and now were accusing Randy of threatening to kill the president, just so the family would be killed or scattered and Terry Kinnison could get their land. There really was nowhere to hide.

As they always did when trouble began, Randy and Vicki—ever the good executive secretary—sat down to write a letter, official and precise even in her feminine, cursive handwriting.

Thursday, February 28, 1985

To Whom It May Concern:
We are the victims of a smear campaign of our
character and false accusations made against us to
the Federal Bureau of Investigation and the
United States Secret Service by some local
residents who have a motive for my decease.
These local residents are named:

They listed the Kinnisons and three others.
She wrote the names of the FBI agents and other
officials they'd gone to for help and noted that they might
have to defend themselves from attack. She wrote that
there were witnesses to a conspiracy against them and said
the Secret Service was building a fraudulent case against
Randy because they didn't like his beliefs. Years later,
when a reporter uncovered the old affidavit, it seemed to be
an eerie prediction.

My accuser set me up as a criminal member of the
Aryan Nations. They accused me of having illegal
weapons. They accused me of saying I was going
to assassinate the President of the United States
and the Pope. Very possibly, a threatening letter
was sent to the President with my name or initials
forged. My accusers hoped the FBI would rush
my home with armed agents hoping I would feel
the need to defend myself and thus be killed or
arrested for "assault on a federal officer."
Fortunately, bad weather (the first part of Feb.
1985), witnesses to this plot and our God, the
Lord Jesus Messiah, King of Israel prevented a
disaster.

Two months later, they wrote the secret service agent who had interviewed them, demanding an apology and a copy of a "forged letter" that didn't exist. Then, they wrote a quick note to Ronald Reagan, apologizing and explaining what had happened.

> Apparently, some local residents, who wished me ill fortune, carelessly used you and sent you a threatening letter (which I'm confident you never saw) and forged my signature on it hoping I'd be sent to prison so they could harass my wife and squat on my land.
>
> Please let me apologize for their evil in using you to get at me. I'm sorry such a letter was ever sent . . .

Then, the Weavers asked for his help. If someone on his staff could mail the letter to the Boundary County prosecutor, perhaps they could prove who forged the letter. Apparently, President Reagan didn't write back.

The rumors doubled the Weavers' fear and paranoia. As always, contention drew the couple closer together and convinced them that the end was at hand. In fact, it was right on time, according to Vicki's predictions.

Suddenly, everyone around them seemed to be traitors and conspirators. The Weavers responded the way they always did when they felt pushed—going a little further out on the fringe. Soon, Randy decided to check out the group he'd heard about from his survivalist buddy, Frank Kumnick. In the summer of 1986, Randy and Frank drove down the road a stretch to the annual Aryan Nations World Congress.

five

About the same time Randy and Vicki Weaver were finishing their cabin, the Coeur d'Alene office of the Federal Bureau of Investigation was launching a full-scale probe of the Aryan Nations, every agent in the office engrossed in a far-ranging investigation of the racist organization. Of course, at the time—the early months of 1984—the Coeur d'Alene office of the FBI was just one guy, a resourceful eighteen-year veteran agent named Wayne Manis, who had just escaped to Idaho himself, requesting a transfer after a long drug investigation in the Birmingham, Alabama, office.

A holdover from J. Edgar Hoover's bureau, Manis looked more like a country-western lounge singer than a G-man. He was a retired Marine with wavy brown hair, halfway to graying. But it was his wardrobe that was so incongruous: flashy western clothes and long coats, gold jewelry and dinner-plate belt buckles. His methods were decidedly nonbureau as well; he preferred common sense to wiretaps, and although he could skulk and spy with the best of them, he knew the best way to find out if someone was breaking the law was to ask them.

It didn't take Manis long to see where his attention was needed. Ten miles north of Coeur d'Alene stood the twenty-acre, wooded compound of the Church of Jesus Christ Christian and its political arm, the Aryan Nations. The church had been founded by Richard Girnt Butler, a leathery-faced former Lockheed Corporation engineer and a disciple of Dr. Wesley Swift, who was an early leader in the Christian Identity movement. Immediately, Manis called for a meeting with Butler, sat down across from him, and just asked what was going on.

In the 1960s, while the Ku Klux Klan was getting all the attention, the real growth in the racist right movement was being made on the religious end, specifically, the hybrid religion called Christian Identity. It was basically the same doctrine Randy and Vicki Weaver had arrived at, an unlikely blend of strict evangelical and Mormon theology, American nationalism, and a defunct movement called British Israelism, which taught that white Europeans were the lost tribes of Israel. If that were true, the Jews had to be something else: impostors—in fact, the very seed of Satan, sown through one of Eve's sons when she was seduced by the serpent, or perhaps through one of Noah's children.

In ten years, Richard Butler had risen to a sort of distinguished professorship in the white supremacy movement, one of a handful of grand old men who passed the hatred and fear along to a new generation of skinheads and state prisoners. By 1983 he claimed six thousand members of the Aryan Nations in the United States, three hundred of them in northern Idaho. The goal of the Aryan Nations was simple: to establish a white homeland in the northwest, preferably great slabs of Washington, Idaho, and Montana. On any given Sunday, as many as forty people might fill the swastika-decorated church to hear Butler—grandfatherly in his blue suit with the Aryan Nations arm patch—preach his Christian Identity beliefs. And, every month, it seemed, another family with such views moved to North Idaho, some drawn by Butler's church.

But as he listened to the Reverend Butler and heard

from people in the area, Wayne Manis became less concerned about what was going on at the Aryan Nations compound and more worried about what was going on outside it.

A bout two weeks after Randy and Vicki Weaver discovered their mountain retreat, a splinter group of Aryan Nations members was meeting in a corrugated steel barn just on the other side of the Selkirk Mountains, in Metaline Falls, Washington. Home to fewer than 300 people, Metaline Falls was a close cousin to towns like Naples and Bonners Ferry. In late September 1983, it was also the home of a disillusioned white guy named Bob Mathews, who gathered eight other men—Aryan Nations members and supporters—in the metal barn he'd recently built, on whose concrete steps were etched a swastika and the words: "White Pride, White Unity, White America." Inside the barn, Mathews, another escapee from mainstream America, was laying out his plan for the violent overthrow of the Zionist Occupied Government.

The men repeated an oath to "deliver our people from the Jew and bring total victory to the Aryan race." They swore, as Aryan warriors, "to complete secrecy to The Order and total loyalty to my comrades." They called themselves the "Bruders Schweigen," German for "Silent Brotherhood." About a month later, members committed their first crime in the name of the great race war, robbing an adult bookstore in Spokane of $369. The group quickly became more proficient criminals and began counterfeiting money to finance their revolution. Still, by the end of 1983, no one outside the group even knew there was a gang of criminal white supremacists.

By June 1984, word was gettting out. Members of the Bruders Schweigen robbed a bank and hit an armored car in Seattle, getting away with $500,000; others bombed a synagogue in Boise, Idaho; others murdered a man they considered to be a security risk in North Idaho; and some executed Jewish radio talk show host Alan Berg in Denver.

The group reached its zenith on July 19, 1984, when Mathews and eleven other members of the Bruders Schweigen robbed an armored car in Ukiah, California, escaping with more than $3 million.

In many ways, 1983 and 1984—when Randy and Vicki were just getting settled on Ruby Ridge—were the high point for white separatists and supremacists in the Northwest. The Aryan Nations was attracting welcome publicity for its cross burnings and for a rally it staged in Spokane's Riverfront Park. The 1983 and 1984 Aryan world congresses at Hayden Lake brought big crowds and Butler's church was at its peak of influence in the far right. Sentiment against the government was growing, and scores of people with radical right-wing beliefs were making their way to North Idaho from other parts of the country. Meanwhile, this group of Aryan warriors—the Bruders Schweigen, or The Order, as they were also called—was raking the countryside, preparing for the great race war.

A fringe member of the Bruders Schweigen had been caught passing a phony ten-dollar bill in Philadelphia, and—keyed by Wayne Manis's investigation and the Philadelphia man's testimony—the FBI began to unravel the group. Dozens of FBI agents were called to the Northwest, and the investigation of the Aryan Nations—specifically The Order—became the number-one priority in the country for the FBI.

In November 1984, the first member of The Order was arrested, after the informant told the FBI where Mathews and another member were hiding. After a gun battle with twenty FBI agents in Portland, Oregon, Mathews got away, but another member, Gary Yarbrough, was arrested, and The Order continued to crumble. A few weeks later, on December 7, four more members were arrested, and federal agents surrounded a house on Whidbey Island, in Washington's Puget Sound. Inside, Mathews was waiting for them. They negotiated, pleaded, and fired back and forth for thirty-six hours. Snipers, helicopters, and bullhorns did no good. Mathews would only leave if the Zionist Occupied Government guaranteed him a white homeland. Finally, with negotiations going nowhere, federal agents decided to finish the standoff. They fired three flares into the house, hoping to smoke Mathews out. It didn't work. A frustrated Wayne Manis watched the house burn to the ground. They found Mathews burned to death, curled up in a bathtub. Afterward, there was little talk of Mathews's civil rights, except among the radical right.

By January 1985, the FBI was rounding up the rest of the Bruders Schweigen and Manis's office was packed with dozens of agents tracing all the leads in the case. Tips were stacking up in the FBI office in Coeur d'Alene. Sometime that winter, a report was floated onto Wayne Manis's desk about a white separatist who had no known connection to The Order, but who was making a lot of noise in Boundary County. So Manis tagged a young agent who'd been sent from Indiana to help and passed along the name of a new figure in their far-reaching investigation, a woodcutter from Iowa named Randy Weaver.

K enneth Weiss had only been an FBI agent for three years when he was transferred from Indiana to Coeur d'Alene and assigned to the biggest investigation in the country, The Order. With so many agents involved, one of Weiss's tasks was to go through a long list—names of people with some minor connection, sometimes no connection, to the larger case that was unfolding. One of those names was Randy Weaver. The disgruntled neighbor Terry Kinnison had gone to the Boundary County sheriff with his allegations against Weaver, and the sheriff had quickly called the FBI.

Four feet of snow covered the Idaho mountains in the winter of 1984–85, and if Randy Weaver was going to be a threat to the president, he was going to have to figure out some way to get down from his mountain. And if federal agents wanted to talk to him, they were going to have to figure out some way to get a message up to him. They finally found a friend of the Weavers to snowmobile up with the word, and in February, Randy and his wife came down and met Weiss, another FBI agent, two Secret Service agents, and a sheriff's investigator at the sheriff's office in Bonners Ferry. Weiss had already interviewed Kinnison, who maintained that Randy was a threat to any federal employee and had even talked of killing the president. He'd said Randy's mentor for his far-right-wing beliefs was a guy named Frank Kumnick. Weiss also interviewed Sam Wohali, who said Randy was obnoxious but probably wasn't going to kill Ronald Reagan unless the president hiked up onto his land and started hassling him in the name of the New World Order. Now Weiss sat down with Randy and Vicki Weaver and asked point-blank, "Did you ever intend to assassinate the president or other government officials?"

Weiss and the other agents were subjected to Randy and Vicki's tale of backbiting and feuds and land problems and setups and plots fomented against them by a broad conspiracy of neighbors. "I got no time for Aryan Nations preachers," Weaver said. His beliefs were "strictly by the Bible." He said the Bible gave him the right to kill, if necessary, to defend his family, but that federal authorities were welcome on his land, despite all the rumors about him.

By the time he left Bonners Ferry that day, Weiss had concluded there was no reason to interview Randy Weaver again, and the case was basically closed. Back in Coeur d'Alene, Manis was keeping tabs on scores of radical right wingers in North Idaho, and the briefing on Randy went into that "mental hopper."

"We were aware of him," Manis said later. "He was a very outspoken separatist who lived in a remote cabin and made it very public to the community his views about the government." But Manis found no strong connection between Weaver and the Aryan Nations, let alone the dangerous Bruders Schweigen. He filed Randy Weaver's name away with those people who had radical views but weren't breaking the law. Yet.

• • •

*L*ike the FBI, the people of Idaho fought back against the racists. In 1982, after a Los Angeles woman with three racially mixed children was harassed by a member of Butler's church, some community members quickly formed the Kootenai County Task Force on Human Relations to provide a TV counterweight to the neo-Nazis and white separatists.

After convincing the North Idaho business community that the Aryan publicity was bad for tourism, the group became influential, holding meetings and demonstrations and constantly pointing out to the news media how small the Aryan presence was in northern Idaho. Their biggest accomplishment was helping to convince the Idaho legislature in 1983 to pass a law making it a felony to intimidate anyone because of their race.

At the Aryan Nations compound, Butler and his followers saw the "Inhuman Rights Task Force" differently. Their pressure on the business community made it nearly impossible for Aryan Nations members to hold down jobs and dissuaded people who might otherwise have come to Butler's white homeland. Suddenly, there was a stigma attached to being a member of the Aryan Nations.

But the real pressure was coming from law enforcement. And it wasn't just the FBI. Competitive police and sheriff's departments had their informants, the FBI had its informants, and the Bureau of Alcohol, Tobacco, and Firearms its own informants. At some meetings, half the participants were reporting to some agency or another.

And the federal officers that set out to investigate the growing radical right weren't the same ones who had battled the Ku Klux Klan in the South. The FBI, for instance, was no longer Efrem Zimbalist, Jr., and his competent G-men, cleaning up on bank robbers and radical leftists. In the wake of Watergate-era revelations about the FBI's violation of civil rights—especially in its investigation of Martin Luther King, Jr.—the chief federal law enforcement agency was hamstrung by new laws that limited their infiltration of groups like the Aryan Nations.

And so, the FBI had been caught off guard by The Order. Afterward, the bureau wasn't going to be surprised again, and so it began using informants, usually other criminals whom they turned and convinced to spy on their friends.

In 1983 and 1984, the radical right was the biggest concern of the ATF as well. In just twenty-five years, ATF had gone from 1,000 agents whose prime mission was to break up whiskey stills in the South to an armed force of 4,300, charged with keeping tabs on the skyrocketing number of weapons in the United States. Guns were the currency of doomsday cults, white separatists, and tax evaders, and increasingly, ATF—a federal agency founded to deal with contraband cigarettes and booze—was bumping up against the fringe of American society.

And its informants were bumping up against the FBI's informants. ATF spies gave reports on FBI spies, neither knowing what the other one was doing. At one meeting of three white separatists, two of them were federal plants. The growing competition between the FBI—a Justice Department agency—and the ATF—under the Treasury Department—spilled over so that neither side was telling the other what it was doing.

But there were results. Arrests were made left and right. The cases were dreams for federal prosecutors. Just say the words white supremacist, detail some of the bizarre beliefs, and you were halfway to conviction. American juries, like the public at large, seemed to understand that such dangerous people needed to be dealt with, and they looked the other way. In a nation desperate to purge itself of its racist history, law enforcement enjoyed the sort of freedom to investigate and prosecute white supremacists that it normally had only with drug gangs and the Mafia. The weapons they used were the same: criminal informants, undercover agents, surveillance, and broad charges like racketeering, sedition, weapons violations, and tax fraud—legal sleights of hand that broke the silence of conspirators and often culled convictions from what might have been weak cases otherwise.

• • •

*I*n April 1985, The Order defendants—twenty-three disgruntled white men and one woman—were indicted on conspiracy and racketeering charges, in all, sixty-seven separate counts, the kind of elaborate web usually reserved for the Mafia. Twenty-two Order members had been caught by then, and they were chained and bolted to the floor during the sixteen-week trial, which ended with an impressive record: twelve guilty pleas, ten convictions—twenty-two for twenty-two. As with the Mafia and drug lords, turning one Aryan against the others was the best weapon in prosecuting them. In the end, more than half the defendants testified against their "brothers." It was a valuable lesson for law enforcement and for the third-chair attorney in the case, a young assistant U.S. attorney named Ron Howen, a driven, no-frills federal prosecutor with a reputation for rejecting almost every plea bargain that landed on his desk. After The Order trial, white supremacists became Ron Howen's metier.

A few months later, Howen was prosecuting Elden "Bud" Cutler, the security chief for the Aryan Nations, who was arrested and charged with hiring a hit man to kill Thomas Martinez, the FBI informant who broke The Order. Unfortunately, the hit man Cutler hired to behead Martinez was an FBI agent who had infiltrated the group.

Howen handled the prosecution. A young attorney named David Nevin, who was raised on the other side of a black neighborhood in Louisiana and who grew up abhorring racism, was Cutler's defense attorney. But to Nevin, Cutler's beliefs—no matter how awful—were beside the point. This was a classic case of entrapment, sending an FBI agent to a simple man and coaxing him into paying to have someone killed. But Howen laid out the intricate kind of case he'd participated in with The Order, tying Cutler's beliefs into the conspiracy to kill the informant. Still, Nevin thought he had a chance, until the government played a tape of his client looking at a doctored picture of Martinez that purported to show him after he'd been decapitated. "Goddamn," Cutler said. "You guys really did it." It didn't help Nevin's case that his client wanted a copy of the picture. So Howen won another one, sending Cutler away for twelve years. "Usually, my clients don't commit their crimes on videotape," Nevin said. "It makes it tough to present a defense."

Howen wasn't done with white separatist prosecutions, but he wouldn't face David Nevin again in such a case for seven years, when they'd face off in *U.S. v. Randall Weaver and Kevin Harris*.

*I*n August and September 1986, five small bombs rocked the trendy resort town of Coeur d'Alene, including one that damaged the home of the Reverend Bill Wassmuth, one of the leaders of the Kootenai County Human Rights Task Force. Other bombs slightly damaged the federal courthouse and three businesses.

Within a few weeks, four people were in custody, charged with counterfeiting, bombings, and murder. Like a bad action movie, they called themselves The Order II. They were much less organized and efficient, and authorities bragged that they'd been stopped much earlier. The key to the case had been a confidential informant.

Richard Butler couldn't keep track of all the informants running around his compound. He had to worry about outright FBI agents, ATF agents and cops working undercover, private investigators and thrill-seekers working on behalf of law enforcement, and real Aryans who had gotten in trouble and had been convinced that testifying about their friends was better than spending the rest of their lives in prison.

It was the latter breed of informant who broke The Order II. Again, Ron Howen handled the prosecution, drawing out the same sort of elaborate racketeering indictment that had been successful with The Order. And again, Howen used the beliefs of the group to create an unmistakable web of criminal acts, each committed to further a broad conspiracy—namely, to overthrow the government of the United States. It was a detailed, masterful prosecution and the blueprint for how to convict white separatists.

Yet these people sprouted like chokeweed, and as soon as you arrested one, another seemed to take his place. Law enforcement responded by sending in informants and snitches, who returned with more bad news. There were indications everywhere that there was a movement to finish the job The Order had started. For instance, the bombings themselves seemed to have sprung from the collective anger at the 1986 Aryan Nations World Congress.

*I*t was like a racist All-Star game. Randy Weaver and Frank Kumnick walked through the gates of the Aryan Nations in the summer of 1986 to find almost every racist right-wing leader who wasn't in prison. All the giants milled about the Aryan Nations' pristine Hayden Lake compound: William Pierce, author of *The Turner Diaries,* the tract that had been used as an outline for The Order and would be used again more than a decade later for the bombing of a federal building in Oklahoma City; Tom Metzger of San Diego's White Aryan Resistance, a role model for skinheads across the United States; Bill Albers of Modesto, California, Imperial Wizard of the American Knights of the Ku Klux Klan; Thom Robb, national chaplain of the KKK; John Ross Taylor, known as the elder statesman of Canadian anti-Semitism; Terry Long, a forty-year-old "high Aryan warrior priest" trying to establish a white homeland in Calgary; and the two religious voices of the movement, pastors Robert Miles of Michigan and Richard Butler. Also there were the wives of the dead or imprisoned martyrs: Order members Mathews and Richard Scutari, and James Ellison, founder of The Covenant, The Sword & The Arm of The Lord in Missouri.

There were informants at the congress, too, poorly dressed guys from the FBI, the ATF, and local police and sheriff's departments. Most of the informants didn't even know who the other informants were, and they milled throughout a crowd of a couple hundred people, taking mental notes of who seemed ready to take over where The Order had left off.

Everywhere, groups of white men stood around, talking about where The Order had gone wrong. Some said they had gotten too violent too fast; others suggested they had wasted too much time robbing armored cars and printing up phony tens. Randy milled about, meeting all sorts of people, like Rod Willey, a computer technician who'd joined the Aryan Nations four or five years earlier. Kumnick introduced the two, and Randy told Willey he was there "to observe the goings-on and learn about the Aryan philosophy."

Despite the crushing defeat of The Order, the Aryans at the 1986 world congress weren't backing down. The last member of the group captured, security officer Richard Scutari, wrote from prison: "I offer no apologies, except for having failed to meet our goals. Learn from our mistakes and succeed where we failed. The Bruders Schweigen has shown you the way." It served as the unofficial motto for the weekend.

Dozens of people bought $7 medallions that promised: "Should you fall my friend, another friend will emerge from the shadows to take your place."

People milled about the congress, eating $1 "hamburgs," listening to the speakers, or checking out the bookstore ("*Nazi Polar Experiments*—$5.95"). Many people wore camouflage and holstered handguns strapped to their waists. Several KKK members wore black robes, others wore Confederate flags, and one young man wore a black concert T-shirt that announced "Adolf Hitler World Tour, 1939–1945." After skipping the event in 1985, the Aryans even had an old-fashioned cross burning, which the tastefully called a cross "lighting," torching a fifteen-foot cross in the woods near Hayden Lake and praying around it.

The congress made a special outreach toward families and children. The pastor Robert Miles, wearing a hooded black robe, told fourteen young boys they'd be good night riders for the KKK. The Klan started for fun, Miles said.

"They took their pillowcases and sheets and painted all kinds of designs on them." A sign on a nearby wall showed an Uncle Sam figure and was labeled "Aryan Sam sez: I WANT YOU . . . Join in support ARYAN FREEDOM FIGHTERS."

But the ease with which The Order had been broken and its members convinced to turn against the others chilled everyone. All weekend, the congress-goers watched each other suspiciously, wrote down license plate numbers, and were careful about what they said and to whom they said it. The public address announcer cautioned: "It is likely there are federal informants and agents among us."

six

At 11:15 on the cold, blue morning of January 20, 1987, a barrel-shaped motorcycle gang member named Gustav Antony Magisono—Gus to his friends—let out a small burp as he stepped out of his red Nissan Sentra and walked across a small, grassy park in downtown Sandpoint. It was a typical winter day in North Idaho, where the wind careens off the deep-water lakes and is channeled up granite valleys until it has built up enough icy speed to slam every screen door this side of Canada. The sky was clear, and the ground frosted with a light snow as Gus looked for Frank Kumnick, the guy he was supposed to meet at a small park behind a downtown motel. Subaru station wagons with ski racks maneuvered the snow-lined streets of Sandpoint, stragglers from a winter carnival that had taken place that weekend on nearby Schweitzer Basin's seventy-eight inches of packed powder. Magisono couldn't care less about skiing. He was in Sandpoint on serious and hopefully illegal business. He was a biker in his early forties, just under six feet tall, round and balding, with a gray beard and a cocksure way of talking about himself that made it clear:

He knew the right people. A weapons dealer and security expert from the East Coast, Magisono had gotten involved with motorcycle gangs and found he agreed with their white supremacist views. The summer before, he'd even gone to the Aryan Nations' summer conference, where he listened to the incredible list of speakers and hooked up with a few guys who believed the way he did. Now he was finding himself drawn into the movement and taking the next step, trying to get involved in a group that would carry on the work of The Order and The Order II.

That's where Frank Kumnick fit in. Magisono had talked on the phone with Frank, who said they ought to meet to go over some important business. Magisono had known Kumnick only a short time, but he seemed to be a player, and he talked of having some ambitious plans. Magisono stood in the park alone, wondering where Kumnick was. Just then, an old Jeep Wagoneer pulled up, and Kumnick leaned out and waved at him. When the Jeep stopped, Kumnick got out of the passenger seat, moved to the back, and Magisono took his place up front. In the driver's seat was a guy he remembered meeting at last July's Aryan Nations Congress, a thin, friendly guy with intense eyes and a midwestern accent, Randy Weaver. Kumnick was a couple inches and thirty pounds bigger than Weaver, who was jumpy, with darting eyes. Magisono hadn't expected to see Randy here, and right away he wondered what was happening. Since the summer congress, Magisono had had some dealings with Kumnick but none with this other guy, and all three men were wary of each other because it was just so damn easy to get caught up with a snitch these days.

From the backseat, Kumnick leaned forward, not quite as friendly as before. "How are you?" He asked if Magisono remembered Randy Weaver.

Magisono looked over at Weaver. "Hey, guy, how are you?"

"Pretty good."

Magisono turned around and talked with Kumnick some more, but Weaver seemed quiet and nervous, and so he turned to him again. "How you been, stranger?" He was beginning to feel comfortable with Kumnick, but he still wasn't sure what Weaver was doing here. For the things they were going to talk about, Magisono knew trust was the key on all sides.

"Didn't you celebrate Martin Luther Coon day?" Kumnick asked.

"I took a big piss for him," Magisono said. He complained about the King celebration in Spokane the day before, with its parade and civil rights speeches. Hell, even the governor had come to Spokane, to make a symbolic statement about the Aryans just over the state line. The others agreed it was a bullshit holiday.

"Yeah," Weaver said. "This country has got its priorities just really screwed up. Yeah. Celebrate a communist, sex-pervert nigger. Son of a bitch."

They all laughed.

They considered talking at the restaurant in Connies Motor Inn, but Kumnick didn't want to because it was so crowded. Magisono suggested driving down to the Edgewater, a hotel on the shore of Lake Pend Oreille, talking in its coffee shop and then going for a stroll through the big city park there.

Weaver drove along Sandpoint's square of one-way streets, past antique stores, specialty shops and restaurants, and across Bridge Street and the wide, shallow Sand Creek, into the parking lot of the Edgewater, a two-story resort at the northwest corner of the lake, sidesaddle to the round, sandy point for which the town was named. Magisono got right to business. He wanted to get involved in something big; he wanted to be around guys who could make something happen. "Are you ready to get things . . . ah . . . rolling?"

Kumnick said yeah. Weaver said nothing. Kumnick and Magisono talked some more about the dangers of getting an organization together. Especially in light of what had happened with The Order I and II. Kumnick seemed spooked by something.

"Gotta question for ya," Kumnick said as they waited in the Wagoneer. "Can you stand an electric scan?"

Magisono was surprised. "An electric scan?"

"Yeah."

"Sure."

"I've gotten really careful," Kumnick said. He said the scanner would tell him if Magisono was carrying a gun or was wired with a transmitter or tape recorder. "As small as a pin, it will pick it up."

Weaver contributed, "If you're figuring out lately, everybody's falling for one reason."

They thought *he* was a snitch? Magisono wondered if they were trying to scare him, but, in a way, he understood. Hell, he was just as careful in his business dealings. "Boy there's a lot of shit going on, I'll tell you that," Magisono said. "It's made me real, real cautious."

Kumnick wanted to talk about the electronic scanner some more, but he didn't pull it out yet. "I picked it up at an auction for a dollar. The same thing in a security place will cost you a hundred and ninety-nine dollars."

After a few minutes, they walked into the Beach House, the restaurant adjacent to the Edgewater. Just ahead of the Tuesday lunch business, they got a table with a view of dark Lake Pend Oreille, where the January wind was lifting whitecaps onto the soft sand beach. The big biker excused himself and walked down the carpeted hall to the bathroom. Inside, he locked the door and, trying to remain calm, said into the tiny transmitter draped over his shoulder and taped to his chest, "He wants to do a scan on me!"

*T*here was no Gus Magisono. It was a cover dreamed up by Kenneth Fadeley, a thirty-nine-year-old private investigator from Spokane. He'd worked in marketing for STP—the gasoline additive—before getting into the business of confidential informing in 1983, when a friend who happened to be a Spokane cop was killed during an investigation of motorcycle gangs. "I took it pretty personal," Fadeley said. Unlike most informants, Fadeley didn't have a criminal record and only did it because of his friend, because of the excitement and because of the little money he made. He also found that his personality was suited to undercover work, to adopting a new identity and rooting out criminals.

Working for a few hundred dollars here and there, he posed as a bank executive to stop a rash of robberies in Spokane and staked out stores that had been hit by repeated burglaries. Once, as Fadeley was watching Johnny Carson on television in the back of an appliance store, he heard a noise and confronted the burglars in the back of the building. One of them raised a metal bar, and Fadeley shot and wounded him. A few weeks later, he helped track down a guy who had robbed a bank with a hand grenade.

But he was at his best infiltrating the Aryan Nations. In 1984, while investigating a church that was stockpiling automatic rifles, Fadeley met a Georgia-friendly agent with the Bureau of Alcohol, Tobacco, and Firearms named Herb Byerly. Two years later, with federal law enforcement worried again about white supremacists in Idaho, Byerly had talked the informant into attending the 1986 world congress, the first one since The Order trial, which had decimated the ranks of the violent young bucks in the white supremacist movement. The 1986 congress was to be the radical right's signal that it wasn't going away. At the Aryans' Hayden Lake compound, Fadeley talked with survivalists, white separatists, constitutionalists, the entire spectrum of far-right activists, even knights of the Ku Klux Klan. He jotted down license plates and met men who would later be arrested for a spree of bombings in the Northwest.

Fadeley's assignment that weekend was to listen to the leaders who spoke and to gather intelligence about what they might be planning. He was also to get to know the radical right wingers at the conference, and so Fadeley introduced himself to a friendly guy who said his name was Frank Kumnick. They were walking across the grounds when Kumnick pointed out a smallish guy standing near a group of Ku Klux Klan members.

"Hey, there's Randy," Kumnick said. "Come on over, I want you to meet an individual who will always cover your backside." Kumnick introduced the two and told Fadeley that Weaver had been in Special Forces and was a really good guy. Weaver didn't seem involved in anything, and so Fadeley didn't think about him anymore. But Kumnick continued talking a big game, and after the Aryan Nations Congress, the ATF agents told Fadeley to continue getting close to him.

A month later, Fadeley and Kumnick hiked up a wooded peak called Katka Mountain, near the Canadian border. There, they talked about raising money for a white uprising by kidnapping kids from the exclusive private school where Kumnick worked. The Rocky Mountain Academy was a boarding school where tuition could be as high as $3,500 a month. Filled with disciplined teachers and administrators, it catered to East and West Coast wealth, and at times housed heirs to the Du Pont and Howard Hughes fortunes. Rocky Mountain offered a place for successful people to send their kids, some of them to get them away from drugs, bad friends and relationships, or sometimes just the city. In the woods, they dried out, chilled out, and—the theory went—couldn't find trouble even if you spotted them a convertible and $1,000.

Kumnick later wrote that the plot to kidnap children from Rocky was all Fadeley's idea, a trick to entrap him. "When I asked why, he said 'for money,'" Kumnick said. He said he was suspicious of "Gus Magisono" immediately and knew he was being set up. He said he only talked about that plot and the others to shine Fadeley on and to see if Fadeley slipped up.

But Fadeley insisted Kumnick was serious about kidnapping the children. "He told me Barbara Walters's kid was there. He told me that Clint Eastwood's kid was there," Fadeley said later. "They were going to target the Jewish kids and anybody that was well-known." According to Fadeley, Kumnick wanted to use the money they got from the kidnapping scheme to finance terrorism, and he was especially interested in Jacqueline Guber, Walters's daughter, who attended the school in the 1980s.

Fadeley said Kumnick wanted to hide the children on

Katka Mountain while they waited for the ransom money. He also said that Kumnick planned to bury food, water, weapons, and other supplies at the base of some trees on the mountain before the kidnapping. They would find the supplies by using key chains that beep at the sound of clapping hands. But Kumnick later denied the whole scheme.

Fadeley was afraid Kumnick was serious, and so he told Byerly in the fall of 1986 and Byerly called some FBI agents in Spokane, who then called the school. Kumnick was fired a few weeks later by school officials who said his van had been seen at an Aryan Nations gathering.

Hopeful that his cover was still good, Fadeley requested another meeting with Kumnick, so they could talk about forming a group to further the white cause, to continue the work of the Bruders Schweigen. But now, in Sandpoint, Kumnick had brought Randy Weaver and was threatening to run an electronic scan on him. Fadeley figured that Byerly was nearby, perhaps a block away, in an unmarked, gold Dodge Diplomat and hopefully he understood how much danger his undercover informant was in. Fadeley knew Kumnick was a big talker, and he suspected he'd brought Weaver along to impress him with the fact that he had at least one colleague. He also knew that if he was going to find out what kind of crimes Kumnick, and possibly Weaver, might be capable of committing, he just had to go along with Kumnick and hope his cover stayed solid. And if things went bad, he sure as hell hoped Byerly wasn't far away.

"I was about to bust a kidney," Fadeley told the men when he rejoined them at the table. They made small talk and ordered three cups of coffee. Randy was quiet in the restaurant while Kumnick and Fadeley— the man they still believed was Gus Magisono—talked about how difficult it was to find anyone to trust. Federal informants were everywhere. Fadeley knew a few other guys who could help them with the group they wanted to form, four ex-military guys that he'd checked out himself.

Of course, they would need weapons. Randy was still distant, still wasn't talking much about their plans, so when Kumnick excused himself to go to the bathroom, Fadeley tried again. This team they were trying to form would never work if they couldn't build trust in one another.

"We don't know each other at all," Weaver said.

"No."

Always friendly, Randy offered a little bit of information about himself. "I base everything I believe on the Bible," he said.

Fadeley said he was raised Catholic but wasn't practicing anymore, that he trusted his values and was loyal to people like Kumnick. The conversation turned to what they believed.

"Your kids go to public school?" Randy asked.

"You bet."

"The Bible says don't take your hate out on your kids, but raise them up," Weaver said. "So when you send them to school, you are giving up your responsibility."

"It's out of my hands," Fadeley said.

"I'm just telling you things I read and study." Randy wasn't trying to preach, but people should know why they believe what they do. "And when the time comes, it's not my responsibility to share brotherly love and my food that I save for my kids with your kids because you didn't . . . And it's comin' down . . . It's goin' down the tubes. And you'd better be prepared to survive. Pray that you be counted worthy."

Kumnick came back to the table for the end of the sermon he'd heard several times from Randy and Vicki Weaver. The talk switched to organization, to forming a group, and again, Randy just sat there, quiet. Kumnick was worried still. These other guys that the biker knew, they didn't know Weaver and Kumnick's last names, did they? Because that could be unwise. No, Fadeley said.

They paid for their coffees and walked out to the Wagoneer again, Weaver in the driver's seat, Fadeley next to him, and Kumnick in back. Randy drove across the parking lot to City Park, a rounded, three-block point of grass and beach that was the tourist center of town during the summer. But on such raw winter days, the park was empty and Randy maneuvered the Jeep to the end of the lot. Here, even though they were smack in the middle of downtown, no one could see them.

Fadeley was becoming nervous, and he suggested they go for a walk.

"You know what sounds like a winner to me?" Weaver said. "It's nice and warm in here. I'm gonna stay right in here."

"That's fine," Fadeley said, wondering how far away Byerly was.

"I ain't got all that blubber on me like you guys do." Weaver laughed.

Fadeley laughed uncomfortably. In the backseat, Kumnick pulled out his electronic scanner, a black gizmo that fit nicely in his thick palm.

Fadeley sat with his left arm draped over the Jeep's bench seat and Kumnick swept the device over Fadeley's arm, stopping just inches away from the wire that ran over his chest. With his right hand, hidden from Kumnick and Weaver, Fadeley fingered the .22-caliber pistol holstered on his ankle. His hand strained to wrap around the handle, but he realized he wouldn't be able to get the Velcro strap off the gun in time. He was screwed.

Instead, Fadeley grabbed the device out of Kumnick's hand and turned it over in his own hand before Kumnick could finish scanning him. "Is that a stud finder?" Fadeley held it in his hand and pretended to be impressed, although he was trying to figure out what to do next.

"Yeah," Kumnick said. "A stud finder will do a lot more than find studs." Fadeley wanted to see if it would work, so he ran it over Kumnick's chest, and it produced a long beep. Kumnick smiled, reached inside his camouflage shirt, and pulled out dog tags. Written on the tags were the words "Liberty or Death." They talked about other things the stud finder might react to. "Hmm. Glasses?" Kumnick teased.

"Yeah, frames?" Fadeley said.

"You think so?" Weaver asked. They're trying to scare me, trying to intimidate me, Fadeley thought. Or worse.

Fadeley slowly set the stud finder down on the bench seat between himself and Weaver.

When the informant turned back around, Kumnick had a pistol, a small .22-caliber gun, black, with a pearl handle. Fadeley saw that it was pointed at his head.

"I'll be damned," Fadeley said. Then he laughed nervously. "I'll be damned," he said again. And then, for Byerly's benefit: "A little derringer."

"Where'd you come from back East?" Randy asked.

"New Jersey. Eighteen miles from Philadelphia."

The Wagoneer sat alone in the parking lot, basketball hoops on one side, the cool, dark waves of Lake Pend Oreille slapping the sand on the other. Inside the car, the questions were about Fadeley's background, where he was from, what he believed. Kumnick kept the gun pointed at Fadeley's head, but the informant played along, laughing and talking tough until Kumnick set the gun down in his lap, still cocked, and finally relaxed the hammer and slid the gun back into his coat pocket. Fadeley's hand loosened on the holstered pistol strapped to his ankle.

The conversation turned back to Philadelphia and Fadeley's childhood. "I grew up fighting black kids on a daily basis. I grew up fighting Puerto Ricans," he said.

They'd finally hit on a subject Randy was interested in. "Yeah, I run into them in the service. Mexicans are bad, but Puerto Ricans make them look tame, I think."

Kumnick said the problem was that white men didn't stick together anymore. Once again, they were talking about loyalty. Kumnick and Fadeley talked more about this group and how to form it. Fadeley hoped they trusted him finally.

"We have a problem," Kumnick said. "And if something isn't changed where we show some new leadership for the patriot movement, it's going to be dead." He said that's where he disagreed with the men "to the south," Richard Butler and his aging neo-Nazis at the Aryan Nations compound.

"The next group coming up—and this is what I consider ourselves," Kumnick continued, "we're going to be new leaders in the new age." Weaver was still quiet. Kumnick asked how committed Fadeley's men were.

"Three of the four are very, very dedicated."

"Okay. I'll tell you what we do."

And then Kumnick seemed to forget his trepidation, and he laid out his plan for the next white uprising. This was the whole reason Fadeley had come down here, to hear where the vanguard of the white terrorists wanted to strike next in their war against the Jew-controlled government. Among people like Kumnick and Weaver, Fadeley believed, there was the seething potential for danger. Members of the Order I and II had robbed armored cars, murdered people and bombed buildings, and here was Kumnick saying they'd gotten off to a good start but had grown sloppy.

Finally, the scattershot Kumnick got around to his own contribution to the white revolution:

"Okay, I'll tell you what we do," he said. "It's so simple, it will blow your mind. You get a quick set epoxy. . . . Go out to all the banks and shoot epoxy into the locks. When they're closed. Hey, come that following morning, there won't be a whole bank in town that will be able to open for four or five hours until they have, you know, pulled all the locks and gone through them."

"Super Glue," Fadeley said simply.

"Yeah. Right." But Kumnick wasn't done. He had other plans. They could somehow mess up people's television reception. And maybe cut some power lines. And then, his masterstroke. Boundary County is home to acres and acres of hops fields used by Anheuser Busch to make beer.

"Well," Kumnick says. "All the hops burn. . . . Hey, when you go into the tavern and they ain't got no beer, shit . . . I mean, they are totally gonna be upset."

Kumnick said he knew what had gone wrong with The Order. "They got too fast, they didn't think this thing through and, uh, they absorbed too many people too quick and boy, you're gonna get a ringer every time."

These were the crack soldiers, the storm troopers of a white revolution that Fadeley had driven north to root out.

• • •

Fraternity pranks weren't the only plans Kumnick had for the revolution. He talked about burning a house that was being repossessed by the IRS and spoke fancifully about shooting the agents who came up to investigate it. But later he backed away from those plans, saying the weather wouldn't be right to ambush IRS agents. Weaver was quiet and didn't say much whenever the talk returned to forming a terrorist group. And Kumnick just came up with bizarre plans that sounded more and more like hazing than guerrilla civil war.

"You know what I would like to do," said Kumnick. "I would like to catch some agents [and] let them go naked. I'm serious man. You know, like take everything away from them, including . . . their money and all that. . . . Hey, they're gonna have a hard time explaining it, you know. You talk about embarrassment. They would wish they were dead."

Kumnick was a big, jittery man who talked nonstop and occasionally took off for the woods when he got an inkling that the heat was about to come down on him. Some people believed he was a federal informant himself. That Tuesday, he talked like a sprinkler, spraying words all over the inside of the truck about what he could do. "I told the sheriff the other day, you know, I says, you know, just playing around, you know, you guys ever come to my property, you'll probably step on, you'll probably step on a nail, and I shit on that nail at least four times, you know, you're gonna die anyway. You know?"

"Yep," was all Fadeley could say.

Fadeley tried to bring Kumnick back to reality a couple times. "I realize we are starting out small, which is what we should do," he said.

"Right," Kumnick said.

Fadeley said their biggest problems were finances and communication.

Kumnick said he was sending for some military radios. "This is what makes the difference between a real fighting unit and those guys that are just a bunch of jokers, see?"

They sat in the Jeep and talked, and Fadeley was

painfully aware of the gun in Kumnick's coat pocket. He tried to appear as snitch-conscious as the other two men. Kumnick said they had to be careful of the guys who would talk "when they put the electrodes in their testicles."

Again the talk turned tough, and Fadeley wondered if he should reach for his gun. He played along with Kumnick but wanted to make sure Byerly knew how to find him if this started to go badly. The conversation "doesn't go any farther than this Wagoneer," Fadeley said clearly into the transmitter. "It is a Wagoneer, isn't it?"

Kumnick talked again about the need for security within the organization. He said that if someone decided to leave, "they're leaving permanently. Because if they think they're gonna go blow a whistle . . ."

"The old story," Fadeley said.

"Right, right," Kumnick said, recalling that The Order I even had to dispatch one of its own. "That's where The Order did one thing right. They did get rid of somebody, and they ain't found 'em yet, you know?"

Weaver still didn't contribute to the conversation about forming this group, and so it was just Kumnick and the guy they thought was Gus Magisono, talking in circles about security and snitches.

"What's your last name?" Randy asked Fadeley.

"Magisono."

"What is it?"

Fadeley repeated it and said, "Italian."

"Right," Randy said. "But you know there are black Greeks and white Greeks and there are black Italians and white Italians." Randy said you couldn't trust all white people either, but he wasn't worried because Yahweh would show him who to trust.

"Right now," Randy said to the stocky informer, "I don't know about you."

"If I start to find wires on the midget, he ain't goin' home," Kumnick added.

They were bluffing, testing him to see if he called in help or ran away. Again, Fadeley had to try to play it cool, to act as tough as them. "That's right," Fadeley said.

"Your ass is dead if you want to know the truth," Kumnick said.

"And the same with you," Fadeley answered.

"I'd put the .22 in your ear and pull the trigger," Kumnick said.

Fadeley coolly called their bluff again. "And the same with you."

The trust seemed to rise and fall as they chattered on about other separatists they knew and such. Finally, after more than an hour in the Jeep, Randy started it up and said they should get going. They drove back across the bridge to downtown Sandpoint and an antique store where Kumnick's wife was supposed to be waiting for them. The solidly built Kumnick—who might've passed for a brother of Kenneth Fadeley—walked inside to check, and for a minute it was just Weaver and Fadeley alone in the Jeep. Randy confided that he didn't agree with Kumnick that organizing a group would help the cause.

"Frank is a good friend of mine, and I'd back him up to anything in a minute, but, like I told you, according to the Bible, things are goin' down the tubes and . . . it don't matter, and we can think, well, let's do this or that. Frank . . . ain't gonna change what's comin'. You can get yourself in trouble trying to change it." The Bible, Randy said, is clear that you should trust no man.

"Frank let me come along today, for whatever reason," Randy said.

"Frank wanted you to come along today to get you an overall impression of me."

"That's probably true," Randy agreed. "I know he wanted me to scope you out and see what I thought of you."

"Sure."

Frank's wife had gone on to the pizza parlor already, and so they piled back into the Jeep and headed on again, a little more comfortable with each other.

Six blocks from the Edgewater, Papandrea's Pizza was the consensus choice for best pizza in Sandpoint, and the end of the lunch rush of skiers and downtown workers was finishing up as the three men slipped inside the door. Frank's wife, Mary Lou, was already there and Vicki showed up as soon as they'd ordered. She'd spent the day at the bookstore, trying to find a *Smith's Biblical Dictionary*. "You know, one that's not abridged." Vicki liked to go to the bookstore when Randy went into town, which wasn't often. "Yeah, see, I might be sitting here till three, so I sit and read."

The men engaged in the bizarre small talk of white survivalists: the new .223 rifle, the new Russian-made shells, this guy who was a snitch, that guy who lives in "Niggerville, Florida."

Kumnick and Fadeley talked more about trust, and they agreed that the time in the Jeep had made each of them more comfortable. Now they got down to business. Fadeley said he was ready to start putting some money up if Kumnick could really provide weapons. But, again, there were no specifics from Kumnick, just a bunch of loose talk.

After a while, the informant turned to Randy, who had eaten quietly with his wife at the end of the table. "After hearing us talk about different things, without making a split-second judgment, do you feel comfortable?" Fadeley asked.

"Yeah."

"Okay," Fadeley said.

"You guys wanna get somethin' goin', that's fine with me," Randy shrugged. "I don't. I haven't heard nothin' . . ." He didn't finish the thought.

Earlier, Kumnick had talked about torching some house near Randy's place that the IRS was about to foreclose on in March, but when Fadeley brought it up again, Kumnick wasn't interested anymore. He said the weather in March probably wouldn't be good enough.

But Randy was interested in that subject, and he agreed with the others that IRS agents in Boundary County could spark a violent confrontation.

Some people were especially hard-core, Randy said. "They go in there and they can kill IRS. Of course, they are damned fools that might do it. They have no brains whatsoever."

They talked more about the guy who lived in the place near Randy's and his trouble with the government. "I'll be honest with you," Randy said. "If they try to throw him out, I hope he kills half of them."

"Yeah," Fadeley said.

But things could get quickly out of hand, Randy said.

"When people stop sayin', 'No more,' hey, it doesn't all of a sudden happen," Randy said. "You can't just bring four, three or four agents. You have to bring the whole army."

Fadeley wanted to get going, and he made one more attempt to get Kumnick on the record with something more than vague, unlikely plots and racist talk. At that point, he didn't figure Randy Weaver was of much interest to his ATF contacts.

"What's our next step?" he asked.

"Surprise resistance," Kumnick fired back.

"Okay." But still, the talk was all philosophy, with Kumnick and his wife doing all the talking. Even when Randy tried to talk, Kumnick cut him off. The talk of jail worried all of them, but Kumnick said he was cool, that he'd go away if he had to.

"I go back to Scripture," Randy said, "where it says in there: 'Some will be destined, and some will be destined to die. For this is the patience of the saints.'" At the end of the table, his pretty wife agreed with him.

Again, the conversation came around to this group they should form, and again, it was Kumnick and Fadeley doing the talking, comparing themselves to the Vietcong, the Afghan rebels, and the Contras.

"We're only a little bit of a ring," Kumnick said. "Put us all together, we'll be a wave."

They drove back to Connies, and Fadeley said he'd walk to his car from there, since it was a small town and it might not be good for them to be seen together. The others agreed. Fadeley tried once more to get some specifics out of Kumnick.

"Why don't you do this," he said. "Why don't you send me down a little list . . . of maybe five . . . things you'd like to see the group accomplish by March, February, March, by say, the first of April."

"Right."

"Or three things," Fadeley said.

Kumnick said "the glue thing" would be one of them. And then, maybe burning down the house the IRS wanted to repossess.

"Take care," Fadeley said.

"I will."

Fadeley watched them pile back into the Wagoneer.

"I got a long walk ahead of me," Fadeley said.

The Kumnicks and Weavers were still sitting there, and he spoke into the transmitter. "They're watching me, so stay away. They are watching me, so stay away!" Finally, the Jeep pulled away, and Fadeley began walking, his boots clicking on the sidewalk. He looked around for the gold Dodge that Byerly would be driving. He was beginning to realize how close he'd come to drawing when Kumnick pulled the gun on him. For some reason, Kumnick had only run the stud finder across Fadeley's left arm. Man, that was close.

Fadeley was beat. He'd spent the afternoon in a Jeep with these nuts, a gun against his head. As far as he was concerned, no matter how loopy their plans sounded, these guys were dangerous, especially Kumnick. Weaver was a weird, racist zealot, but Kumnick was clearly the one who needed his attention. As he turned another corner, Fadeley allowed himself to relax. "Don't see 'em anymore." He sighed. "Lord."

seven

Dust rose off the driveway like fog on a river as the summer of 1987 dragged on, hot and dry as any in memory. David and Jeane came for their yearly visit in late August, bringing more boxes of clothes and books for the kids. Sara, who was eleven, dove into the books, but that was no great shock, because she was so much like Vicki her grandparents expected her to be a good reader. Sam was the surprising one. A couple summers before, you couldn't force a book into his hands, but by 1987, he read everything he got his hands on, and it wasn't long before well-worn paperbacks filled up the kids' bookshelves and gathered in piles under their beds: *Trixie Belden*, *The Black Stallion*, *Heart of the Blue Ridge*, *Saddle and Ride*, *Eight Cousins*. Sam loved adventure stories and books about warfare—*The Story of the Green Berets* and *Naval Battles and Heroes*.

Soon, Sam was reading and memorizing encyclopedias and Sara would get frustrated when they'd get into an argument over some piece of trivia and Sam would always be right.

Vicki's first date for Armageddon had passed without any trouble, and back in Iowa her family got the feeling she was mellowing a bit, maybe even finding her place in the rough life of northern Idaho. David and Jeane's visits had become nice breaks for both families, and as usual Jeane and her daughter stayed up all night canning fruits, vegetables, and herbs, and talking about all the nieces and nephews.

David and Jeane were amazed at the amount of work Vicki did. Besides teaching and taking care of three children, she kept the house repaired and served as doctor, cook, and caretaker for the whole family. In just a few days, she canned fourteen quarts of green beans, seven quarts of pears, and seven quarts of fruit cocktail—peaches, pears, red grapes and honeydew melon—all of it while bent over a stoked woodstove in the searing August heat. She and Sara planted, picked, and canned all manner of vegetables, not to mention mustard greens and nettles. Vicki studied books about medicinal and edible plants and herbs, and her knowledge of gardening made her the equivalent of a country doctor who, in her chatty letters back home, prescribed teas and roots for the ailments of people in Iowa.

That summer, Sammy especially enjoyed his grandparents' visit and couldn't wait for them to return the next year, so he could "show Grandpa around my mountain again." Samuel led the smiling, bowlegged David up to mountain streams, where the nine-year-old proudly fished for four- and five-inch brook trout. Sam would jerk them out of the water, slip 'em off the hook and slit 'em up the belly, clean them with his thumb, and then they'd go back and panfry them for dinner. When David asked for a walking stick, Sam ran into the draw near their house and brought back a light, smooth birch limb, then another and another, until David had six to choose from. It became a game for all the kids, finding the best, smoothest, lightest, strongest stick for the aging but energetic Grandpa David.

When he wasn't walking with the boy, David always found work to do around the cabin, like helping Vicki rig up her washing machine, a gas-powered monster from the

1920s, with a kick start and an engine that looked like it could run a helicopter. With nothing to cook on but the woodstove, the house heated up like a toaster, and so David built Vicki a screen door and hung it up so they could leave the door open without feeding half the flies in Idaho. He fixed the plumbing, set up the water system and—as he'd done to his own house forty years earlier—soldered a sidearm onto the woodstove to heat water for a shower. David was the kind of guy who couldn't be still. Besides, anything was better than sitting around, listening to Randy and Frank Kumnick talk. Randy had gotten a construction job nearby, working on a log cabin that was being built. But too often, it seemed to David, he sat around smoking and talking with Kumnick about all these wild ideas that David didn't really want to hear about.

David Jordison never much cared for Kumnick. He would get to talking about this conspiracy or that plan and he was just the kind of guy to get Randy all riled up and into trouble. On the drive back to Iowa, he told Jeane: "Pete should stay away from that guy."

After David and Jeane left that summer of 1987, the family got ready for the barter fair, one of the kids' favorite events, a gathering of every social stratum in the woods: the loggers, the second-generation hippies, and the other Christian survivalists. That year, Sara sold cookies and popcorn balls.

The three Weaver children were doing well, David and Jeane thought. Those first years, Vicki kept pretty strict hours for their school, starting it in the morning and not releasing them until the afternoon. Although Randy and Vicki said the kids didn't need anything but home schooling, they told all three children that, if they really wanted to, any of them could go into town and brave the Beast's public school for a while.

"All you got to do is learn to read, write, arithmetic through maybe the eighth grade," Randy told them. "And then you can make it on your own." He didn't guess any of his kids would make it in public school, though, because they were being taught to question authority the way he did, and he figured they'd just get kicked out for arguing with the teacher.

None of the kids wanted to go to public school anyway. They were happy where they were.

The children also became more involved spiritually, although Sara said her parents didn't push her into the beliefs. "Don't take my word for things," her father would say. "Look into it yourself." They never felt brainwashed or trained in any way, Sara said. They were just a family.

And they were pretty happy, enjoying the world more and obsessing less about its end. Vicki had even taken a job refinishing antique furniture. It had been one of her many talents growing up, and now she dove back into refinishing and completed three good pieces in just a few weeks for a local refinisher. "Maybe when the kids are raised, I'll have my own shop," Vicki wrote her mother in November 1987.

*I*n the fall of 1987, after Vicki's date for Armageddon passed, the Weavers looked hard for a place to house-sit, to get off the mountain for the winter. Eventually they found a place, but they had to rent it, a cute red house with a river rock foundation, just off the highway, with a telephone and a satellite dish. They were living in civilization. The family seemed to be emerging from their paranoia, and Vicki's parents hoped they had come out of the darkness. The Weavers even agreed to pose for pictures, and back in Iowa, Vicki's sister and brother couldn't believe how much the kids had grown. Sara was almost as tall and pretty as her mother, Sam was a little man, and Rachel was growing up even cuter than they'd remembered her.

When David and Jeane visited in 1988 and 1989, while the Weavers lived in the house at the bottom of the hill, they had the best times they could remember. Jeane took all the kids into town for some shopping, and they hit a few garage sales, looking for the denim skirts that all the Weaver girls wore. Later, they drove up a road that rivaled their own mountain driveway for inaccessibility, until they came to an open field and a wall of huckleberry bushes. They all piled out and began picking huckleberries. The kids found a small lake to swim and fish in while the adults picked buckets full of the bright purple berries. Tourists had discovered the Northwest's huckleberry pies, and so when they took the huckleberries to the local fruit stand, the grocer gave them an incredible deal, trading a few gallons for a pickup load of apples, plums, and peaches. Vicki and Jeane stayed up all night canning.

It was around that time that David helped Randy and Vicki build an eight-by-fourteen movable outbuilding about the size of a bedroom and shaped like a tiny barn. Vicki said they would use it for a guest house, a place for people to sleep when they visited the mountain. The house was also used as a retreat for Vicki when she was menstruating. God had shown her that she was unclean during her periods and that she should separate herself from the family and pray until the time had passed. In Iowa, when Julie heard about the menstruation shed, it reminded her of the baths Vicki used to take in Cedar Falls. She wondered if, deep down, the shed was just a way for Vicki to get away from the family and get some rest.

Nothing came of the Secret Service investigation; the Wohalis moved; and even the problems with the Kinnisons were turning out okay. Terry and his wife had filed a $9,000 lien against Randy and Vicki, trying to collect the money they claimed to have put in and the work they'd done on the barn. But the Weavers hired Everett Hofmeister, a Coeur d'Alene attorney who represented, among other people, Richard Butler, the leader of the Aryan Nations. With Hofmeister as their attorney, the Weavers countersued.

Terry Kinnison and his family had moved to Alaska, and he wrote the court a half dozen times asking them just to drop the whole thing.

"I do not have the money to pay any more attorneys," Kinnison wrote.

I did not [do] one thing wrong. I am guilty of
nothing. I do not wish nor have ever wished to
fight anyone. . . . I can no longer take the stress of
all this. So just give the Weavers the lien. But Sir,
I will not pay the man or any one individual
involved any amount of money. I will not. I have
given all I will give. So give them the lien papers
and let it go. . . . And may the God of Abraham
rebuke them for what they have done.

Unfortunately for Kinnison, the God of Abraham didn't
decide the case, and since Kinnison didn't show up for
court, Randy and his family were awarded $1,000 in
damages and another $1,100 in costs and attorneys' fees.
Eventually, the money was garnished from a small land
settlement that was still going to the Kinnisons.

Even though their racism was progressing and Randy
had become interested in the Aryan Nations, it seemed to
the Jordisons that Randy and Vicki had come through some
long, dangerous time. Now, Vicki's family thought, as long
as Randy stays out of trouble, everything will be okay.

Randy was in town again, and he was ready to talk.
There was no Sambo's in Naples, Idaho, so he did
the best he could, hanging out with friends at their
houses or at the restaurant at the Deep Creek Inn. He went
to a couple more gatherings at the Aryan Nations, became
friendly with some of the members, bought a belt buckle
there, and talked about the movement with Frank Kumnick
and others.

Randy had also begun to talk about fixing the messed-
up system. He'd been in conversations where guys like
Frank said that revolution was the way to repair things, but
he and Vicki took a more biblical approach, figuring God
would let them know if violence was needed. Yet Randy
was also living in the world again, and so he and Vicki

began asking God whether one man—Randy—couldn't try to fix the world, at least the part of it in Boundary County. So he went to a few public meetings, including one held by the state game department where Weaver reportedly stood up and said that if he couldn't shoot grizzly bears, and one of his kids got mauled by a bear, he might come back and shoot the people at the meeting.

In 1987, Boundary County formed a Human Rights Task Force like the one in Kootenai County, after white separatist Robert Miles suggested he might move his 2,000-member Mountain Church from Cohoctah, Michigan, to Bonners Ferry. At one meeting, after the group showed a Canadian documentary about the Aryan Nations, a gully-cheeked man stood up and addressed the task force. Randy Weaver said the film was all wrong. He'd been to the Aryan Nations, and the people here were misunderstanding their message. He said Jews were running the world's economic system, and groups like the Aryan Nations were only trying to help America get back on its feet. The Human Rights Task Force, Randy said, was denying whites their civil rights. It was a classic Randy speech.

Later, he said he'd been unimpressed with the task force. "The only thing they're concerned with is white racists, that's what bothers me," he told a newspaper reporter. "They don't show any movies on black racists. . . . There are lots of Indian racists in this county."

That spring, Randy figured out how he could change the system and maybe keep his family in food and clothes for good. He announced his candidacy for sheriff, running on the Republican ticket. He figured he was a good candidate; a former Green Beret, he'd been involved in investigations in the army and had never had a criminal record. In fact, it seemed he'd been planning to run for quite a while. He told friends that one reason the family moved down off the mountain was so Randy could do a better job campaigning. But he didn't have a lot of money, and so he handed out business cards with "Vote Weaver for Sheriff" written on them. On the other side was written "Get out of jail free."

He was like the college radical who runs for student body president, promising to abolish student government. Randy said Boundary County had turned into a business, feeding itself on the people by taxing and fining them until their needs were secondary to the Beast, to feeding and keeping the revenue-monster going. The philosophy of Randy's campaign was a long-standing right-wing thought: that the sheriff was the only legitimate law in the county and the place to begin repairing the screwed-up government.

But if Randy's candidacy was right-wing dogma, it was also good-old-boy neighborliness. He promised to enforce only the laws the local people wanted, so that if locals voted to allow drunken driving, he'd enforce it that way. Seat belts and motorcycle helmets were stupid, he said.

"I question whether they should pass any laws to protect the individual from themselves," he told a newspaper reporter. "I believe in what I call Scriptural socialism: take care of your neighbor."

His candidacy got the attention of the *Spokesman-Review* from nearby Spokane, Washington, the largest newspaper in the area. And so one of its reporters, who was based in Sandpoint, called Weaver at the little house on the highway where he was living and interviewed him. But after a wide-ranging interview, Randy got mad and said that he'd assumed the reporter was just chatting with him and that he'd never given permission for an article to be written. The reporter said he'd identified himself at the beginning of the conversation and that, clearly, he was interviewing Randy for a story. Randy said the media adhered to the One World Government and would twist his words and make his candidacy some sort of race issue, when that wasn't it at all. "If you want to make a racial issue out of this thing, they're going to bring busloads of blacks here to prove a point," he said. "The people up here don't want that. I don't want that."

Furious, Randy said that if his quotes were used in the story, he would call his attorney and sue the newspaper. The next day, the story ran on an inside page, under the headline "Boundary Sheriff Candidate Opposed to Mixed Marriages."

"I have been to the Aryan Nations World Congress to see what's going on," Randy was quoted as saying in the May 21 issue of the *Spokesman-Review*. "I've also been to the Boundary County Task Force on Human Rights. Both have good things."

Randy said he wasn't a member of the Aryan Nations, but he shared their beliefs that the Constitution was being subverted and that interracial marriage was against God's will. Still, he said, "some of my best friends are of other races." He said he also shared the group's belief in a white-dominated America and that the Constitution had to apply mainly to white people "when you consider that it was white folks that wrote the Constitution and some of those folks owned slaves.

"I don't believe in slavery," Randy added, "but, religiously speaking, I don't believe in mixing the races."

Sheriff in such a large, sparsely and strangely populated county was often a difficult job to fill. In 1983, the sheriff of Boundary County had gone to Alaska for a vacation and just never returned. His chief deputy assumed the job because there was no one else, and then, five years later, he quit, too. But in 1988, there were five people eager to jump into the spot, three Democrats and two Republicans. On the Democratic side was a part-time school bus driver, a bartender, and Boundary County's first-ever bailiff. Running for the Republicans were Lonnie Ekstrom, the chief investigator for the sheriff's office, and Randy Weaver.

Ekstrom said Randy's views of law enforcement were impossible. "He's much more radical that he lets on," Ekstrom said.

In a county of 7,000 registered voters, 1,500 people voted—most of them for Democrats. Randy got 102 votes, which some folks took as a pretty good estimate of the number of white separatists and constitutionalists in their county. Randy lost the Republican primary to Ekstrom, who had 383 votes.

*T*here was a dispute between Randy Weaver and Steve Tanner over $30,000 that Tanner said belonged to him. It was one of those strange mountain deals that no one wanted to talk about above a whisper. But some people in the Naples area just knew what they saw—the Weavers suddenly renting this nice house on the highway, driving a new Ford Ranger pickup, quickly buying and selling other rigs.

Tanner called his friends in the area—including the Raus, who lived below the Weavers' cabin—and told them about the dispute. Once again, Randy and Vicki Weaver were at the center of a Boundary County feud.

Some of their former friends theorized that, once the Weavers decided someone wasn't one of God's chosen, they had the right to do whatever they wanted to that person. Everyone was either on their side or was part of the great conspiracy. Their severe beliefs didn't leave any room in the middle. People figured the Weavers were so desperate to provide for themselves and prepare for the Great Tribulation they could justify any action. Whatever the truth, the fact was that almost all the initial friends the Weavers had made when they came to Bonners Ferry were now enemies.

*I*n the summer of 1989, the whole family was planning to go to the Aryan Nations World Congress. Ever since they'd arrived in Idaho, the Weavers had been drifting toward Christian Identity, the racist religion of the skinheads and the Aryan Nations. They called God "Yahweh" and Jesus "Yashua" and eventually dropped the word Christian entirely from the description of their beliefs, saying they were Identity believers. If you looked up the word "cretin" in the dictionary, they pointed out, you'd find a reference to the word "Christian." They'd picked up books and literature at the Aryan Nations before, and Randy and Vicki had found some things to help in their latest project—researching old Hebrew words (like

Yahweh) and trying to find the holy words that had been replaced in the English language by pagan words. Humans, for instance, weren't white people, they were hue (color), mans (people). Aryans were the true people chosen by Yahweh. Sammy picked up his mother's love for language and devoured the Bible, while Sara studied the way her mother prayed and understood that their religion was different from everyone else's because of the way it was revealed directly to them by the Creator.

Even though they agreed with many of the tenets of Richard Butler's church, there were differences. For instance, the Weavers celebrated the sabbath from Friday at 6:00 p.m. to Saturday at 6:00 p.m., while the Aryans held a traditional Sunday celebration. There were other reasons Randy had resisted joining the group. He told friends that Aryan Nations was filled with ex-cons and others he didn't want to associate with. Yet he liked going to the world congresses every summer, listening to the speakers and the philosophical debates, and keeping up with the bluster of the movement and tossing in his own opinions.

The 1989 world congress continued in Richard Butler's recent theme of trying to recruit younger people. The focus was on skinheads, the young soldiers who would step in to take over from the old men of the movement. Randy, Vicki, and the kids had brought a friend, a visitor from the Deere plant in Iowa, and they all camped at the edge of the Hayden Lake church compound. They stayed in tents and in the back of Randy's pickup, barbecuing and drinking beer with other families and with some skinheads from Las Vegas whom they'd met.

Later, the skinheads remembered being impressed by the Weavers. Here was a strong, God-fearing group of people who understood Scripture and were as tight as anyone they'd ever seen. But someone complained that Randy had given the young skinheads beer, the kind of thing the Reverend Richard Butler frowned on, especially since his compound was being watched so closely by law enforcement.

Even though the Weavers were having a nice time

camping, the congress itself wasn't as dynamic as those of the past, certainly not like the star-studded gathering of 1986. The widow of Gordon Kahl—a North Dakota tax protester and Posse Comitatus member killed by federal agents after a shootout and standoff—was there, but otherwise it was the same old rhetoric by leaders who seemed more and more tired. The pressure from law enforcement finally seemed to be getting to Butler, and people talked among themselves of the need for new leadership. The best part, for Randy, was always seeing friends he'd met at earlier conferences. In fact, toward the end of the gathering, Randy looked up and saw one of the first guys he'd met, the biker Gus Magisono. Gus hadn't been at the last Aryan Congress, but this year he had an important job. Butler had assigned him as the security officer for the widow Kahl.

"Hey, Gus, how you been?"

"Good, Randy. How about you?"

The informant Ken Fadeley said he'd been too busy to come in 1988. They talked a little and Randy introduced "Gus" to some friends from out of town who'd come for the congress. The biker asked where Kumnick was, and Randy said they'd had a falling out.

"Give us a call sometime, Gus," Randy said.

He was supposed to be watching some of the Aryan leaders here, not a fringe associate like Weaver, but Ken Fadeley figured he'd keep in touch with him, just to keep an eye on what he and Kumnick might be doing.

"Yeah," Fadeley said. "If I get a chance, I'll come up and say hi."

*I*n August, Ken Fadeley drove his Nissan Sentra through Naples, a couple miles north on the old highway, where he pulled off the road and into the driveway of the little red rental house. Fadeley was impressed. The house was nice, the yard well kept, and a new Ford Ranger pickup truck sat in the driveway. Even though Randy had complained about money at the world congress, it looked as if he was doing okay. It was a beautiful summer day, and the mountains behind the house framed it with lush green.

"It's good to see ya," Randy said.

"This is a pretty nice place."

"Thanks."

As they started for the house, Fadeley was surprised to see Frank Kumnick step outside. He thought there was a rift between Kumnick and Weaver.

"Well, hi Gus," Kumnick said. "How are you?"

Inside, Fadeley felt some tension between the Weavers and Frank Kumnick. From the kitchen, Vicki Weaver offered to cook Fadeley anything he wanted. She made some chicken soup and they small-talked, Randy and Vicki returning to their favorite subject: Scripture. They had some cookies and coffee and talked about how unimpressed they all were with Butler's gathering that year. "The same old rhetoric," Randy said.

Those old guys always repeated themselves, they agreed.

Then, out of nowhere, Kumnick and Randy got into it, arguing about where the Weavers had gotten the money for the new truck and the rental of the little house and the other things they'd been buying. Fadeley thought it might even turn violent.

Randy couldn't believe it. He was pissed off, and Fadeley backed him up. "It's nobody's damned business," he said.

"Exactly," Randy said. As the argument heated up, Vicki began crying, curled up in a chair in the living room, saying she was tired of all the rumors.

Finally, Kumnick left. Fadeley walked him out, and they made an appointment to meet at a restaurant in Bonners Ferry later that day. When Fadeley returned to the house and told Randy that he was meeting Kumnick later, Randy said Frank wasn't trustworthy anymore.

"Frank's too radical," Randy said. He said he didn't want Frank to know what he and "Gus" were talking about. Frank and his wife had been their closest friends in Idaho, Randy said. And now he couldn't be trusted. "He's gone off the deep end."

The Weavers had lost another set of close friends, in fact the oldest friends they had in Boundary.

Before Fadeley left, he testified later, Randy launched into some bizarre speech about how he was being prepared for something by God, being set up to do something big for the white race. And then the children came running inside, saying the dog was about to have puppies, and so Fadeley made his way out, although he and Randy set up another appointment, this time to talk business.

After the August meeting, Fadeley talked with Byerly about where to go next in their investigation of white separatists. They had some fourteen hours of tape on Kumnick, but the only crime they had on him was a sawed-off rifle that Fadeley said Kumnick sold him. Since all the craziness with the kidnapping plot and the talk of Super Glue and stripping federal agents, Kumnick was keeping a low profile.

But something was going on at the Aryan Nations. Federal authorities were aware of the dissatisfaction with Butler's leadership and knew the legal pressure had driven committed warriors away from the Aryan Nations. In fact, the ATF was becoming more interested in some people just over the border in Montana. Specifically, the bureau wanted to investigate a constitutionalist named David Trochmann and a Ku Klux Klan member named Chuck Howarth, who had settled in Montana in 1987 after getting out of prison on an explosives violation. Richard Butler's top two lieutenants eventually would leave him to join Howarth, at a church they promised wouldn't have the swastikas that covered Butler's compound. "Adolf Hitler is dead and the Third Reich is gone; it's history," Howarth said. "All you've got now is seventy- and eighty-year-old men sitting around talking about history. It has nothing to do with the movement and problems we face today."

The Montana leaders knew the movement was destined for bigger and more widely accepted things. Later, the Trochmann family formed the Militia of Montana, the prototype for militia movements around the country.

In 1989, authorities were investigating tips that

Trochmann and Howarth might be gunrunning, a violation with which they were never charged. But at that time, it was up to Fadeley to figure out exactly what they might be up to. Randy Weaver had talked about knowing those guys, Fadeley told Byerly, and maybe he might lead the informant to Montana.

On October 11, Randy and Kenneth Fadeley met again at Connies, around 9:00 a.m. After taping the first meeting with Weaver, Fadeley hadn't worn a wire since. Again, he wasn't wired, but he made careful notes of the meeting.

They made small talk, had a cup of coffee. Randy asked if Fadeley had been busy.

"Very much so," Fadeley said. He'd sold all of his "product"—his guns. "How are you surviving?"

"Just that, Gus." Then Randy said times were tough and that they were moving back up to the mountain because they couldn't afford rent anymore and that the world was sliding again. In the Soviet Union, for instance, Gorbachev was having trouble, and it looked like the hard-liners might dump him any day, sparking a war with the United States. "It's all goin' down the tubes," Randy said. Fadeley had met with Randy only a few times, but almost every time he heard that the world was goin' down the tubes.

Fadeley brought up the Montana trip. Maybe at the end of the week, he and Randy could take a little drive over, check in with Mr. Howarth.

Randy said that was fine.

They talked some more about guns, which ones were in highest demand among the arms dealer's mysterious clients. Fadeley said the biggest sellers were .223s, .308s, and shotguns, especially the shotguns. Twelve-gauge mostly.

Then, according to Fadeley, Randy said he was ready to do business with the gun dealer. He said that he could get his hands on Remington model 870 shotguns—the standard duck hunter's gun—and that he could saw the barrels off five shotguns a week if there was a market. Randy claimed it was Fadeley who asked *him* to do business.

They walked out of Connies and stepped behind the restaurant. Randy got into his truck and pulled a gun case from behind the seat. He opened it, pulled out the kind of shotgun they'd been talking about, and pointed to a spot on the barrel.

According to Fadeley, Randy said, "I can cut it off to about here."

The government informant pointed to the gun and said, "About here." Or, "About *here?*" Fadeley insisted later at trial that it was a question.

Weaver said it was an instruction.

Either way, they were set. They made a plan to meet on Thursday, exchange the guns and the money, and drive to Montana to meet Randy's friends. Randy seemed relieved. "I need to make some money, Gus. Hey, if this works out, maybe I can keep feeding my kids."

"It's a struggle, ain't it?"

Back in the office, Fadeley made his report to Byerly, and they talked about using Randy to get to the new leaders in Montana. As soon as Randy got them to Howarth and Trochmann, Fadeley said, they would be done with him. He was a nobody, certainly not a leader in the movement or a target for law enforcement. But any big criminal investigation is made up of dozens of smaller ones—tiny bits of seemingly unrelated information, intelligence from everywhere, guys rolling over on their buddies—until the whole case suddenly comes together. Randy Weaver had a role in this investigation, getting them to Montana. In 1989, as they prepared the Montana plan, the ATF didn't even have an active file on Randy Weaver. So that day Byerly typed his name into the computer to see if he had a criminal record or a file elsewhere at ATF. Nothing came up.

Two days later, on October 13, in the ATF office in Spokane, Herb Byerly tapped out the phone number of Randy Weaver's house and then handed the telephone to Kenneth Fadeley. A woman answered.

"Hey, Vicki, this is Gus. How you doing?"

They talked a little, and then Randy got on the phone. Fadeley said he couldn't go to Montana like they'd planned because his mother had suffered a stroke. He looked up at Byerly, who'd told him to come up with some excuse because they couldn't get air support to cover the dangerous trip to Montana that day. They put it off until they could get a plane, and Fadeley had to make up the excuse about his mother. In truth, she had been dead for three years.

"I have to catch a flight down to see her," Fadeley told Randy. "She's in real bad shape." Now they talked about "chain saws"—the code word for guns that Fadeley had picked up at the Aryan Nations the summer before. When they wanted to talk about gun lengths, they substituted the chain saw's bar length. Fadeley said that if Randy could really produce such short chain saws, he had a buyer.

They made plans to meet at Connies again on October 24. "Uh, maybe you could put something together and bring it with you," Fadeley said.

They repeated the drill, meeting again at Connies, and driving, this time to City Park, the same place they'd gone that icy day in January 1987, when Kumnick had pulled a gun and a stud finder on the big, affable informant. After going without a wire during the meeting when the gun deal was actually set up, Fadeley wore one for the transaction itself.

Weaver was suspicious again, but Fadeley tried to put him at ease, saying the guys he was dealing with would never cause him problems. "I've been in this business far too long to get fucked over now."

Weaver showed him two shotguns, one pump-action, the other a single shot, both about five and a half inches shorter than the law allows. For the guns to be legal, Randy would have to apply to the government and pay a $200 registration fee. Of course, he had done neither.

Randy said he cut the guns himself, "sitting under a shade tree with a vise and a hacksaw. When I get my workshop set up I can do a better job."

"That's good quality," Fadeley said. "Oh yeah, that's beautiful. How 'bout I pay you three hundred for both?"

"I'm going to have to have three hundred on the pump," Weaver said. The single shot would cost $150. Fadeley promised to bring him the rest of the money later. They transferred the guns and returned to Connies to talk some more. Randy said he needed the money badly because he was having such a tough time feeding his family.

"There's money to be had in it, you know," Fadeley promised.

"That's what the good name of the game is." As far as the guns, Randy said, "Personally, I hope they end up with a street gang."

"I got an idea where some go. . . . It's just, number one, there's no need to know."

"No," Randy said. "I don't want to know. . . . You know what my biggest problem is?" Randy asked.

"Yeah?"

"My biggest problem is going to be running around and picking up here and there and . . . going out and making buys."

"Well," Fadeley said. "That's why . . . I want it to be worth, worth your time. You figure four or five a week?"

"Yeah. Or more."

"Pretty good," the informant said. "That's paying for all your running around."

At the end of the meeting, they talked about the Aryan Nations. "What do you hear from down south a little ways?" Fadeley asked.

"I haven't heard anything from them," Randy said.

"I've never felt comfortable with them," Fadeley said.

"I don't like them," Randy said. They were back to the same conversation they'd started two years before, when they sat in Randy's Wagoneer with Frank Kumnick and agreed that "the movement" needed new leadership.

And then the meeting ended. Randy Weaver went home to his family and Kenneth Fadeley drove off to meet Herb Byerly and give him the two sawed-off shotguns he'd just bought for the U.S. government. By the time this case was over, they would be the most costly shotguns in the world.

eight

Gaunt and tired, Randy Weaver walked up to the passenger door of Kenneth Fadeley's red Nissan Sentra, which was parked in the usual meeting place, along the grassy strip behind Connies Motor Inn in downtown Sandpoint. It was just before 9:00 on the morning of November 30, 1989—another slate-colored winter day in the Panhandle. Fadeley and Randy were supposed to drive across the Montana state line to a little town called Noxon, where they were going to meet Chuck Howarth and the Trochmann family. This time the informant was not alone. ATF agents were everywhere, four of them parked in two cars a few blocks away in Sandpoint, one waiting to pick up the tail in Noxon, and a pilot on standby in Spokane. This was to be the major attempt at getting inside the Montana movement and the beginning of a new phase of Kenneth Fadeley's undercover mission. It was the opportunity for ATF to get inside a group that even the FBI had failed so far to infiltrate. And once they got to Noxon and Fadeley got inside, this portion of his investigation would be over and Randy Weaver would've served his purpose.

"This is a fancy car," Randy said as he slipped into the passenger seat.

"Well, actually, it was a hell of a good deal," Fadeley said. "This thing got wrecked."

"I can't go to Montana," Randy blurted out almost immediately.

"How come?"

"Oh, I'm busy. I got somethin' goin'."

"Can't even run over there for a short time or nothin'?"

Randy shook his head. "I really better not, Gus."

Fadeley worked him for a while, saying Randy had promised and that it would just take an hour, that he'd driven all that way. But Randy wanted to talk business. He wanted to sell more guns. He and his family had moved back up onto the mountain and Randy was broke and getting desperate. He wanted to sell four more single shots and a double-barrel. Then, maybe after the deal, Randy would have time to go to Montana. Or at least he'd think about it. Randy got out of the car and into his flatbed pickup, and Fadeley followed him to a mall at the other end of town, where they parked next to each other. Fadeley got out and they stood between the two vehicles, their breath steaming as they dickered over the price of the guns.

"Let's sit in the car and talk about it," Fadeley said. "I'm freezing my ass off."

Inside the clean new Nissan, Randy went over the math again: four single shots at $150 each, that's $600. A double-barrel at $300 and the $150 still owed on the first transaction. Randy wanted $1,050.

Fadeley wasn't going to give a grand to a guy they didn't need anymore. He said he didn't have the money and blamed it on his "contact," the guy who was eventually getting the guns. "He only sent me a hundred dollars to give ya on that last deal," Fadeley said.

The informant counted out four twenty dollar bills and two tens—most of the money still owed from the last deal—and put them in Randy's hand. Randy put the money in his billfold, slid it back in his pocket, and lit up a cigarette.

"I'll tell you what, Gus," he said, drawing out the smoke. Then he started in again about family, wondering why the biker had never introduced Weaver to his wife and kids. Randy talked about this all the time, Fadeley realized, as if showing your family was some kind of lie detector.

"If we could meet your family, that would make things a lot cooler, you know?"

"We can set that up," Fadeley lied. He was not about to drag his family into this, have his wife and kids pose as the family of a racist biker and gun dealer. Fadeley ran the conversation back around to Montana. "Would it hurt if we just met this guy and had a cup of coffee?"

Randy said he wasn't comfortable.

"I don't want you having any kinky feelings about me," Fadeley said. He reminded Randy that this was a business they were involved in. "We're both in this to make money." Fadeley made it clear that Randy initiated the gun deal. "That's pretty much how you've approached it with me, and you said, 'Gus, I wanna go to work for you,' that's what I figured you wanted to do. I think I was right on that, wasn't I?"

"I'll be honest with you," Randy said. "I'd like to feed the family, but it's more than that with me." He said that's why he had to meet Fadeley's family, to see if the biker was clean.

The informant backpedaled. "Let me do this, let me get with the wife, and let's see what we can come up with."

Randy took a breath. "I had a guy in Spokane tell me that you are bad."

Fadeley went numb. "Who told you that?"

"I'm not gonna give you this guy's name, but I talked to the wife about it, Gus, and let me tell you something. That's all I care about is my family, and if I go to prison, or anything, I'm gonna be pissed off. That's all I gotta say." And then Randy said that Fadeley initiated the gun buy. "You approached me and offered me a deal."

Fadeley realized he could be losing three years of work here, and he scrambled. "I would like to know who the fuck told you that. He's lying through his teeth 'cause I'm not a badge, and if I was a badge, then I suppose I'd be wired, and you're welcome to check me for a wire." After the close call with Kumnick two years before, the ATF wasn't wiring Fadeley anymore. Instead, the tape was fixed to the underside of the dashboard, and it continued to run as the informant tried to convince Randy he was okay.

"Whoever told you that was a fucking liar," the ruddy-faced informant continued ranting. "Why, go back to Frank. Frank did some business with me. You see Frank walkin' the fuckin' street.

"If you want to believe someone else that is walking around paranoid that I'm a . . . fuckin' pig, it's been nice doin' business with you," Fadeley said. "Have a nice life. In the long run, it all pays the same, and you figure how many conferences I've been to. I've been to your house! Stupid Frank sittin' there in your Jeep or whatever the hell that was, with a goddamn stud finder and a gun on my head. . . . I mean, let's get real. Dismiss those ideas right there."

Fadeley's tirade seemed to work, and Randy muttered that maybe he was wrong. "You know, I didn't really believe this guy, but I don't want to take chances, you know. When somebody says something like that, it makes you think."

Randy tried to make Fadeley feel better by admitting he'd been mistaken for a cop before, too. "Didn't it bother you, knowin' that I ran for sheriff?" he asked.

"You lost, didn't ya?" Fadeley asked.

They both laughed.

"You only got twenty-five votes, didn't ya?"

"I got more than that."

Fadeley put an end to the subject. "Will you do me a favor? You get ahold of whoever this clown is in Spokane, and you tell him to shove it up his ass, and tell him it'll be a good idea for him to do it. Because if I find out who he is, I'm gonna shove it up his ass for him."

They talked again about whether this meeting had been set up for the Montana trip or as a way to exchange more guns.

"I wish you had brought some money," Randy said. "I came down here to make a living. . . . And I'm sorry, you know, I couldn't go to Montana. I gotta get home. I got somethin' goin' on up there and, uh, the next time that I tell you I'll go with ya, you know, I'll make sure I'll go with you. No bullshit about it."

And then Randy got out of the car.

"All right," Fadeley said.

"Thanks," said Randy.

Fadeley watched him climb into his pickup and drive off. As he turned his car around and headed off to meet Herb Byerly, he wondered just how much Randy Weaver really knew.

O nce again, Randy and Vicki had new best friends. Bill Grider was a 6-feet-3-inch ironworker, parcel-service driver, and softball player from Detroit who, a few years earlier, had followed his wife, Judy, and adolescent son, Eric, to Moscow, Idaho, a town at the edge of the state's flat farmland, 120 miles south of Sandpoint. "I was running for my life when I left Detroit, physically and spiritually," said Judy Grider, whose vibrato speech was a thick soup of Bible references. In Moscow, the Griders opened a cleaning-and-maintenance business for apartment buildings, but Bill hated Moscow, which was also home to the University of Idaho and people who were too liberal, pretentious, and artsy.

So, in June of 1989, Bill and Eric went north, looking for a place to fish and—eventually—to live. They were fishing the Kootenai River, a cold, rock-bottomed ribbon of clear water, when they met Randy and Sam, who were fishing the same stretch of river. When he heard Randy's views of the Bible, Bill couldn't wait to tell his wife. Like he figured, when Judy met Vicki Weaver and heard her views about the Identity religion, something clicked into place, and Judy Grider realized the Father—whose name she was learning was Yahweh—was stirring her to action.

"We've always been into the Scriptures, always

searched for truth," Judy said later. "I didn't know Identity. I didn't know there was such a thing until I got to Idaho. I was excited when I heard about it. It made more sense. It just all sort of fit together."

Bill had been raised a Presbyterian, Judy an Episcopalian. But when they moved to Naples in late 1989, they became Identity followers and the Weavers' closest friends.

That fall, Randy and Vicki and the kids moved back up to their cabin in a caravan of loaded pickup trucks that ground up the dusty road past Wayne and Ruth Rau's place. Close friends of Steve Tanner, the Raus had heard about the dispute over $30,000 with Randy. They ignored the Weavers when Randy and Vicki drove past and waved. And the Rau children—who had always been good friends with Sammy and Sara—ignored the Weaver kids when they waved from the cabs of the overloaded pickup trucks.

While they were living down the hill, the Weavers had allowed the Raus to run a pipe from their spring down to the meadow, but there wasn't enough water for two families, and so when the Weavers got home, the Raus lost the use of the spring. Randy said the Raus tried to make a claim on the spring. The Raus said the Weavers just cut their pipe with no explanation. When the Raus' dog disappeared, they blamed Randy, and when it showed up two days later, they never apologized. The antagonism seemed to have its own momentum, and soon the two families were at war.

But the Weavers had an ally. The Griders had moved up to the meadow and squatted in an old house on Arthur Briggs's land, near the trailer where the Weavers had first lived. Briggs didn't live there anymore, and the land had been repossessed by the IRS, which had virtually no use for it. The Griders spent much of their time up the hill at the Weavers' cabin, studying the Bible, yelling at the Raus, and target shooting with the family at cans and at a "Weaver for Sheriff" sign.

The Weavers who retreated back up to Ruby Ridge were more militant and frightened than the family that had moved down the hill eighteen months before. Randy's yearly trips to Aryan congresses had deepened his racism and had made the family some severe friends—like Aryan Nations member Proctor James Baker and his wife, Katy—whose stories and experiences confirmed everything the Weavers had been saying about the government conspiracy. And, strangely, the family began to swear more often. Pagan words like Christian were the true profanities, the Weavers believed. Words like fuck, shit, and goddamn were simply the rough-hewn language of pioneers.

Along with the Griders, the Weaver family's Christian Identity beliefs became even more militant. Impressed by the skinheads' commitment at the Aryan Nations Congress and moved by a Bible passage in Jeremiah in which warriors shaved their heads in mourning for Israel, Randy and Samuel Weaver and Bill and Eric Grider shaved their own heads.

Sammy and Sara, along with Eric Grider and another boy from Naples, wore swastikas and marched with their guns in front of the Rau place, calling the Rau kids niggers and chanting, "What do we want? White power! When do we want it? Right now!" Whenever the Raus saw Randy, he was wearing a holstered pistol. The Weavers and Griders fired guns at night and dumped garbage on the road in front of the Raus' house. The Raus told the sheriff, but when he questioned Weaver, Randy claimed the Raus were harassing him and had let the air out of his tires. It was another bottomless feud.

For more than a year, the three families yelled back and forth and always seemed on the verge of a fight. The Raus accused the Weavers of stealing the pipes for their water supply, and later federal agents found the equipment at the Weaver house. But the constant gunfire was the worst for the Raus. They were afraid the Weavers were going to finally snap and kill them.

Once, during a yelling match near the Rau property, Ruth found herself standing near Vicki, whom she always admired for being such a hard worker, whom she always suspected of being smarter and gentler than Randy.

"What did we ever do to you?" Ruth asked suddenly.

The acrimony disappeared from Vicki's face, and Ruth found herself looking at a woman with—more than anything—deep fear and bruised feelings. "You wouldn't let your kids play with our kids."

Kenneth Fadeley wasn't going to stop working Aryan cases just because Randy Weaver was suspicious. A few months after his last meeting with Randy, he met with two Aryan Nations members—a Montana Ku Klux Klansman and the group's most colorful member—a 300-pound, former professional wrestler named Rico Valentino. An engaging mix of bravado and flamboyance, Rico had shown up suddenly in 1987 and had become an occasional bodyguard for Richard Butler and one of the biggest financial supporters of Aryan Nations. He wore dark sunglasses and a pound of jewelry, overtipped every waitress he met, and made a beef stew so tasty, it was said to be the chosen food of Aryan warriors. Fadeley had tagged Valentino as one of the more dangerous members of the Hayden Lake church, certainly a guy to watch. When he wasn't flying around the country, promoting professional wrestling matches, Valentino was hanging around the Aryan Nations compound, flashing wads of bills and teaching martial arts and wrestling holds. So, that night in early 1990, Fadeley met Rico and the other man at a restaurant in the Spokane Valley and listened to see what kind of things they might be planning. After the meeting, as Valentino drove away, Fadeley wrote down his license plate number. But when he ran the plate, he found it odd that it was registered under a different name.

In March, Fadeley returned to the Aryan Nations compound and told its leaders he had been gone because he'd had a heart attack. The Aryans welcomed "Gus" back, and he quickly got involved in the preparations for a local skinhead conference. His assignment was to write an article for a brochure that would go out to the young men. His topic was commitment.

Toward the end of March 1990, as Fadeley sat at a picnic table on Butler's pastoral church grounds, the Aryan security chief, Steve Nelson, began walking toward him with a man named Proctor James Baker and another man. Baker held a video camera, and when they reached Fadeley, Nelson told him to say hi to everybody.

They set the camera down, and Nelson sat facing Fadeley. Baker moved in behind him. These were two of the men ATF agents had suggested he watch—the severe, thirty-five-year-old Nelson and the fifty-seven-year-old Baker, guys Fadeley had identified as two of the most radical members of the group.

"Gus," Nelson said, "the license plate on your car doesn't match your vehicle. How could that be?" Apparently, he wasn't the only one writing down license numbers.

"I've been driving a car that belongs to my dad," Fadeley said. "I'll have to ask."

"Gus, do you know who Kenneth Fadeley is?"

Oh boy. They were all alone on the grounds of the Aryan Nations, and these guys were jamming him. With his peripheral vision, Fadeley tried to see what Baker was doing, but the older man had moved in behind him and so Fadeley stared straight into Nelson's eyes.

"Yes," he said. "I know him."

"Well, is he a close relative or something like that."

"Something like that."

Nelson said they had two addresses for "Gus." And then he read off Fadeley's address and his dad's address. The informant had never given either address out. Did they have a source in law enforcement? Had they been following him?

"Are you a member of any law enforcement agency?" Baker asked.

"No."

"Is your father in law enforcement?"

"No."

"Are you a member of the Jewish Defense League?"

"No."

"Why don't we have an address for you?"

"You do. A post office box."

"No, I mean a residential address."

Fadeley, trying to stay calm, said he was staying with friends and that he didn't need a bunch of Aryan Nations members ruining his friendships.

"I need to be able to trust you, Gus," Nelson said. "Right now, there are too many circumstances here for me to be able to do that. I'm going to have to ask you to leave."

Given their other options, Fadeley didn't think it was such a bad offer. As he walked off the compound toward his car, the informant figured that, as far as the Aryan Nations were concerned, Gus Magisono was officially dead. At the ATF, Byerly looked at the two cases that had come from the Fadeley operation—Kumnick and Weaver. They were beginning to realize Kumnick was all talk and wasn't connected. But Weaver . . .

*I*n May, just two months after kicking Fadeley out of the Aryan Nations compound, Steve Nelson, Proctor James Baker, and another man, twenty-nine-year-old Robert Winslow, were arrested by the FBI and charged with conspiring to blow up a gay disco in Seattle. Nelson and Winslow were on their way to Seattle when they were popped. Baker was at his house in Hayden Lake, with his wife, Katy, when FBI agents showed up and searched the house. An FBI informant had taped one hundred hours of conversations with the men, in which they talked about the plan and even set off a practice bomb, packing it into a coffee can filled with gravel and nails. After it detonated, spraying a field with fragments and embedding nails into trees, Baker said, "Think what that would do to a roomful of people."

At first, Richard Butler refused to believe that his close associate and bodyguard, the former wrestler Rico Valentino, was an FBI informant. He said Valentino had gone with Nelson and Winslow to Seattle to record an

English version of "The Panzer March," one of Hitler's favorite songs. When the others were arrested and Valentino didn't come back, Butler had to acknowledge that he had been duped again.

Valentino claimed to be a born-again Christian who went undercover to fight evil. He was paid $100,000 by the FBI for the three years he worked undercover and flashed the money around, tithing more than any other member and buying a 550-gallon water tank for use during the Aryan Nations World Congress. He paid to roof and floor a bunkhouse for visiting skinheads and bought two guitars for church services. He even paid $3,300 for the large metal "Aryan Nations" sign that covered the gate of the compound. He taught martial arts to skinheads and got to know every member of Butler's groups.

Defense attorneys for the bombers said Valentino had set their clients up, encouraging them to act out a macho, right-wing fantasy that they would never have tried if the informant hadn't been coaxing them into it and tossing around his substantial cash. They argued that the informant had spent three years inside the Aryan Nations without finding anything, and so he was eager to set someone up. Baker's attorney, Everett Hofmeister, wondered how his client could have been involved when he didn't even drive to Seattle with the other men. The key physical evidence tying him to the bombing, a pipe bomb, was given to him by Valentino, Baker said. And a piece of pipe found in his house was going to be used to vent a fuel line in his new home, not for a bomb. Why, Baker had even allowed the FBI to search his house.

But the prosecutor in the October 1990 case was Ron Howen, the assistant U.S. attorney from Order I, Order II, and the Bud Cutler trials—a guy who was making a career out of Aryan prosecutions. He brought expert witnesses who testified that one section of pipe in Baker's house was capped on two ends, with a small hole drilled in it—perfect for making a bomb. Howen presented a detailed, steady case—thirty-eight witnesses over seven days—that focused on the men's racist beliefs and on how those beliefs translated into criminal acts. On October 19, 1990, all three were convicted of possessing explosives and plotting to use another bomb in what they called a "kill zone" around the gay bar. The key witness was Rico Valentino.

After running his phony license plate, Fadeley had figured Valentino for an informant. And he guessed Valentino had made him for an undercover operative as well. Their covers were mirror images of each other, Fadeley realized—big, flashy, tough-talking guys who had the money for the wild plans these guys talked about. Fadeley's only question now was who had burned him. His money was on Valentino, either to prove his own value to the group or to keep ATF from the embarrassment of having an undercover informant involved in the bombing plot. But there was another person who could've blown his cover: Randy Weaver.

*O*n Ruby Ridge, the disco bombing case scared and angered Randy. All the other Aryans who had ended up in prison were just names to him, guys who might have been set up, but also really might have been criminals. But Randy knew Proctor James Baker. PJ was one of the guys from the summer conferences, an old auto mechanic, not a terrorist. They were at the Bakers' house just before it was searched. The arrest especially hardened Vicki, who later wrote that it taught her a lesson.

"Remember my friend whose husband was 'framed'? . . . He spent two years in prison. He got home last December half a man. He physically doesn't look well and half his memory is gone. '*They*' gave him a drug to destroy his mind and told him it was heart medication. He called my friend, his wife, three times in one day and couldn't remember talking to her. She asked him if he was taking any medication and if so the name of it. He gave her the name and she went to a druggist and asked what it was used for. It destroys the mind (memory) and what's gone will *never* come back!!" Vicki said she would not let that happen to her husband.

In the meantime, things had gotten so bad between the Weavers and the Raus that Wayne and Ruth were looking for some way to split the Griders and Weavers up. So when they saw that the IRS was selling the repossessed house that the Griders had been living in—rent free—they quickly bought it. They instructed their attorney to find someone to evict the Griders, and so he hired the county bailiff, a man named Ron Sukenik, who had spent fifteen of the previous sixteen years working in a restaurant.

On June 7, 1990, about 6:00 p.m., Sukenik drove his four-wheel-drive Dodge Raider across the Ruby Creek Bridge, followed closely by two Boundary County sheriff's cars with their lights on. They drove to the Griders' house and stopped in the driveway. Sukenik stepped out and stared at a flag that he recognized as the Aryan Nations banner—a Z on its side with a sword through it. A red truck sat near the house with a plywood sign—"Yahweh is white power!"—leaning against it.

"That's far enough." Ten feet away, Sukenik would later testify, Randy Weaver appeared with a Mini-14 semiautomatic rifle. Bill Grider stepped out from behind a berm with a long pistol holstered on his hip and then Vicki Weaver came out, her own revolver sheathed on her waist.

They were fanned out in front of him, Sukenik testified. In his short career as a bailiff and court security officer, Ron Sukenik had never been on this end of a gun before and he tried to stay calm. He spoke to Randy, who lowered the gun to the ground.

"I'm not here to see you," Sukenik said. "I'm here to see Bill."

Randy began preaching about the IRS being an illegal entity, a tool of the federal government and the Jews. As Randy spoke, Sukenik allowed himself a deep breath. From the sheriff's car, one of the deputies engaged Weaver in conversation about his beliefs, and he turned his attention away from Sukenik long enough for the bailiff to explain to Vicki what he was doing up there and to pass the papers along.

Vicki walked over, took the papers, and gave them to

Bill. "They're throwing you out of your house," she said simply.

"I gotta go," Sukenik said. He and the deputies backed away slowly and left. A few days later, Sheriff Whittaker saw Grider in Bonners Ferry and told him if he didn't move out, Whittaker was going to move him out. So Bill and Judy moved down the hill and once again the Weavers were alone.

A green Forest Service pickup truck bounced and rattled along Randy Weaver's driveway and stopped in a cloud of dust below his house. The first thing the two men saw as they approached the house was a Confederate flag, stirring in the light breeze above the cabin. It was a clear, sunny afternoon, June 12, 1990—just five days after Ron Sukenik's eviction of the Griders.

Sara, who was thirteen, came running out first, her black hair pulled back in a tight ponytail, her long skirt clean and neat. She looked like any teenage girl except for the World War II gun belt slung around her waist and the semiautomatic pistol sticking out of the covered holster. "What do you want?" She was polite but businesslike.

Ten-year-old Sammy came out next, wearing jeans and a T-shirt, with a hunting knife holstered to his hip and a short, butch haircut.

Herb Byerly leaned out to talk to the kids. Even after twenty-two years with the ATF, it was unnerving to see children walking around with guns. Byerly could see Randy Weaver's truck wasn't there, and so he and the other agent, Steve Gunderson, figured Randy wasn't home. "Hi," Byerly said. He asked what road they were on. "We're from the Forest Service. Have you seen another crew up here?"

"No," Sara said, "I haven't."

Byerly and Gunderson took another look around, then backed up, drove down the ridge and onto the old highway, where they spotted Randy's red flatbed truck outside one of the rooms at the Deep Creek Inn. Byerly recognized it from the meetings with Fadeley in Sandpoint, and so the ATF agents parked across the street and waited for Randy to come out.

Since Fadeley's cover had been blown, the ATF needed someone else who could tell them what was happening with the Aryan Nations, and especially the separatists in Montana who were under investigation. Randy seemed like a perfect candidate. He was ex-military and had run for sheriff in Boundary County, so they figured he might, at least, have some respect for law enforcement. He had the trust of the Aryan Nations without any ties to them and, clearly, he didn't agree completely with their views.

Byerly and Gunderson waited until they saw a woman come out of the motel to smoke a cigarette. Then they drove through the gravel parking lot and across the street, rolled down their windows, and asked the woman if Randy Weaver was inside. Byerly said they wanted to talk to him privately. The woman figured them for local Forest Service employees and went back inside.

Randy came out in a black leather jacket decorated with two Nazi SS lightning bolts, a dress shirt, jeans, and an Aryan belt buckle that caught Byerly's attention. Randy walked up to the truck, smiling, and stood near the open window. Vicki stood well behind him on the porch, watching, just out of earshot.

The agents introduced themselves, and Randy stepped back. "Don't say anything," Byerly said. "Let me finish what I have to say." He told Randy they had evidence that he'd manufactured and sold sawed-off shotguns. He offered to play a cassette tape of the gun buy and showed Randy Polaroid photographs of the guns. He said they'd presented the case to the U.S. attorney's office and—while they didn't have a warrant yet—there was a good chance it would go to a grand jury and Randy would be indicted on federal weapons violations. But there was something he could do.

Gunderson, a former sheriff's deputy in Montana, was good at these kinds of pitches to blue-collar and country folk and so he took over for his partner, trying, in a friendly but foreboding way, to lay out the different ways this case could go. He called it telling someone what their future held: "Number one, Herb has got a case here, and he thinks it's prosecutable, and he thinks you can be found guilty. But here are all your options: You could cooperate. You could wait until you get arrested. You could go to a jury in North Idaho and you might be able to beat the case and you would never go to jail. But you'd have to go through the process of being initially arrested, getting initial appearance, getting bonded, and all that. That's a big hassle. Or you could decide to work with us on it. Doesn't mean you'd have to actually get somebody arrested for us. It just means you go out of your way to cooperate with us, and then you can evade all this other stuff that could occur."

"You can assist yourself," Byerly finished, by providing information on other Aryan Nations members, particularly the group in Noxon. They wanted him to be a spy. Byerly wrote the ATF address and phone number, and "Herb" on a piece of paper and handed it to Randy. "Come alone, tomorrow, to the federal courthouse in Spokane at 11:00 a.m."

Randy had listened quietly, but when the agents were done, he said he wouldn't be a snitch. "You can go to hell," Randy said. He turned and began walking back to the motel. "Vicki, Vicki! Come on out here!" The agents drove away.

At the time, Gunderson thought they'd played it just right. It was a reasonable offer, and even if Randy didn't take it, he wasn't going to fare that badly on a simple gun charge, not with a jury likely made up of gun-owning, antigovernment Idaho citizens. It'd be probation or a minimum security work release facility at the very most. It wasn't until later that Gunderson realized Randy didn't see it that way and that Gunderson was making a mistake treating Weaver the way he treated the normal criminals he dealt with. He'd never seen Randy's expression on any of the gunrunners and drug dealers he usually busted. Those guys understood just *how easy* it would be to get out of

this. Randy didn't seem to get it. "Usually, I'm dealing
with someone who knows, deep down inside, that he's
committed a crime," Gunderson said. "But with Randy's
beliefs, I don't think he felt he'd done anything wrong in
the first place. He just looked at us with these vacant
eyes. . . . It was going right over his head." It would not be
the last time federal agents would misjudge Randy Weaver.

R andy and Vicki couldn't believe it. It was just like
Proctor James Baker, like Randy's supposed threats
to President Reagan! Why couldn't they leave the
Weaver family alone? He was being set up! Gus had come
to him, had shown him where to cut the guns. They
decided Frank Kumnick must be a snitch, too. Well, Randy
would rather go to jail than be a snitch! The Weavers raced
up to the cabin, gathered the children, and explained what
had happened. Vicki cried, and they prayed that the Creator
show them what to do in the face of such oppression and
deceit. And then Yahweh guided them. About 10 p.m.,
Randy and Vicki assumed the familiar roles, Randy raging,
Vicki writing. She addressed the letter to "Aryan Nations &
all out brethren of the Anglo Saxon Race." In it, Vicki
explained how the ATF agents had approached Randy and
had tried to make him a snitch and how she and Randy and
the children were ready to stand for truth and freedom. She
challenged "the Edomites" to bring on war. When they
were done, Bill Grider took the letter to the Aryan Nations
compound.

> If we are not free to obey the laws of Yahweh, we
> may as well be dead!
>   . . . We have decided to stay on this mountain,
> you could not drag our children away from us
> with chains. They are hard core and love the truth.
> Randy's first thought was to let them arrest him to
> protect his children—but he is well aware that
> once they have him the Feds will send agents to

search and destroy our home, looking for "evidence." He knows his children—they won't let that happen to their mother.

Let Yah-Yashua's perfect will be done. If it is our time, we'll go home. If it is not, we will praise his Separated Name! Halleluyah!

The conspiracy theories flew like snow in a November storm. The government was trying to separate the family and indoctrinate the children! The Raus and Tanners were trying to steal their land, just as Kinnison had tried! It was a plot to kill all the Green Berets! Frank Kumnick had been a plant, a government agent whose mission was to introduce Randy to the informant! The Jews, working through the Masons and the government, were finally making their move on those who knew Yah's truth!

Two, three, four at a time, the Weavers' friends trekked up the drive-way and witnessed the family's ever-deepening fear and their resolution not to be bullied.

"If we are up here when they come, we will stick by you," said Judy Grider, who had taken on Vicki's prophetic tone, her stern, matriarchal leadership, and even her denim-skirt wardrobe. "They'll encounter two families with the strength of Yahweh."

But there was another problem in the Weaver cabin. The money from Iowa was long gone and—whatever had happened with Tanner—that money was gone, too. The family was broke. Randy had made a few hundred bucks on the shotguns he bought and sold, but that business didn't last long. They cut cords of wood for elderly people but, despite all that time in the woods, never came across a deer, and so Vicki worried that they wouldn't have enough protein that winter. They sold off some of their possessions and picked up an occasional eighty-dollar check from their settlement with Terry Kinnison. But the family was worse off than it had ever been. And so, when they stopped paying their property taxes, it was partly because of politics, partly because they just couldn't afford it anymore.

The tax bill was piling up, Randy was under investigation for gun charges, and the family was out of money; they were backed as far as they could go. And they responded the way they always did when they felt the pressure of the world against them. They became even more radical. Now it was time to make a stand, time to reject the lawlessness of Babylon.

Angry summer faded into fall and winter, and ATF agents contacted Wayne and Ruth Rau and asked for their help. In December, an indictment was handed down for Randy Weaver's arrest on the gun charge, and a letter went out to the Weavers' attorney. The Raus agreed to take a radio and contact the agents whenever they saw Randy and Vicki Weaver drive down the hill.

On the ridge top, Randy was torn. "Maybe I should turn myself in, so nothing happens to the kids." But Vicki was convinced that as soon as Randy left, the government would confiscate their land, send the children to AIDs-infested mental hospitals or foster homes, and—when the Great Tribulation began—the Weavers would be spread throughout the Beast's institutions and prisons. They took a vote; none of the kids would allow their father to turn himself in. Sammy, especially, was unbendable. He would not have his mother subjected to the horror of losing their land.

But by winter nothing had happened, and Randy and Vicki were still low on money. They set to work cutting firewood and selling it in town, rushing while they still had decent weather because the December storm clouds were moving in and each load could be their last.

nine

*O*n the first Sunday of December 1990, dark, bloated storm clouds crashed into the Selkirk Mountains and left fifteen inches of snow on the granite and softwood of Ruby Ridge. On top, the snow slid off the Weaver cabin's metal roof and piled three feet against the doors, isolating the family as perfectly as they could ever hope. The wind riffed through the plywood walls, while the kids huddled around the woodstove, reading action books beneath kerosene lamps, playing Scrabble and Parcheesi, and trying to learn chess from their father. Vicki read Scripture, wrote letters, and mended rips in the family's snowsuits, until, after the storm, she and Randy broke a trail through the snow from the ridge top to the meadow below and rode snowmobiles to the county road, where their pickup truck was parked. They drove the truck to the store, filled their backpacks with milk and a few other groceries, drove back to the base of the mountain, climbed aboard the snowmobiles and rode back to the top of the hill. Such trips to the grocery store were day-long events, during which the kids took care of the dogs and other chores and waited with their holstered pistols until Randy and Vicki returned. Even after the snow passed two feet, the couple went into the woods most days, trying to find a cord or two of dry firewood to sell. But their chain saw wasn't working very well, and the tough winter slowed them down considerably. They had a few older customers, but during a North Idaho winter, trees are about the only thing in ready supply, so most folks cut their own wood. Vicki got word of a couple of refinishing jobs waiting for her in the spring, but one of the furniture shops was going out of business because some government agency had ruled that the chemicals used to strip the furniture in the shop's big tank were too dangerous. "Our wonderful tyranny in action again," Vicki wrote. The Weavers were doing so badly, they considered selling the flatbed truck, probably their last possession of any value.

Two more snowstorms battered the ridge until, on January 17, the weather cleared a bit, and Vicki and Randy started up the snowmobiles and rode them down the hill, past the Raus' house. Just below the meadow, they switched to the pickup and continued down the snow-packed road until they reached the bridge leading to the old highway. A pickup truck and camper was broken down on the bridge, wedged sideways just on the other side of Ruby Creek. A blond, shaggy-haired guy was bent over the truck's open hood, his head buried in the engine, while his mousy wife stood next to him, shivering in jeans, a shirt, and no coat.

Seven years in Boundary County hadn't shaken the Iowa from Randy and Vicki, and they stopped the truck before the bridge. Randy opened his door and hopped out. The young woman walked toward Vicki's side of the truck, and so Vicki opened the door and stepped out into the snow to greet her. "What's the matter with your truck?"

"Oh, it's broke down," the woman said. "We were trying to get it off the road and you were the first people who came along."

Randy kept walking toward the open-hooded pickup. From behind, he could see the guy—early thirties, scraggly jeans, flannel, surfer hair—still messing around under the hood. "Ya broken down?" Randy asked. Fifteen feet from the truck, Randy slowed, and the long-haired guy spun around in the snow and jammed a 9-mm pistol toward Randy's face. "Federal agent, you're under arrest!"

The county sheriff, Bruce Whittaker, jumped out of the back of the camper with three ATF agents, including Herb Byerly. From the woods, agent Steve Gunderson, in full snow camouflage, kept the scope of his AR-15 trained on Randy Weaver's chest.

"Get on the ground!" yelled Lance Hart.

Vicki was the responsibility of agent Barbara Anderson. "Turn around!" yelled the petite woman with no coat. "Get down!" Vicki turned to run away, but Anderson ran a few steps, pushed her, and Vicki fell face first into the snowbank.

There was yelling and confusion as Hart wrestled Randy to the snow-crusted ground, put the gun against his stomach, and waited for the other agents to help subdue Randy. A short newspaper story about the arrest said Randy "offered no resistance," and he wasn't charged with resisting arrest.

But during the struggle, Hart testified later, Randy had slapped at the agent's gun and then had reached for his own coat pocket. When another agent got there, he reached in Randy's pocket and pulled out a .22-caliber pistol. They read Randy his rights and handcuffed him. Byerly ran to where Vicki lay on the ground, and they cuffed her hands behind her back, too, grabbed her by the arms, and stood her up. She was mumbling something that Herb couldn't quite make out—the steam slipping from her lips—until he realized it was chanting, some sort of prayer in a language he'd never heard before. Barb Anderson ran her hands all over Vicki, who stared off haughtily—horrified and angry to be slammed into the snow and then body searched by the agents of Babylon. There was nothing on Vicki. Back in the truck, her purse contained a small .38-caliber handgun, which the sheriff gave back to her as soon as everything calmed down. They took the cuffs off Vicki. But the agents pushed Randy into a car, and as Hart walked by, Randy nodded at him. "That was good," he said. "But you'll never fool me again."

Vicki cried as they drove her husband away.

When it was over, Byerly thanked Anderson and Hart, a former navy pilot who had been an ATF agent since 1986—when he'd begun growing his hair out. By 1991, it was long enough for a dishwater ponytail, and Lance had become a specialist in undercover operations. Young and muscular, with stooped shoulders and surfer looks, Hart met the most important requirement for undercover work—no matter how long you stared at him, he just didn't look like a cop.

The ATF had arrived the day before, and Byerly had given Ruth Rau a two-way radio, telling her to call when Randy came down the mountain. That morning, Hart and Anderson had waited with the others in Naples until Ruth Rau called and said the Weavers were coming. And then they moved in. Byerly felt good that they'd arrested such a dangerous, stubborn man without any bloodshed.

It was a good ruse, but it didn't do much for Randy and Vicki Weaver's distrust of the government.

In Coeur d'Alene that afternoon, Randy stood in front of the booking camera, tight-mouthed and stunned. He refused to sign his name. That night, when he complained about the conditions in his jail cell, one of the jailers said, "Wait until you get to prison."

Shackled at the wrists and ankles and wearing an orange jail jump-suit, Randy Weaver walked in small, chain-rattling steps down a hallway with Byerly on one side and Barbara Anderson on the other. They escorted him into a long, narrow courtroom on the second floor of the federal building in downtown Coeur d'Alene—one of the buildings damaged five years earlier by the bombings of The Order II. At the end of the off-white room, in front of windows overlooking shimmering Lake Coeur d'Alene, an attorney and part-time magistrate named Stephen Ayers waited to hear the case. Since there was no federal judge in Coeur d'Alene, it was the magistrate's responsibility to try first appearances and misdemeanors for federal prisoners in Coeur d'Alene.

Anderson left for a moment and came back with coffee for the three of them. Randy held his Styrofoam cup with both hands, because, with his wrists cuffed, it was the only way he could drink it.

The charges were read, and Ayers entered a not-guilty plea for Randy, who leaned over the railing and talked with his wife and with Wayne Jones, the security chief from the Aryan Nations, who was there to vouch for Randy's character.

After only one night in jail, Randy—like many federal prisoners—didn't have an attorney yet. But what upset Byerly was the fact that the U.S. attorney from Boise, Ron Howen, had declined to fly up for the arraignment. Byerly had called Howen—the U.S. attorney's expert on white separatists—and told him about the arrest. He told the prosecutor that Randy and his kids were armed and that, if the government didn't request Randy be held until the trial, there was a good chance he would jump bail. But federal attorneys didn't travel to Coeur d'Alene often, and Howen didn't think they had grounds to keep Randy in jail. So the first appearance continued without an attorney for either side.

Ayers set a $10,000 unsecured bond for Randy's release, meaning he'd have to pay the bond only if he didn't show up for court or if he violated the terms of his release. Ayers added the conditions that Randy find a job if he didn't already have one, that his travel be restricted to northern Idaho, and that he check in with probation officers constantly. He was to give up all firearms or destructive devices and stay away from alcohol and drugs.

Ayers asked Randy if he had a preference about an attorney, and when Randy said he'd like Everett Hofmeister, the magistrate said he'd do his best to appoint him.

"Yeah," Randy said, "in fact, if I can't have him

appointed, we will try to hire him. My wife might even have to hock her ring."

"All right," Ayers said. "Well, you need to understand that . . . if you're found guilty of this charge, you will probably be required to reimburse the government for the cost." Ayers later admitted he had made a mistake.

Ayers read the pretrial report, which concluded that Randy's only real asset was his property, which was assessed at $20,500. "And you own that free and clear?" he asked.

"Yes sir," Randy said. But he said the land was probably worth only half that. Now Randy was confused about the bond. "You're telling me that you're recommending that I could be released with this ten-thousand-dollar deal over here. I don't understand, secured bond? I can't get no ten thousand. I don't understand."

Ayers said Randy didn't have to put up the money, but if he didn't abide by the rules, then he forfeited his bond and the land could be sold to make up the $10,000. Again, Ayers was mistaken about how an unsecured bond worked.

What if he got pulled over for having dirt on his license plate, Randy asked. "Would I blow the whole ten grand there?" Randy asked. "Would they throw me in the slammer?"

Ayers said probably not.

Randy also was confused about the gun issue. "How far does that go? My kids own firearms."

"You can discuss it with your attorney," Ayers said, "but, you know, in my view, I think you'd probably be better off not having those things in your house. You'd better put them with somebody else."

But what if he needed a gun for protection?

"Look, Mr. Weaver," Ayers said, "even if you are in fear for your life, you cannot possess a firearm."

"I'll get rid of them," Randy said. And then he promised to show up for court, and he signed the paperwork that indicated he would show up in Moscow, Idaho, a month later, for trial.

•   •   •

Randy and Vicki raged and prayed all the way back to Ruby Ridge, gathered the children, then raged and prayed some more about what they should do. The treachery and deceit were almost more than the couple could bear. The lawless highwaymen of ZOG had no legal warrant, didn't identify themselves, seized Randy's legal gun, and usurped the power of the county sheriff. Vicki wasn't charged with a crime at all, and yet they treated her like a gangster.

The vision that came from Yahweh coincided with the family opinion: they were to stay on the mountain forever. When it was settled, Vicki wrote a letter to the U.S. attorney's office and then another one to "all my Saxon brothers and sisters in the Aryan Nations." Vicki wrote that Israel was outnumbered millions to one and that they should never give up. "Don't they realize that one with Yahweh is the majority? He shed His blood for us, can we expect to do any less."

There was more proof of the One World Government's plans when the notification of Randy's trial date arrived from the U.S. attorney in Boise. Although Randy had been told to report to court on February 19, the notification listed the trial date as March 20—something they later tried to pass off as a simple mistake. Perhaps they were trying to confuse him or get him to show up on the wrong date, so they could throw the leg irons on him again and say he'd missed his trial date. Their lying treachery was so transparent.

Finally, Vicki wrote a letter to her mom, back in Iowa. She tried to reassure her, promising that the family was okay and that the arrest had been unlawful. The government had no intention of giving Randy a fair trial, Vicki wrote, and the Weavers were not going to obey such liars. "The conditions of his probation would be impossible not to violate," Vicki wrote. "We are staying home. We have been shown to separate from lawlessness and warn our people of the coming troubles." Vicki sealed the letter and sent it down the hill with Judy Grider.

In Fort Dodge, Jeane walked slowly away from her

mailbox, planted along the straight rural road with a hundred others, like rows of short saplings. She carefully opened the envelope, pulled out the letter, and read Vicki's clean cursive. Its tone was strange and—apart from the usual Yahweh and One World rantings—it was like a good-bye note.

"We have all loved you and pray that Yashua bless you for all the things you've done over the years to help us," Vicki wrote on the last day of January 1991. "The past seven years have been trial and testing for me. I know what I need to do to take care of my family under hardships. . . .

"I'll keep writing as long as someone can come up. We have the peace that passeth understanding."

*I*n the February 7 mailbag at the U.S. attorney's office, there were letters and packages addressed to most of the thirteen lawyers who worked in the office, but only one was addressed to the "Servant of Queen of Babylon." Like he did with hundreds of other pieces of mail, the young clerk opened the envelope, glanced at the writing, stamped the letter with the date, and set it in one of the piles, this one for Maurice Ellsworth, the U.S. attorney for Idaho. The clerk walked into the bright offices on the third floor of the federal court building and slid the letter, along with the others, into the U.S. attorney's mailbox. The U.S. attorney's secretary came by, grabbed his mail, walked into the office, and set it on his desk. And that's where Maurice Ellsworth found one of the strangest letters he'd ever seen. It was addressed to Ellsworth, who was listed as the Servant of the Queen of Babylon. The letter said the stink of lawless government had reached Yahweh and Yashua. "Whether we live or whether we die," the letter read, "we will not bow to your evil commandments."

It was dated February 3, 1991, mailed from a P.O. box in Naples, Idaho, and signed "Mrs. Vicki Weaver."

The attachment that Ellsworth was supposed to pass up the chain of command was more alarming. There was more of the Yahweh Yashua stuff, a Bible quote from Jeremiah, and—this is what worried Maurice Ellsworth—a quote he soon found out was from Bob Mathews, the martyred leader of The Order. "A long forgotten wind is starting to blow. Do you hear the approaching thunder? It is that of the awakened Saxon. War is upon the land. The tyrant's blood will flow."

Ellsworth got strange antigovernment letters occasionally—with suggestions for Japanese trade or Russian foreign policy—but this was unlike anything he'd ever seen. Ellsworth picked up the telephone and called the supervisor of the U.S. Marshals Service in Boise, the agency charged with protecting federal buildings and employees, as well as arresting fugitives. A deputy marshal came to his office, and Ellsworth handed him the letters. He said they appeared threatening and he'd like to know what this was all about.

The marshals quickly found out Vicki Weaver was the wife of a guy who'd recently been arrested. They found out that Weaver had been a difficult arrest, that his kids were armed and even slept with their guns, and that Randy had vowed not to be arrested again. Weaver's court date had been changed from February 19 to February 20. His attorney was notified and he tried to let his client know, but Randy wouldn't answer his letters. So, on February 20, a failure-to-appear warrant went out for Randy Weaver, and the case was assigned to the U.S. Marshals Service, the agency charged with bringing in fugitives. In Boise, the chief deputy marshal, Ron Evans, sent a letter to his boss in Washington, D.C., writing that Randy Weaver had the potential to be "another Bob Mathews and his homestead another Whidbey Island standoff."

*D*ave Hunt wasn't worried. In fifteen years as a deputy U.S. marshal, Hunt had seen plenty of guys like Randy Weaver. How many times had he gone to a government auction where some guy's life was sold out from under him, all because the dope decided to skip his trial date or not pay his taxes? Guys like that lost their whole lives over $600, over principles and ideologies that have no basis in reality, no logic. Hunt didn't understand that blind devotion. He was a guy who was ruled by sense.

A former Marine with two tours as a guerrilla warfare trainer in Vietnam, Hunt had arrested more than 5,000 fugitives in his fifteen years as a deputy marshal. Yet he'd never fired his gun in the line of duty. He was world-weary and gruff, a guy who stooped and shambled along persistently and who could bring anyone in with enough reasoning and cajoling. His best weapon was his tenacity. He just wouldn't give up. A big, gentle Baloo-looking guy, the other marshals saw him as the hardest ass in the service, simply because he wouldn't give up.

At first, Hunt didn't think Weaver was any more dangerous than any of the other tax protesters and government bashers in the West. But his boss, Chief Deputy Evans, had a bad feeling about the case from the minute he saw the Queen of Babylon letters. Evans had been the chief deputy in North Dakota in 1983, during the shoot-out and standoff with tax protester and Posse Comitatus member Gordon Kahl. Wanted on a simple probation violation, Kahl got into a gunfight on a rural highway near Medina, North Dakota, and two lawmen were killed. Three other officers were injured, along with Kahl's son. The sixty-three-year-old Kahl fled to Arkansas where, four months later, lawmen found him hiding in a house near the Ozark Mountains. When the local sheriff walked into the house to arrest the fugitive, Kahl shot him. A federal agent outside the house also fired, again hitting the sheriff, who, before he died, managed to squeeze off a shot that killed Kahl. Agents outside didn't know Kahl was dead, and so they fired tear gas inside, blistered the house with gunfire and, finally, dumped fuel into the house and burned it to the ground. The case made Kahl the first modern martyr among the radical right wing.

Evans saw the same potential for violence in this case. By February 6, a clerk had typed "very uncooperative" in Randy Weaver's Marshals' file and someone scribbled "crackpot" on the case file. By February 20, after Weaver's initial court date passed, his description went out over the National Crime Information Computer as: "Aryan, carries firearms." Evans and Hunt waged a friendly debate over how dangerous and how stubborn Weaver was. From the beginning, Hunt felt the pressure from Evans to bring in the tactical guys and go after Weaver—to rush the cabin. But the deliberate, hangdog Hunt proceeded the way he always did in such cases, learning the fugitive's habits, tracing his family, figuring out where his money came from, trying to get inside his head.

He and the other deputy assigned to the case, Warren Mays, went over the case file and read the biblical passages that Vicki and Randy quoted in their letters. From his desk in the marshals office on the seventh floor of the federal building in Boise, Hunt looked out the window at Schaeffer Butte, which towered over Boise. Marshals duty in North Idaho was considered the worst assignment in the country because of cases just like this, stubborn antigovernment types whose failure to appear became more serious than the actual crime they'd committed. Such cases, once they started going badly, could pick up their own momentum, like a stone rolling downhill.

There wasn't a lot of personal stuff on Hunt's desk, a few family pictures and a photograph behind him, a picture of the bloody car door of a law officer who had gotten too excited during an arrest and had accidentally shot himself in the leg. It was a good reminder to proceed cautiously and a warning of how badly things could go.

*N*aples, Idaho, looked like a town that was in a constant state of evacuation. In March of 1991, it was essentially a dying rail yard and lumber mill, a tattered school and a general store, all clinging to an old highway in a glacial valley—the town's gaps filled by trailers and single-story houses in need of new paint. Down the road a piece, Budweiser was on tap at the North Woods Tavern, where the occupation of every third customer was "handyman"—some guy looking for work on a road crew or a dairy or a Christmas tree farm, anything to pay the property taxes while they finished the corral and got the roof on the log cabin. The town wasn't incorporated; there might have been 300 people in a mile radius of the Naples General Store and five times that many with Naples addresses who lived on the veins of dirt that cut into the foothills and mountains around the town.

In early March 1991, Dave Hunt and Warren Mays drove up to Sandpoint and checked into Connies Motor Inn, the same place Weaver used to meet Kenneth Fadeley. The deputy marshals drove north on Highway 95, eased onto the old highway and into Naples. Their mission consisted of the patient drudgery of fugitive chasing: find the Weavers' friends and ask what it would take to get them down. They spent five days in northern Idaho and talked to people who knew the family, to a gun-shop owner who'd sold Randy some guns and to the people who ran the general store and post office—who told Hunt that the Griders had been picking up the Weavers' mail.

They drove to Bonners Ferry and talked to the sheriff, Bruce Whittaker. When they were done interviewing him, Hunt asked if the sheriff could help them find Bill Grider.

"Yeah, that's him over there." Whittaker pointed across the street to a grocery store parking lot where Hunt saw an imposing, muscular guy, 6 feet 3 inches tall, with a bushy mustache, loading groceries in a battered green pickup truck.

Bill and Judy Grider were suspicious, but they listened.

Hunt said he was trying to solve the problem peacefully and would the Griders please take up a message. "Tell him he has to surrender to the courts. I don't care how he does it. He can do it through Hofmeister, he can do it through the local sheriff if he has more faith in that, he can do it directly to the court, or he can surrender to me. I don't care how he does it. I just want to get this thing resolved."

Hunt and Mays met the Griders the next day at the house they were caretaking—the red house on the old highway that the Weavers had once lived in. Bill and Eric Grider had shaved their heads since the day before.

They were in mourning, Bill said.

Thirteen-year-old Eric wore a white T-shirt with a skull and "White Power" written on it. He pulled from his wallet a worn piece of Scripture—Jeremiah 7:29—and began reading: "Cut off thine hair, O Jerusalem, and cast it away, and take up a lamentation on high places. . . ."

Grider said he'd passed Hunt's message along and he gave the marshal three letters from Weaver—handwritten copies of the two that had already been sent to the U.S. attorney and a third—with the same defiant, poetic language.

"You are servants of lawlessness and you enforce lawlessness. You are on the side of the One World Beastly Government," the letter read. "Whether we live or whether we die, we will not obey your lawless government."

That apparently was a no. Already tired of this case, Hunt knew this was no way to carry on a dialogue.

"Why shouldn't I just go up there on the mountain and talk to him?" Hunt asked

"That's not a good idea," Bill Grider said. "Randy's ready to meet his Maker." He told Hunt that Weaver's kids were up there and that the whole family was pretty strong in their convictions and their belief that they were the target of a conspiracy. "Let me put it to you this way. If I was sitting on my property and somebody with a gun comes to do me harm, then I'll probably shoot him."

Hunt got along well with the Griders though, and he had the impression that he was building trust. He shared a beer with Bill, and young Eric was fascinated to hear the stories of Aryan heroes that Hunt had arrested or transferred, especially David Tate, the Order member who shot and killed a Missouri state trooper in 1985.

Hunt went back to Boise, filed his notes, and kept working the telephones, talking to anyone who knew the Weavers, anyone who had a suggestion about how to get them down. He even called Vicki's parents in Iowa. They were very nice and tried to be helpful, but it was clear they had no control over Randy either.

Alarmed by Evans's constant concern about the Weaver

case, Hunt kept detailed notes, and after his first trip to northern Idaho a portrait of Randy Weaver emerged:

- He was paranoid, maybe not clinically, but in practice. Friends said he saw deputy marshals behind every tree and connected unrelated incidents as part of an ongoing plot to defame and maybe even kill him. Whenever Randy did settle down, his wife seemed eager to stir him up into his irrational fears again.
- He was not in control. Vicki was. Randy wasn't even clear on the doctrine he followed, according to Hunt's interviews. During Bible study sessions, Randy would get so worked up, he'd lose his place and have to catch Vicki's eye. "Where do we believe on that now?" She'd set him straight, and he'd go off again. From the beginning, Hunt figured that if he could get to Randy when Vicki wasn't around, he could begin real negotiations. Without his wife, Randy would fold like an empty shirt.
- He was very sociable. He seemed like the kind of guy who only wanted to hop in his pickup truck, drive to a buddy's house, sit on his porch, and bitch about the government. More than anything, Randy seemed to want to be a preacher, to have a following of people who agreed with his views. After all, this was a guy who—just a couple years earlier—had run for public office. He didn't seem suited to isolation.
- They were overestimating his military training and the danger that might pose. Randy had never been in Vietnam, although he didn't seem to mind people thinking he had been. And he was no explosives expert, just an army grunt with some Special Forces training, which—despite the wild rumors—didn't teach you to make bombs out of household items like soap and matches. Hunt completely discounted the theory that Randy booby-trapped his mountain.
- He was lazy and quite possibly a coward. He hadn't held a job for a long time, and even some of Randy's

relatives said he was "as lazy as they come." Vicki, who had grown up on a farm, seemed better equipped for the privation and solitude of Ruby Ridge. But one thing about Randy bothered Hunt more than anything else: his cowardice, the fact that he didn't hesitate to place his children right in the middle of this danger.

Something else struck Hunt about the Weavers. He'd done a little work on The Order and Order II cases, and he recognized three elements to this kind of people: racial, antigovernment, and religious beliefs. Randy and Vicki Weaver were especially dangerous, because to them, it was almost all about religion. And Hunt knew that when someone thinks they've been ordered by God to do something, they're going to do it.

"Let's go work some other cases," Hunt told his boss, Ron Evans, after a couple of months with the Weaver case. There was no hope for negotiations right now. Randy was so paranoid and his wife so controlling, he wasn't going anywhere. He was parked on that mountain, daring the government to make a move. "Then fine," Hunt said. "Let him sit up there. I don't care." Maybe if they gave it some time, the Weavers would calm down, and they could figure out a way to start negotiations without the kids right in the middle. Hunt had his own kids. He thought of the family on that craggy point, with no running water—too hot in the summer, too cold in the winter—no company, all jammed into a plywood cabin. He wanted to pound a sign into the base of Ruby Ridge: "U.S. Penitentiary, North Idaho."

Vicki was pregnant. In Iowa, her family couldn't believe it. Randy gets in trouble with the law, and Vicki gets pregnant? It made no sense. It was insane. The first part of the pregnancy was awful. Vicki was anemic and so nauseous in the morning, Sara and Rachel had to climb the narrow stairs to the sleeping loft and serve her meals in bed. But Vicki couldn't stay down and even when she was sick she worked, starting an inventory of her maternity clothes. She'd saved her nursing bras and had three jumpers and plenty of skirts, and she decided to put a panel in another skirt and make a couple of cotton dresses. "Good thing big shirts were popular for a while," she said.

Her parents sent care packages—a curling iron for Sara, belts for the family, mixed nuts for the kids, and pantyhose for Vicki. She wrote them back a short, nervous letter ("I don't really want to put anything on paper") and told them not to bother sending checks, because they couldn't leave the mountain to cash them.

After such letters and the trouble with the law, Vicki's parents couldn't take it anymore. They had heard only Vicki's side of Randy's arrest, so they called the sheriff and got the other side. In April—four months before their regular visit—they wrote Vicki that they were coming out immediately to try to talk the family into coming down from the mountain, so that Randy could turn himself in and Vicki could have the baby in a hospital.

Vicki was outraged. Sometimes Bill Grider delivered the mail, and Vicki sent down notes to Judy with him. This one said her parents "had been in contact with [the sheriff] and are on their way out here. They say how nice he is and we can trust him. I'm furious!" Going to court, she said, would be "going to our own hanging."

"I now have to contend with more unbelief and deception," Vicki wrote. "Such is my job. I should be glad to be back to work. But I wish the Master would not tarry."

When David and Jeane Jordison arrived, they were especially worried about Vicki, who had worked herself raw all winter and had been wiped out by the early part of her pregnancy. Her hands were calloused from constantly knitting wool socks and mending other clothes. They worked hard on the cabin that spring, fixing the pump that David had started the summer before. For the first time, the Weavers had running water.

But the Jordisons had no luck reasoning with Randy and Vicki. They were adamant. They were going nowhere. David and Jeane gave up and drove back to Iowa.

While the Jordisons and others urged Randy to give up to protect his family, he said that's what he was doing. If he surrendered, they would run Vicki and the kids off the land, and they'd be homeless or worse. Vicki was prepared to stay on the mountain forever. Her parents gave her 200 pounds of potatoes, and Vicki had canned gallons and gallons of berries and fruit that year. She had an endless supply of eggs and several chickens to butcher. For years, Vicki had been buying and storing corn, wheat, beans, rice, and everything else for when they were permanently cut off from the world. She still had the dehydrated fruit, peanut powder, milk, yeast, and vegetables that she'd brought from Iowa, and she'd gotten new stuff, so that commercially packed survival foods lined the root cellar and the kitchen. She made every kind of herb tea and had dried and stored a variety of herbs, nettles, and greens. After all that work, she was furious when she heard rumors that she was starving or abusing her children or keeping them from seeking medical attention.

"I have been given the wisdom of also using and collecting wild herbs for medicine," Vicki wrote her mom, encouraging her to ignore the rumors. "Don't ever let the feds fool you into thinking they can starve us out."

More groceries arrived with Vicki's parents that April, with the Griders and with Kevin Harris, who brought sixty pounds of potatoes, one hundred pounds of flour, three gallons of cooking oil, forty boxes of macaroni and cheese, twenty-five pounds of sugar, four sacks of apples, ten pounds of oatmeal, six dozen eggs, and ten loaves of bread.

When the marshals didn't come after Randy right away, the family went back to their normal lives, playing board games, target shooting, reading, and working in the gardens. Yahweh was testing them in many ways, though. For the third year in a row, the cabin was swarmed by stink bugs, spiders, and ants. The family smashed them and sprayed them and blocked their holes but still the insects came.

They had plenty of more welcome company that spring: Kevin, the Griders, Vicki's friends Jackie Brown and Katy

Baker and some other Ruby Ridge neighbors. On April 23, they had a great birthday party for Samuel—his twelfth— with his favorite meal: roasted turkey hot dogs, chocolate cake, potato salad, chips, and dill pickles.

They also got an "intelligence letter" from someone named Simon, who wrote a melodramatic, staccato note filled with intrigue and One World conspiracies: "In the name of Yahweh, I am just beginning to understand your letter of last summer," the letter began.

> Community has become aware of a considerable number of individual and family "plants" in the area, plus a high number of "agents." . . . Electronic devices in place under satellite system in the area of border counties. The Federal Emergency Management Act signed by Bush in January. Constitution has been set aside by the stroke of a pen. The people have not been informed. This is High Treason. Much Bloodshed Anticipated. . . .

Such information only made Vicki and Randy more confident that they were doing the right thing. "All is well and no threats," Vicki wrote to her family. "This situation will be resolved some day. I'm not desperate, but peaceful . . . [but] I have no way of knowing for sure who is a friend and who is not. We live in a time of great deception."

Ron Evans was right. After a few weeks, Dave Hunt realized the Randy Weaver case was going to be more difficult than he thought. He figured there were only a few solutions: a direct, tactical attack on the house, possibly with stun guns or rubber bullets; a ruse like the ATF had used; or steady surveillance until Randy separated himself from his family, when marshals could rush in and arrest him alone.

But Hunt didn't agree with Evans's other proposition,

that they would have to use a tactical solution to get Randy out of the cabin. Hunt thought there was too much potential for injury to marshals and to the Weaver family. And he still held out hope for negotiating a settlement. He'd sent another message through Bill Grider and had gotten his first almost rational response. Randy would come down if the ATF admitted it was wrong and that it had set him up; if the government returned his pistol; and if the sheriff gave him a written apology for telling people that Randy was paranoid. Those conditions were impossible for Hunt to agree to, but at least he had Randy talking. Hunt also asked Grider to try to get the kids to come down, but he didn't like the answer he got to that proposal.

"If the kids can't live in peace on the mountain," Bill said, "they don't want to live."

So, with the children still up there, Hunt dismissed the tactical approach. The ruse was a possibility, but Randy wasn't likely to fall for something like that twice. Surveillance made sense, but—after looking at aerial photographs—Hunt estimated twelve marshals would be needed to watch the cabin from all angles at all hours, and he didn't have that kind of manpower. They were stuck.

From the beginning, Ron Evans had wanted to bring in SOG, the Special Operations Group, a military-style elite marshals force used for raids and especially tough federal fugitives. The SOG marshals were in the best physical condition and had the best equipment: high-tech optics, surveillance cameras, portable computers, and guns—a dizzying arsenal of state-of-the-art weapons.

But Hunt resisted calling in SOG. He was an enforcement marshal, one of the guys responsible for the day-to-day, often boring, sometimes dangerous work of bringing in most fugitives. Enforcement deputies have always believed they were doing the purest marshals work and were mildly dismissive of the hot dogs who spent thousands of dollars to stake out some house and then bust down the door with an army of marshals. It wasn't that he had any animosity for the intense, cocky SOG marshals—

but Hunt wasn't about to call in the cavalry until he had exhausted every peaceful, commonsense solution. Yet, as the case dragged on, Hunt knew the SOG guys could provide technical assistance on surveillance, and maybe even some fresh troops, so he made a formal request for their help.

Evans flew to Camp Beauregard, Louisana, and met with the SOG commander, deputy commander, and several team members. They went over aerial photographs, reports on the Weavers' beliefs, and photos of the family before coming up with a plan. SOG members would fly to northern Idaho and assess the situation. They planned to come around June 20, Vicki Weaver's birthday, when they hoped Randy would wander far enough from the house for them to separate him from the rest of the family and arrest him. The SOG team handed the case file over to a psychologist, who produced an assessment of Randy and Vicki Weaver from the file. Vicki, the psychologist reported, wanted to keep the family together so much that she might even kill her children. His assessment of Randy concluded:

> In my best professional judgement [sic], Mr. Randall [sic] would be an extreme threat to any police officer's attempt to arrest him. Further, Mr. Randall has indoctrated [sic] his family into a belief system that the end of the world is near and that his family must fight the fences [sic] for evil that want to take over the world. I believe his family may fight to the death. If Mr. Randall is captured by your force, I feel the remaining members of his family will use all force necessary including deadly force to regain Mr. Randall's freedom.

The first SOG team arrived June 17, and Hunt went up with them to Naples, but they didn't need his help. They crept around in the woods for a couple days and then flew back to Louisiana, promising to send Hunt their assessment of what it would take to end the stalemate.

All summer, Hunt waited for their report. In the meantime, he worked other cases, took one more trip up to Naples, and sent other marshals up for background work. Hunt thought about posing as a surveyor to arrest Randy but figured he'd wait for the SOG assessment before he did anything. He thought about cutting off the Weavers' water supply but didn't want to do anything that could escalate the problem. Maybe, Hunt thought, I ought to just drive right up there and talk to him face-to-face. But he realized he'd avoided violence all those years by gathering as much information as he could before acting, and so Hunt decided to wait on this case until he could do some surveillance.

Finally, the SOG assessment arrived, and Hunt dove into it to find their recommendations on surveilling the cabin. There were none. The recommendation was to try to get Weaver away from the cabin and, when that didn't work, to use armored vehicles to get him out. But even the SOG commander didn't seem too thrilled about a tactical solution. He wrote that Weaver was dangerous and maybe suicidal and that there were likely booby traps and bombs set up all over the compound. The Weaver case, the SOG chief said, was the worst fugitive situation he'd seen in twenty-three years as a marshal.

May and June were as peaceful a time as the Weavers had ever spent on Ruby Ridge. In mid-June, while the SOG marshals watched the cabin with binoculars, the Weavers worked in their gardens. Vicki estimated she was about six months pregnant, and since she was past her morning sickness, she began preparing in earnest for the baby, collecting clothes and food. She planned to use cloth diapers, but Yahweh instructed her that she might want some disposable diapers on hand for the winter months, when it was more difficult to wash diapers because most of their water froze.

On June 20, Vicki's birthday, Sara made her a German

chocolate cake but rain clouds moved in and made it too
wet to barbecue chicken. The heavy rains had fallen all
spring and summer, rotting out half the green beans in Sara
Weaver's favorite garden.

Other friends visited and picked up mail at times, and
Vicki wrote that Judy Grider didn't like these other
visitors. Among them was Jackie Brown, an owl-eyed
waitress who had become close friends with Vicki and who
ran errands for her. Jackie's husband, Tony, liked Randy
and enjoyed debating him—they didn't agree politically—
but he wouldn't visit the cabin because he didn't condone
Randy's stand. So Jackie came alone, once a week.

Katy Baker drove up that summer, too—a couple of
times with Rodney Willey, the computer technician at
Hewlett Packard who had met Randy at one of the Aryan
Nations summer conferences. During one visit in June,
they arrived in an orange van, loaded with food and some
toys and books for the Weaver children. As always, the
Weaver kids met the van with their guns and then
allowed it to pass. They had a nice visit and chatted
about gardening, fishing, and the marshals' surveillance.

"Planes are flying over all the time," Randy said during
that visit.

Ruth Rau was taking down license plate numbers as
they wound up the mountain road, then passing them on to
the marshals. Soon after the June visit, Rodney Willey got
a call from the U.S. Marshals Service, asking if he'd pass
on some messages to Weaver. He agreed and, on July 9,
met with Weaver's attorney, Everett Hofmeister, Richard
Butler, the head of the Aryan Nations, and the deputy
marshal from Moscow, Idaho, Jack Cluff.

Butler and Hofmeister gave Willey letters to pass on to
Randy, urging him to come down and settle his case in
court. Jack Cluff's message was verbal.

Willey drove up to the cabin that same afternoon. This
time, he was by himself. Again, the kids responded with
guns and dogs but were happy to see it was Rodney. They
showed him the deck the family had recently finished and
talked about wood cutting and the weather. Then Rodney

told them why he had come. Inside the cabin, he sat in the clean, cozy living room with the family and handed the letters to Randy, who opened them and read them aloud. Rodney passed along Jack Cluff's message too, that he wanted to end this thing and avoid a confrontation. Cluff said he'd meet Randy anywhere, that he wouldn't use handcuffs, and that he could probably get the failure-to-appear charge dropped. And Randy had a good chance of beating the gun charge, Cluff said, or at least getting off with only probation.

"But I've been set up," Randy told Willey. "There's no sense going through all the hassles because I'm innocent anyway."

The visit turned into a vigil. Rodney sat up all night with the Weavers, from sunset to sunrise, praying and talking and going over everything that had happened to the Weaver family. Randy and Vicki held hands and told how the informant had asked Randy to make him some shotguns; how the ATF agents had approached him and said that if Randy didn't spy for them, they would get him "one way or another"; how the feds arrested the couple, shoving Vicki down in the snow and searching her even though she'd done nothing; how the judge had said Randy would lose his land if he was convicted of the crime; how they'd sent the wrong court date to mess with his mind.

"They'll kill him in prison," Vicki said. And then the family would lose the cabin, the government would take the kids away, and Satan would have won.

Rodney tried to convince them to give up, reading passages from the Bible that he hoped would convince Randy and Vicki that he might get a fair trial.

The kids interrupted occasionally, tossing in some point from history or Scripture. Rodney was especially impressed with twelve-year-old Samuel, who knew the names of all the presidents, had memorized encyclopedias, the Constitution, and much of the Bible. He seemed to know the date of every event in Western history from the Roman Empire to the present.

They talked philosophy. Randy was quite a speaker,

Willey thought, but this was where Vicki really shone. She said the Babylonian system had taken over America, through infiltrators who seeped into every level of government, through the Masons, the Jewish Defense League, the Illuminati, and the Trilateral Commission. Vicki and Randy spoke about how the Zionists were in control and were trying to get their tentacles into every part of American life, to create their One World Government. Most of it was familiar to Rodney. For the last decade, he had been a regular every Sunday at Richard Butler's Church of Jesus Christ Christian, on the Aryan Nations compound. But many of the Weavers' beliefs differed from the Aryan teachings. The Weavers relied more on the Old Testament and on messages from God—personal visions and biblical insights that governed their lives more than any doctrine.

They talked all the way around the situation, piling on more and more evidence of the ZOG conspiracy and the slim chance that Randy would get a fair trial. They cried and pleaded in prayer, and finally the whole family talked about it in a sort of informal polling, making sure they all agreed with the stand Randy had taken.

With the sun coming up, Randy said he would not talk to Richard Butler, Everett Hofmeister, or Jack Cluff. He saw what they did to his friend Proctor James Baker, and Randy wasn't going to go through that. If the marshals came for him, he would fight to protect himself and his family.

Rodney Willey left that morning exhausted but impressed with the family's resolve and intelligence. He was still worried about them, but—in the back of his mind—he wondered if they might not be right.

The next day, when Vicki wrote to her mom, she just said friends had been up. She also wrote she was having trouble with Judy Grider and that she and her husband weren't going to run their errands or pick up their mail anymore. "She is capable of any lie to deceive you, so don't believe any phone call you may receive from her as of the past two weeks," Vicki wrote. "All of my letters have had to pass through her hands & I could never write my suspicions to you."

Once again, the Weavers felt they had been betrayed by close friends. Vicki wrote a curt letter to the Griders, who had been seen talking to federal agents and whom the Weavers suspected of trying to steal their money.

"We appreciate what you have done for us in the past, but things have changed and we have made other arrangements," Vicki wrote. "For your own safety, please do not come to visit anymore unless we contact you first. Thanks. The Weavers."

ten

Dave Hunt drove back to Naples in October 1991, hoping to finally end this thing before it gathered any more momentum. He and the marshal for Idaho, Mike Johnson, had mailed a letter to Weaver in August, urging him to give up. In late September, seven SOG deputies had returned to North Idaho to talk about a direct approach on the cabin. But Hunt hadn't given up negotiating and, despite the pressure from Evans and others, wasn't ready to go tactical yet, so he sent the SOG team home. Still, the persistent enforcement deputy was becoming discouraged. Jack Cluff and Ron Evans had driven up the mountain, pretending they were looking for land. Weaver's kids had come out first, and when Randy finally came out, he stood back, shielded by his armed kids, and told Cluff and Evans to move along. It made the marshals sick. Hunt had this awful feeling Randy was taunting him, sending down these ridiculous letters that left no room for negotiation and that pointed them ever closer to a confrontation.

In October, Hunt and Mays found out someone else was

picking up the Weavers' mail, a thirty-seven-year-old family friend named Alan Jeppesen. The marshals drove to Jeppesen's house in Naples and talked to him in his driveway. They listened to his constitutionalist rhetoric for a while, and then Hunt asked the same question he'd asked Grider.

"Can't I just walk right up the mountain and arrest him?"

"I wouldn't," said Jeppesen. But he promised to tell Randy that Hunt wanted to meet with him face-to-face, talk about his beliefs, and see what it might take to end the stalemate peacefully.

The next day, Hunt and Mays met with Jeppesen again, and he gave the marshals a letter written by Vicki and Randy on a single sheet torn from a notebook. The rantings were starting to get predictable: "The U.S. Government lied to me—why should I believe anything its servants have to say. . . . Your lawless One World Beast courts are doomed. We will stay here separated from you. . . ."

So Hunt tried the Griders once more.

"Bill," he said, "can you help me figure out how to do this without anyone getting hurt?" He offered Grider $5,000 to bring Randy down.

"Keep your money," Grider said. Randy had accused him of being a snitch, and the short-fused Grider said he'd just kill Weaver. "I'll call you and tell you where the body is."

It just got stranger and stranger. Hunt knew he had to get past this circular, religious language and macho survivalist paranoia to find out what was really bugging Randy. He wrote a letter promising that the government wouldn't interfere with Vicki's custody of the children; that Randy's friend Alan Jeppesen would be able to stay with him while he was arrested; and that the government wouldn't take Randy's property in place of the $10,000 bond.

He knew he couldn't send the letter without running it past his boss, Evans, and the assistant U.S. attorney, Ron Howen. So he faxed them the letter, along with five other negotiating points he thought might end the logjam, points like the following: that Randy be given his choice of jails while he waited for trial, and that his family be given liberal visitation rights.

"Dear Mr. Weaver," Hunt's letter began. "I was hoping that you would be willing to discuss with me how we might come to some reasonable solution to this matter."

While Hunt waited for permission to mail the letter, Jeppesen brought down another letter written by Vicki Weaver, this time addressed to Hunt and Mays. It listed six questions:

1. Why couldn't government informants be cross-examined?
2. Why did the magistrate tell them they were going to lose their land if they lost their case?
3. Why was there a concerted effort to set up former Green Berets?
4. Why was the government trying to disarm law-abiding citizens?
5. Why were there no more constitutional local sheriffs?
6. Why was there no arrest warrant or search warrant in Randy's case?

Hunt hadn't heard back from Howen about the negotiating points in his own letter, so he wrote another letter, answered the six questions as best he could ("I can't answer number three because I don't know of any concerted effort to set up for prison or murder all Green Berets") and assured Randy that if he surrendered, he'd be treated fairly.

On October 17, he saw Jeppesen one last time.

"I'm not supposed to talk to you anymore," Alan said. He gave Hunt another letter from Vicki: "My husband was set up for a fall because of his religious and political beliefs. There is nothing to discuss. He doesn't have to prove he is innocent nor refute your slander. Mrs. Vicki Weaver."

"Can't I still meet Randy face-to-face?" Hunt asked.

"No," Jeppesen said. "He won't meet with you."

So the marshals drove back to Boise. This case was really starting to unnerve Dave Hunt. He'd never encountered someone so unreasonable. Then, the day he got back to Boise, he got Howen's official response to his proposed letter. They weren't allowed to negotiate with someone who had an attorney and—even though Randy wouldn't meet with his—a lawyer had been appointed. "I cannot authorize further negotiations or discussions along this line with the defendant or his agent," Howen wrote.

The U.S. attorney's office had scuttled the marshals' one best chance to negotiate an end to the case.

Okay, neither side wanted to negotiate. That's it, Hunt told Evans. "Let's let him sit up there another winter." Maybe the Weavers would tire of their primitive existence. The weather was turning, and Vicki was about to have a baby. They couldn't stay up there forever.

There was no way Vicki Weaver was going to have her twins in a hospital. There was too much risk of catching AIDS, even at a small hospital like the one in Sandpoint. Besides, Randy's army training included delivering babies, and Vicki was in touch with a midwife who had studied to be an emergency medical technician and who was going to try to be there when the babies were born.

Alan Jeppesen was doing a fine job with the groceries. The Weavers had always paid the Griders for that job, but Alan only asked that Vicki knit him wool socks. It took Vicki forty-two hours curled up with yarn and needles under the kerosene lamp just to knit three pairs, but Alan told her that when he showed them to an elderly woman he knew, she mistook the socks for machine-made. Once a week, Jeppesen brought milk, eggs, butter, bread, cereal, bananas, and a package of hamburger. He delivered mail and often bought a half-gallon of ice cream for the kids.

Kevin Harris came up quite often during the summer and

fall of 1991, helped Randy cut tamarack trees for the winter firewood supply, and helped Vicki can vegetables and fruit. She was working hard through the late months of pregnancy and writing a running letter to her parents that she mailed whenever Kevin or Alan went down the hill. David and Jeane were planning to visit, hoping to catch the birth of Vicki's twins. Each summer, Vicki sent them a winter list of things the family needed, and that year she included cracked corn, canned goods, a food grinder, toothpaste, men's underwear, and shoes for the whole family.

David and Jeane left for Idaho in early October, intent on bringing Vicki down the mountain temporarily, so she could at least be close to a hospital. On the way, they met with ATF agent Steve Gunderson in the Kmart parking lot in Coeur d'Alene. They told Gunderson the same thing they'd told Hunt, the sheriff, and everyone else. Gunderson—one of the agents who had offered Randy the chance to be an informant—felt bad for Vicki's parents, who were clearly caught between forces they couldn't control: their family and the laws of the U.S. government.

"We'll pass along what you say," Jeane Jordison said, "but they don't listen to us on religious and legal matters. And when all is said and done, that is our daughter and our son-in-law, and we love them and we're going to support them."

David and Jeane brought a truckful of supplies and helped out with the canning and work on the cabin, but by the time they had to leave, Vicki still hadn't had the baby. They had no luck convincing her to leave Randy. Jeane tried once more to talk sense with Vicki and asked if the family wasn't causing some of its own trouble.

"You know why I have trouble and other folks don't?" Vicki wrote her mom. "I understand these things—most people never study or read and have their heads ruled by the 'electronic toilet.' They aren't dangerous to the tyranny in place. We are—we speak out against it. If we had 'free speech' protected by the Constitution—then why wouldn't we be left alone? Why do you suppose I live where I do? Certainly not to bother anyone else with what I understand. That should be obvious to anyone."

David and Jeane left Idaho the second week of October. On the twenty-third, Vicki went into labor, and the next morning Randy delivered a healthy baby girl in the birthing shed. She had soft, reddish-blond hair and deep blue eyes, and by the time the midwife showed up, the baby was quiet and alert. There was no twin, just a blood clot that Vicki had mistaken for another baby. The midwife, Carolyn Trochmann—married to one of the men from Noxon, Montana, whom Randy was supposed to spy on—said Vicki and the baby looked fine. It was a morning as peaceful as any she'd ever seen.

Choosing the right name was crucial to Randy and Vicki. Since 1978, they'd talked about living on a mountain top with biblically named children. The last two both had "el" in their names, which Vicki had discovered was a more proper name for God. Samuel meant "lamb of El" and Rachel meant "gift of El." Rachel's middle name had been Marie, until Vicki discovered that was a Catholic derivation and therefore pagan. Vicki changed it to Miriam. For the new baby, Vicki and Randy chose a name that translated to "El is my savior." There was no birth certificate, only Vicki's careful entry in her Bible:

Elisheba Anne Weaver, Born Oct. 24, 1991, Roman, 11:15 A.M., 7th month, 15 day Hebrew, Feast of Tabernacles. On a mountain, Ruby Creek Canyon, Naples, Idaho.

*B*y the spring of 1992, Bill Morlin couldn't wait to pitch this story: A white separatist gun dealer holed up on a mountain for a year with his wife and kids, stymieing federal agents, who flew over with airplanes and tried to figure out how to get him down. Morlin was the tireless federal courts reporter for the *Spokesman-Review* in Spokane, an investigative specialist who, for twenty years, had broken many of the biggest stories in the Northwest and who had spent much of the last decade writing about Aryan cross burnings, pipe bombs, and bank robberies. He'd written a couple of small stories about Randy Weaver's arrest and had filed Weaver away as one of the white separatists he needed to keep track of. That spring, he asked the district marshal what was happening with the Weaver case. He was given little information except a coy smile, the kind of look that was often the first tip Morlin got in a really good story. Remembering that the case had been an ATF bust, Morlin tried their office next and was given a little information and the same smile. He ping-ponged back and forth between the offices and played their competitiveness against each other. Finally, Morlin got enough for a story and went to his editors, but they weren't sold on it. Morlin sulked briefly and then figured out how to convince them: It was the picture. He asked one of his federal sources for a copy of the aerial photos they had, then busted into the photo editor's office, plopped down the picture, and said, "Look." There was the cabin, its metal roof glinting in the sun, just a spot in the middle of a deeply forested ridge. You never saw such solitude. It was the perfect symbol of this guy's separatism and the marshals' dilemma. Days later, Morlin was in Naples, interviewing Weaver's friends and trying to get up the mountain himself. The Weavers should talk to him, Morlin told Alan Jeppesen, because he was going to write a story anyway, and this would be a chance to get their side out.

Alan "has come up claiming the reporters want to talk with us," Vicki wrote to her mother. "They won't say what prompted their interest. We said we wouldn't want to talk with them and they said well—they'd make up a story anyway!"

On March 8, Bill Morlin's story ran on the front page of the *Spokesman-Review*, under the headline "Feds Have Fugitive 'Under Our Nose.'"

Morlin's lead read: "For more than a year, Randy Weaver and his family have been holed up on a North Idaho mountaintop, waiting for the federal government he despises to make the next move."

In the story, Mike Johnson described Weaver's cabin as "the closest thing to having a castle with a moat." He said they hadn't gone in because of Weaver's kids. Some Naples residents were glad the marshals were taking their time and said in the story that Weaver wasn't hurting anyone, that "his home is his jail."

A *Chicago Tribune* reporter followed Morlin to North Idaho and his story was picked up by newspapers across the country. In Iowa, Randy's niece cried when she saw the wire story. Suddenly, the Weavers were hot. *Star* magazine wanted to come up. The *Los Angeles Times* wanted to come up. Geraldo Rivera's new television show, *Now It Can Be Told,* wanted to come up. The Weavers said no to all of them. So *Star* rented a helicopter and buzzed the cabin. Geraldo's crew rented a chopper, too, and it hovered around the house like a persistent wasp while Randy flashed his middle finger at them. The next day, the Weavers were listening to their radio when they heard a report that they'd fired guns at Rivera's helicopter.

"The only thing I shot them was the bird." Randy laughed. The crew later admitted they were probably mistaken.

"I guess they've decided to 'try' us with public opinion," Vicki wrote. She said "Jewraldo" and his crew's false report of gunfire were attempts by the government to make the Weavers look violent. "They may still do a hatchet job on us: white robes, burning crosses, swastikas, and skinhead street fights. What's that got to do with 'failure to appear for a frame-up'?" To Vicki, all the attention was just more proof that even if there were some good men in law enforcement, they were being forced to do the Beast's bidding by the Zionist-Illuminati-One-World infiltrators.

"Well, you wanted to know if all the publicity is good or bad?" she wrote her mother. "It's to put pressure on us & the feds who don't think the 'charge' merits force. . . . I've always told you—they will start it—well they have and they'll never give up."

• • •

*O*n March 27, 1992, fifteen people met in the narrow director's conference room on the twelfth floor of the marshals' headquarters in Arlington, Virginia, overlooking the Pentagon and Washington, D.C. The top officials in the U.S. Marshals Service sat down at the conference table, along with a public relations specialist who was there to answer questions about the sudden bad publicity this case was generating. One writer had even suggested that all criminals move to Idaho, where authorities were afraid to arrest them.

Also at the meeting were electronics, surveillance, and tactical experts, all there to figure out a plan finally to end this thing.

At the end of the table sat the marshals service's acting director, Henry Hudson, who had been on the job only a couple of weeks and had already been blindsided, not only by a sticky case, but by the accompanying lousy publicity. A tough former prosecutor, Hudson listened as his top deputies explained how an Idaho woodsman could hide behind his kids for fourteen months and evade deputy marshals who knew exactly where he was.

The U.S. marshal for Idaho, Mike Johnson, spoke first. A thirty-eight-year-old former coroner and county commissioner, Johnson was a gun collector and savvy local politician in Boise, Idaho. The boyish, stocky Republican appointee wasn't liked by some of the lifers who worked under him, the career marshals who did the day-to-day work. But he was in his element in Washington, D.C., especially with Hudson, who was in the same general position he was—an ambitious, young politician in charge of lifetime civil servants. Johnson and Hudson had known each other for ten years, since the two men served together on a highway safety committee. Now, Mike Johnson went over each painstaking step that his deputies had taken to negotiate and investigate this case. They'd followed all the rules and had even called SOG teams out twice. And now, Johnson said, his chief deputies were requesting help from headquarters.

After Johnson finished, the leader of the Special

Operations Group told Henry Hudson how his members had gone out to Naples in June of 1991 and again that fall. He gave the SOG assessment and solution—several tactical assault plans that involved getting Weaver or other members of his family to respond to some noise, separating the family, and moving in to arrest Weaver. But this case was tricky, he acknowledged, and the tactical approach carried a high risk for Weaver's family and the deputies. Still, the tactical option was the cheapest and quickest and SOG members specialized in pulling off such missions safely.

Finally, Arthur Roderick spoke. Roderick was a former SOG member who had specialized in fugitive investigations and then had returned to the enforcement division to become one of its stars. Mediterranean handsome, he had a bachelor's degree in criminal justice, three years as an army MP, three years as a cop, and a reputation as one of the best young deputy marshals in the service. The last thing Roderick wanted to do was go into the woods after a fugitive hiding behind his armed children, but he knew the case was coming his way, so he had studied the problem.

Roderick outlined his three-part plan. First, he and a hand-picked team of deputies would fly to Idaho and be briefed by Dave Hunt and the other deputies already working the case. They would see the area for themselves, do some preliminary interviews, perhaps take one more shot at negotiating through Weaver's friend. Second, they would spend a few weeks on intensive surveillance, with deputies and with hidden, long-range cameras. That would fill the gaps in the case, like whether or not Weaver ever left his cabin. And third, after they knew when Weaver left the cabin or what kind of ruse he might fall for, they'd make the arrest.

There was one other option. Hudson and Mike Johnson excused themselves and went into Hudson's office. They called Maurice Ellsworth, the U.S. attorney for Idaho and asked him to publicly drop the charges against Weaver and, after the fugitive relaxed, convene a grand jury to secretly indict him. Ellsworth said no. It would be unethical.

Back in the meeting, they spent a few minutes going over some of the tactical plans. One of them involved surrounding the Weaver cabin with thirty-three deputy marshals while Sara slept in the menstruation shed, then grabbing the family members one at a time as they came out to visit her. Another plan involved splitting the family into two groups by getting some of them to respond to a noise in the woods. In both cases, the marshals would use nonlethal weapons—beanbag guns, rubber bullets, stinger grenades—to subdue Weaver and, if necessary, his family.

It took Hudson only a few minutes to dismiss the tactical ideas. The risk to Weaver's family and to the deputies was too great. And so Roderick's plan was approved, and he was told to begin immediately. He code-named it Operation Northern Exposure, after a popular television show filmed in the Northwest, featuring a mud-ugly town surrounded by beautiful wilderness and filled with strange people.

At the end of the meeting, Hudson said that when a plan was ready, perhaps it should be shown to the FBI's Hostage Rescue Team, to see if they had any suggestions. Some of the career marshals fought back scowls. In his first month as director of the marshals service, Hudson was underestimating the competition between the marshals and the FBI.

The meeting ended, the men stood up, shook hands, small talked, and made their way out of the conference room. One by one, they filed past a painting of a grizzled Old West marshal wearing a badge, bringing in one bad guy handcuffed to his wrist and another draped over his shoulder. If only it were that easy.

*D*ave Hunt wasn't sleeping. He stayed up too late at night, pacing and sucking down cigarettes, rehashing Vicki Weaver's cryptic letters in his mind, and second-guessing whether he'd been right to proceed so cautiously. He talked with his wife and played with his two children in their rambling, modern house in the highlands of Boise, Idaho. Their house was perched almost as steep as Randy Weaver's, but it was in a neighborhood filled— not with pine trees, boulders, and mountain grasses—but with other nice homes, minivans, and tricycles. Again and again, Hunt read the case file and pored over Revelation, Deuteronomy, and other Bible chapters, looking for explanations to Weaver's behavior, for clues about what he might do next, and for passages that might convince him to surrender. And although he was grateful for the help from headquarters, it ate at him to have one of his cases—one of his fugitives!—get so out of control.

The day after meeting with the director in Washington, D.C., Mike Johnson made the marshals service's final attempt at negotiating the case. The U.S. attorneys Howen and Ellsworth had made it clear they couldn't negotiate the points Hunt had spelled out in his October letters. But Hudson told Johnson to try once more to negotiate and settle the thing. Johnson called Alan Jeppesen and passed the message along: they wanted to know what it would take to bring Randy in. A few days later, Johnson got his answer.

"He said to stay off his mountain," Randy's friend said.

On the last day of March, Arthur Roderick flew to Boise. If Hunt was concerned that he was losing control of the case, Roderick quickly put him at ease, deferring to his knowledge of the situation and the area and involving him in each step of the new plans. Roderick spent a couple days going through Weaver's case files and talking to Hunt, Mays, and Evans. Then he drove north and began gathering his team: Hunt and the other two local marshals and three guys from out of town, two of whom were electronics specialists and the third an emergency medical technician.

The marshals looked for a place to set up a command post and finally settled on a town house–style condominium at the base of Schweitzer Basin, a fashionable ski resort finishing up its last week of the season. They moved twenty-five crates of equipment in—cameras, tapes, high-beam spotlights, night-vision goggles, and guns. Their chalet headquarters was perfectly situated, fifteen minutes from Sandpoint and forty minutes from Randy Weaver's cabin. The deputies were struck by how beautiful the view was from their balcony, where they watched a black bear come by regularly to sun himself in the parking lot.

They set up the advance command post in a vacant cabin on Wayne and Ruth Rau's meadow and filled it with antennae, two closed-circuit televisions, taping equipment, radio equipment, and two generators. The technicians got the cameras ready to place in the woods and the out-of-town marshals did their best to stay low, so Weaver's friends wouldn't spot them. If they were discovered, they decided to pass themselves off as telephone workers or newspaper reporters. At night, using military illumination charts that let them know when the moon and stars would be the brightest, Roderick and his team hiked into the woods around the cabin and scouted for the best places to put the cameras. When they were ready, Roderick sent one of his deputies back to Washington with a message: Phase I of Operation Northern Exposure was complete.

After a peaceful, mild winter, the Weavers sensed the government was getting ready for something big. They were hyperalert and even began suspecting Alan Jeppesen of being an informant ("We think Alan is cooperating with the agents of the One World Beast Government"), and although Alan denied it, Vicki had Kevin Harris start picking up the mail. The $5,000 that Bill Grider was offered had inflated to $20,000 by the time the Weavers heard the story. After living for years on a few bucks from firewood, rug sales and the charity of friends, Randy and Vicki wondered who could resist money like that? Guns always at their sides, the Weavers approached everyone who rattled up their driveway as a possible federal informant.

After refusing to talk to *Star,* Geraldo Rivera, and the *Los Angeles Times,* the Weavers agreed to talk to Mike Weland, an unassuming, down-to-earth reporter for a tiny weekly in Bonners Ferry. Mutual friends brought Weland up, and the family talked with him for several hours. Sara and Rachel played and worked in the garden. Kevin, Randy, and Sammy wore sidearms and were nervous at first—especially the shaven Sammy Weaver, who glared at Weland early on. By the end of the interview, Sam was quoting Scripture and tickling Elisheba, who cooed with laughter on the couch in the bright, tidy living room.

"Our situation isn't about shotguns," Randy was quoted as saying in Weland's story. "It's about our beliefs. They want to shut our mouths."

"We're not Aryans; we're not Nazis," Vicki said. "The reason we are here is to do our best to keep Yahweh's laws. The people who came to this country came to escape religious persecution, but there's nowhere left to escape our lawless rulers."

Randy said he didn't stand a chance in court against paid informants, and Vicki said that even though Randy wanted to turn himself in to protect the family, she and the kids wouldn't let him. Then he offered a sort of terms of surrender, saying the only way the situation would end would be for the ATF to return his .22 pistol and admit they'd set him up, and for the sheriff to apologize for calling him paranoid.

"Right now, the only thing they can take away from us is our life," Randy said in Weland's story. "Even if we die, we win. We'll die believing in Yahweh."

That spring, Vicki's letters home were filled with stories of "agents provocateur," government informants, and at least one plot to steal her baby. "Needless to say," Vicki wrote, "the April exploits of the feds all failed and I've got rid of or discovered 4 snitches because of it."

The media attention brought fan mail, too, seven letters in all, Vicki noted, "from as far away as New York City telling us not to give up, that lots of people know who controls our corrupt government." When someone sent her a news clipping that quoted authorities saying that letters were flooding in from all over the country, Vicki figured the feds were holding more mail back. Even though they wanted to be separated from the world, Vicki and Randy were pleased when a news crew trying to get an interview told their friends that Randy was becoming a Wild West hero, like his boyhood hero, Jesse James. "The original publicity was to force the feds to get rid of us," Vicki said. "I guess it backfired."

The Weavers trusted fewer and fewer of their friends and so Kevin, who had always been the solid, quiet big brother of the family, spent much of the spring and early summer on Ruby Ridge, riding his motorcycle down to fetch the mail and groceries. He was everything for the family: Randy's friend, Sam's buddy, older brother to Sara, Rachel, and Elisheba. He even baked bread and canned vegetables alongside Vicki.

Even with all the intrigue, it was a nice spring for the Weavers. Sara had a suitor, the fifteen-year-old patriot son of Vicki's midwife, whose family visited when they could. The rest of the time, Sara was busy with her gardens and the rugs she wove alongside Vicki. Kevin was building a log cabin in a forested gully below the Weavers' house, and Sam was helping, both young men working from sunup to sundown and then falling in bed, exhausted. Sam was also playing with Striker, his big yellow Labrador mix, who had grown from a nervous puppy into a great watchdog, with a bark so deep it kept a lot of people from getting out of their cars. Just a year before, with Vicki's aging father struggling to get up a hill, Sam had put a harness on Striker, gave the other end to David, and had the dog pull Sam's grandfather up the hill. "His bark is bigger than his bite," Sara wrote, "and he is really just a BIG lovable puppy." Randy watched fourteen-year-old Sam proudly. Not even 5 feet tall—smaller than his dog—he was a little man, braver and stronger than many of the adults Randy knew.

In a letter decorated with pictures of wrapped presents and people in party hats Rachel wrote that she had baked

Sara a birthday cake. She told her grandparents that she was taking care of the chickens. "I pick grass every other day for the chickens. Sara found the first flower of the year on the mountain."

Elisheba was sitting and scooting but not yet crawling. She was cutting some teeth and learning to say "Mom." As always, Vicki was working, taking care of everyone and keeping them strong in their obedience to Yahweh's law and their war against the lying One World Government.

"Are you starting to get the picture yet? Do you begin to see the reasons for the course we have chosen?" Vicki wrote to her mother. "*They* hate having their deeds brought to light and want to destroy *anyone* who exposes them. That includes me and my children.

"The *quality* of our lives is just as important—more important than the length of our lives. The past 14 months we have been a family; rich in love and experiences, stolen from the desires and intentions of our people. They want my family separated and destroyed. Not with my help!! Tomorrow is promised to no man.

"We aren't stupid, nor paranoid," Vicki wrote. "Nobody has to worry about being shot or in danger unless they shoot at us or are aggressive at us."

*T*he ridge just west of Randy Weaver's cabin was code-named "the lumberyard," so that anyone listening to the marshals' radio transmissions on Ruby Ridge would mistake the traffic for loggers. Phase II of Northern Exposure began in mid-April, when the first camera went up in "the lumberyard," three-quarters of a mile from Randy Weaver's cabin. For a week, Roderick's team went out at night, wearing full camouflage and lugging hundreds of pounds of equipment—microwave transmitters, photo lenses, tripods, transmitting control boxes, batteries, and cable—into the woods and hills around Randy's cabin. When the camera was in place, they covered its tripod with a camouflage tarp and, by April 20, it was beaming pictures of the Weavers—toting rifles, gardening, and urinating in the woods—back to the screens in the cabin on Homicide Meadow.

Placing a camera on the north ridge—code-named "the sawmill"—was easier, and between April 20 and May 11, the deputy U.S. marshals taped 118 hours on sixty-seven separate videocassettes, some showing an empty mountaintop, others showing the Weaver family's daily routine of chores, playing, and walking around with guns. Inside the little cabin on the Raus' property, deputy marshals watched the tapes and took notes. Amazingly, it seemed Randy, Vicki, and the kids never left the knoll. Occasionally, Samuel would ride his bicycle down around the bottom of the driveway, but for the most part, the family—especially Randy—never went past the springhouse below their cabin, where the driveway intersected the old switchback road.

There were visitors occasionally, people looking for land they'd bought, a local newspaper reporter trying to get an interview, some friends of the Weavers. By April 25, Kevin Harris was back at the cabin, rumbling up the driveway on a motorcycle, wearing a hat that looked like it belonged to a Greek fisherman, a pair of goggles, and a backpack containing the mail and groceries.

The marshals tabulated how often the various family members carried weapons on the videotapes: Randy 72 percent of the time, Vicki 52 percent, Sara 38 percent, Rachel 31 percent, Kevin 66 percent. Samuel carried guns a shocking 84 percent of the time and was almost always carrying a sidearm at least. They seemed to have a pattern of patrolling the compound, and in the mornings, Rachel carried long rifles and ammunition out to some of the rock outcroppings, which apparently served as bunkers.

Thirty times during those twenty-one days, marshals witnessed what they called "a response" from the family. The noise from a car would echo up to the ridge top, and the Weavers would run to a rock outcropping, hold their weapons, and watch the bottom of the driveway to see if someone was coming up. Vicki or Kevin and one of the kids would go to the driveway to see who it was while someone else covered them from above.

In many ways, the tapes only confirmed what the deputy marshals feared. This was not going to be easy.

The cameras worked well until May 2, when the west ridge camera picked up Kevin Harris and Sammy Weaver staring across to the camera on the north ridge, where deputy marshals were spiking anchoring pins into place for a solar battery charger. That night, after it had stopped beaming pictures, the north ridge camera was stolen. Later, deputies would find the camera burned and buried on the Weavers' land.

While the $110,000 remote-controlled video system was still being put in place, Roderick took about two dozen trips up the mountain, to make sure it was safe for his deputies to continue working. Once, Kevin Harris almost saw him, and when Roderick returned to his pickup truck, he found the air had been let out of the tires. Another time, April 23, he got within a few hundred yards of the Weaver cabin and found himself in the deep brush near the base of the Weavers' driveway. Already that morning, Roderick had seen rain, snow, sleet, and sunshine, and he huddled in the brush beneath a thermal NASA blanket that was silver on one side and camouflage on the other. Just then, a Ford Explorer rumbled up the driveway and stopped at a rock-and-log barricade the Weavers had put out. Sam and Vicki came down the driveway, Sam with a pistol, Vicki with a rifle and sidearm.

Roderick watched a man get out of the pickup, a logger from Oregon who'd bought some land up there and had been up the day before, when he'd talked to Randy. The man asked for Randy again.

"He's busy," Vicki said. The guy laid out maps and documents from the title company on the hood of his vehicle and showed Vicki where he wanted to go, on the access road past their driveway into the woods behind the Weaver cabin. They went over the maps for about twenty minutes before Vicki said okay. As long as they put the barricade back up.

The deputy marshals had long noted that the Weavers let

some landowners and prospective buyers through the barricade and, occasionally, invited them in for cookies and water. Their plan had already taken shape when Roderick watched the Ford Explorer rattle up the rutted road.

Mark Jurgensen was a deputy marshal from Washington State who could fit in with the people of North Idaho for several reasons: first, he had a great beard; second, he was an excellent carpenter who could pass as someone building his own cabin; and third, he had false teeth that he could pull out, making him look like a toothless mountain man. Phase III called for Mark Jurgensen to pose as a man who'd bought land behind the Weavers. Randy was such a social guy, he wouldn't be able to resist a friendly, bearded, toothless guy hammering away just down the road. Eventually, his guard would drop and the deputy marshals would swoop in. Or so they hoped.

Roderick estimated the expense in per diems, lodging, overtime, travel, and rental cars for a dozen deputies—$30,000 a month. Still, once they started the undercover mission, he was prepared to wait as long as Jurgensen needed to make the arrest. After more than a month in the field, Roderick flew back to Washington and met with his supervisor, Tony Perez. Phase II was complete, Roderick said. He was ready to start the third part of the plan.

Perez told him that Henry Hudson wanted to hold off for a while.

Hudson, the acting deputy director, had been described by the *Washington Post* as "a hard-liner's hard-liner" who mowed through five years of dope dealers, corporate heads, and defense attorneys as a Virginia prosecutor. His biggest trophy had been the Pentagon procurement scandal, a web of bribery and fraud by military consultants and managers. "I offer no apologies whatsoever," Hudson said about his tough reputation in 1991, when he left his position to work for a private law firm. Even then, he admitted he'd be back in government, perhaps in a run for Congress or as a state attorney general. Instead, his break came in 1992, when George Bush appointed Hudson to the top post in the marshals service. Now he was going before a Senate confirmation committee.

Roderick was all ready to begin Phase III of Operation Northern Exposure when Perez took him aside and told him to hold off until after the confirmation hearings.

"*I've* had enough!" Wayne Rau grumbled into the telephone. It was August 1992, and Rau was threatening to drive up with his father to Weaver's cabin and settle this whole mess himself. On the other ends of the three-way conference call, Dave Hunt in Boise and Tony Perez in Washington, D.C., tried to calm him down. Rau explained that his water system, which ran creek water to their cabin, was missing, and he suspected the Weavers might have taken it. The marshals had just been sitting on their asses for eighteen months—five months since they came up with all the electronic gear. If they didn't do something to get the Weaver family down, and if they didn't do it fast, Rau said, he was going to sue the government.

Finally, Perez calmed the tree farmer, offered to pay for the pipe, and promised that something was about to happen.

Northern Exposure finally became operational again in early August, just a few days after Henry Hudson was confirmed as the full-time director of the marshals service. In a way, Roderick was glad for the delay. Maybe he'd lost some momentum, but he also hoped the three months had given the family some time to settle down after discovering the north ridge camera, which Roderick suspected had happened. Besides, the break had given Roderick time to get his plan into shape and get his teams assembled.

The marshals service was a small, select organization—ninety-five politically appointed marshals, one for each federal judicial district, and 2,400 deputies. Fewer than half of those deputies actually worked fugitive cases, and just a handful of those volunteered for SOG teams. There were four regional SOG teams, unlike the Hostage Rescue Team of the FBI, which was a full-time SWAT team. SOG members handled their regular marshals duties until some problem developed; then they reported together to the crisis scene. Roderick knew who was in the service, he knew who was in SOG, and he knew whom he wanted for the first part of this mission, scouting the mountain one more time and preparing for the ruse.

In Washington, D.C., Roderick had the first team chosen. He and five other guys would take one more trip up the mountain, find the best places to hide snipers, and familiarize the marshals with the bluff. And then they would put the undercover agent in place. Dave Hunt, who knew the case and the area better than anyone, would be on the team. Roderick needed someone with medical training so he got Frank Norris, a seven-year veteran of the marshals service, a witness protection specialist from the East Coast, who was also a tactical EMT, specially trained to treat combat injuries. The electronics specialist on the mission would be Joe Thomas, a deputy based in Indiana. And, finally, Roderick requested one of the best marshals he knew: Larry Cooper. SOG, in the meantime, agreed to send another ace, Billy Degan.

While Roderick put the team together, Dave Hunt and the deputy who was going undercover, Mark Jurgensen, worked out the details of his ruse. His new name was going to be Mark Jensen. They called the owner of some dry scabland behind Weavers' place and paid him $2,700 to let "Jensen" pretend to buy the land and spend a few months building a cabin on it. He wrote letters back and forth with the landowner and backdated them for the deputy marshal to use as proof. He got a driver's license and a dog license with the phony names, a credit card, and a gas card. He got a government pickup, registered it under his alias, got a parking ticket, and backdated it. By the middle of August, Mark Jensen was ready to go.

He would have to be good because the Weavers were suspicious. "The feds keep sending informants up here as 'friends' trying to get us to show them the corner stakes of our property!" Vicki wrote to her mom earlier that spring.

"They've tried that four times! Do they want to bury something to justify murdering everyone? It is very curious."

The Weavers' response to strangers was automatic. The dog barked, and the family rushed out of the house. Randy usually covered the stranger from rocks above the driveway while Vicki or one of the kids went down to see who it was. For a year and a half, they met visitors that way.

But once, in the summer of 1992, a friend of Vicki's walked up the driveway without alerting the dogs. She thought about calling up to the cabin but finally decided just to walk all the way up. She walked past the rock outcroppings, came to the door, and knocked.

Vicki opened the door, saw her friend, and started crying. She whispered, "Do you know how long it's been since someone has knocked on my door?"

*C*oop walked down the short runway at the Spokane International Airport and smiled when he saw his best friend at the other end.

Billy Degan was what a U.S. marshal was supposed to look like. Six feet three inches tall, lean, with easy blue eyes and wire-rim glasses, Degan had a face that could be friendly and tough at the same time. His close brown hair was barely a half-inch long and was starting to pull back in front like a retreating wave. At forty-two, Degan was still athletic, a ripened version of the muscular offensive end who'd set pass-receiving records twenty years earlier for the University of New Hampshire football team. He was quiet and thoughtful, stoic even, except around old friends like Cooper, who loved his dry sense of humor. He was one of four national commanders for the SOG team and the only one who was allowed to stay in his home office—Boston— after the SOG command was moved to Louisiana. That was the kind of deference with which Billy Degan was treated.

It was a reunion of sorts in the optimistic, sixties-style terminal of the Spokane International Airport. Cooper— and to a lesser extent, Degan—had trained the young Art Roderick almost a decade earlier, and they teased him about how far he'd come since then. Cooper and Degan went even farther back, fourteen years, to the day they showed up separately at the Federal Law Enforcement Training Center and were given dorm rooms right next to each other. They graduated, reported to the service, and were given Special Operations Group training together; every time they traveled together, Billy and Coop shared a room. Of course, Roderick had joined SOG later, and the three men worked cases together now and then and met once a year for training at Camp Beauregard, Louisiana.

Cooper, mustached and solidly built—like a grown up high school shot-putter—had known about the Weaver case for some time. He'd seen some surveillance tapes and knew that other SOG members had come out to help the local deputies.

But he'd paid little attention to the case until a few weeks before when he was paged out of a law enforcement seminar and told to call headquarters. Roderick had already explained the difficult assignment, and now he had a question for Cooper. Would he go?

Cooper had been trying to get away from SOG. A few months earlier, his father had died and Cooper wanted to back away from his duties as an instructor and SOG team member to spend some time with his mother. But the case sounded important, and Cooper looked forward to working with Roderick and, especially, Bill Degan.

The next morning, August 18, the deputies briefed the marshal in Spokane about what they were going to do. Cooper helped Degan with some of the SOG equipment, which he'd had on display for some Boy Scouts in Boston. Degan had it shipped to Spokane, and he and Cooper loaded it into the Jeep and a van they rented. They drove to the condo at Schweitzer and began unloading the equipment. There were a number of guns: an M-16 machine gun, a 9-mm machine gun with a silencer, a short shotgun, and a .308-caliber sniper rifle. Roderick brought other M-16s, and each of the marshals had his own pistol as well.

On Wednesday, August 19, Hunt and Roderick left the condo to talk to some people around town and gather some final intelligence. Norris and Thomas practiced packing the heavy equipment up the ski hill and Cooper went shopping at the army/navy surplus store in Sandpoint. He bought two pairs of camouflage gloves, one for himself and one for Billy. Degan stayed at the condo alone. That night, they watched the surveillance video, talked about the mission, and sat out on the balcony of their condo, which looked out over the parking lot to a pristine mountain lake and the jagged Selkirks. On a similar peak, Randy Weaver waited.

The next morning, August 20, five of the six deputies—all but Norris—drove west to a shooting range in Davenport, 100 miles away. They practiced firing the guns they'd brought and adjusted the sights, in case they'd been bumped during shipping. The hunters and local policemen had never seen some of the guns the marshals fired, and they whistled and shook their heads.

Back at the condo, the deputies checked their equipment one more time and talked about the next day's mission. Norris, the medic, asked Roderick if they wanted a medical helicopter on standby, in case something went wrong. Roderick said not to bother. After the evening briefing, Hunt walked to the balcony of the condo and looked north at the long ridge that ran east from the ski lodge toward giant Lake Pend Oreille. He and Roderick had talked for the last few months about chartering a boat once this was all over and fishing the lake for some hefty Kamloops rainbow trout. Behind him, the other marshals talked about the mission, but Hunt stood in the cool, dry summer air, the last traces of sun painted on the horizon. Roderick joined him on the balcony. Hunt had grown fond of Artie and appreciated the way he deferred to Hunt's experience and knowledge of this case.

"What do you think, Dave?" Roderick asked.

"I don't know, Artie. I got a bad feeling about this one." He explained that the plan sounded good, there was just something. . . .

Roderick cut him off. "We're just gonna go up there and take a quick look and then we'll get out of there and go fishing."

eleven

*B*illy Degan dressed quickly in camouflage shirt and pants, pulled up his olive-colored socks, and laced his black military boots one rung from the top. They'd talked one more time the night before about wearing body armor, but it was going to be a scorcher that day, a long hot one under the relentless August sun. Besides, if things went badly, the vests they brought weren't likely to stop the kinds of bullets Randy Weaver and his family would fire.

This was the way Degan spent much of his adult life, waking up in some out-of-the-way place, putting on cammies, and heading out in the field, preparing to catch the worst of it. It was tough on his wife, Karen, and his two teenage boys, but it was just what he'd always done. A Marine from 1972 to 1975 and still a member of the Reserves, Degan had joined the marshals service in 1978 and immediately volunteered for the SOG team. He was singled out because of his military training and unflappable personality and was always called upon for the most dangerous and sensitive missions. After Hurricane Hugo in

1989, he led a SOG mission to St. Croix, in the Virgin Islands, quelling riots and keeping an eye on the local police, who were randomly looting homes and businesses. For his leadership, he was given the highest honor a deputy marshal could receive, the attorney general's Distinguished Service Award. The next year, he was given the Marshals Service Director's Special Achievement Award for rounding up fugitives during a drug crackdown in Washington, D.C.

Degan knew that Coop was backing away from the Special Operations Group, backing slowly out of the life. Degan was nearing retirement, and it had to be tempting for him, too—spend more time with his wife in the neighborhood where he'd grown up, coach his kids' hockey teams, throw down a couple of beers, and keep an eye on the neighbors' houses when they went out of town. But for now, he was still in the middle of the life, and even though this wasn't officially a SOG mission, it was one of the strangest and most challenging jobs he'd ever been on.

Bill Degan put his Camel Club lighter and a pack of Kool cigarettes into his pants pocket, next to his remaining spending money, three one-hundred dollar bills, a ten and three ones. At the sink, he filled his camouflage canteen with water. His digital watch showed 2:30 a.m.

Degan stepped outside the condo, into thin, mountain air. The night sky was cloudless, the stars in the east just starting to fade before the suggestion of sunrise. The other deputies were coming out of the condo in the same standard-issue camouflage clothing and loading supplies into the two rigs. They loaded green canvas bags filled with cameras, film, batteries, medical equipment, night-vision goggles, ammunition, and all the machine guns. They left the sniper rifle behind.

Cooper had slid a yellow T-shirt over his cammies, and the other guys followed suit, so that locals who saw them driving toward the cabin wouldn't make them for federal agents. They tested their radios one more time, the tiny earpieces in place, mouthpieces in place, the sheriff—the switch and wire that keys the microphone—slung down their sleeves and bound with rubber bands to their wrists. Satisfied, they split up, climbed into the blue minivan and the white Jeep Cherokee, and started on their way, into the heavy darkness that precedes dawn. Roderick eased the white Jeep down the winding mountain road to the highway, the minivan following closely. At the highway, they turned north and drove through pastures and forest, strobed by the moonlit shadows of deep timber and steep hillsides. They left the van at Sheriff Whittaker's house in Bonners Ferry, piled into the Cherokee and headed back through the foothills of the Selkirk Mountains. They crossed the Ruby Creek Bridge in the dark, their headlights catching the root-lined banks of the narrow dirt road.

At the Raus' house, still a mile from the Weaver cabin, the deputy marshals pulled off the road, slid their night-vision goggles over their camouflage masks, grabbed their machine guns, and started out on foot. Through his goggles, Degan's first views of Ruby Ridge were glowing and eerie, like watching a black-and-white television through an aquarium. The wooded brush was thick, and the deputies stuck to the rough access road, following Roderick, who had been up the mountain two dozen times already and who best knew the mission and the bluff they were approaching.

They walked quickly up the eroded dirt road, almost a mile below the Weaver cabin, until they came to a Y, where the trail split into two legs, one veering up a hillside across from the Weaver cabin, the other winding up Ruby Ridge and ending at the Weaver driveway. Roderick, Cooper, and Degan headed straight; Hunt led the other two camouflaged agents up the other trail, to watch the cabin from the other hillside. Roderick's group walked under a canopy of fir and tamarack trees, into a field of waist-high weeds and a thinning forest. For the first time, they could see the crown of rocks behind which the cabin was built.

Roderick pointed to one of the outcroppings, from where the family could see anything coming up the road or the driveway. "That's where they respond to when a car approaches," Roderick said. They crept around the wooded hillside, Roderick pointing out other areas: the fern field, the lower garden area, good places for watching the house, good spots for snipers. The sun was starting to come up and so they took off the night-vision goggles and backed down the hill again. At the Y where the two trails met, Roderick led Degan and Cooper on the trail toward the observation post, where the other three deputies were waiting for the family to come outside so they could photograph them. The sun was climbing in the sky—closing on 9 a.m.—but clouds were blowing in from the southwest, slipping over the Cascade Range and into nearby eastern Washington. As they walked, the deputy marshals talked into their microphones about spots with good vantages of the family, places where snipers could hide later. At the hillside surveillance post, they held up binoculars and spotter scopes to watch the family—like viewing a play from balcony seats a quarter mile away. Sam, a rifle in his hand, patrolled the compound, and later Kevin joined him. Hunt adjusted the enhancer on the 600-mm camera lens and fired off pictures that froze the family forever on their last peaceful morning: Sam and Kevin, rifles at their shoulders, talking in the clearing past the outhouse; Randy, who had shaved his hair off like a skinhead, walking with Striker; Rachel walking to the outhouse. Each family member had a code name that corresponded with smoke jumpers; that way, if anyone found the deputies' radio channel, they might think they'd come across traffic from state firefighters. And so, when Vicki walked outside the cabin and started pacing—like some ghost in her long, white nightgown—Hunt fired off pictures and said into the radio that he saw "the assistant crew chief."

At about 9 a.m., Roderick, Cooper, and Degan backed away from the ridge, where Hunt and the others were hiding, sneaked back down the road to a stand of birch trees, and slid back into the woods to get even closer to the house. Degan moved behind a rock 200 yards from the cabin, and Roderick and Cooper edged in behind a tree fifty yards closer. They were separated from the cabin by a crease in the two hills, a low spit between their vantage point and the cabin's.

"I want to see what the dogs will respond to," Cooper said.

Roderick grabbed a baseball-sized rock and threw it into the wash between them and the cabin, but it plunked down harmlessly and the dogs didn't alert. He tried again. Still, nothing. They spent about twenty minutes there, watching the cabin, and then they backed away from the closest observation post and began walking down the hill. They moved along the tree line, and Roderick pointed out places to hide snipers during the undercover phase of Northern Exposure. By 10:45 a.m., they were done for the day and were walking down the hill to meet the other deputies at the Y in the road when the radio crackled.

Vicki Weaver stood in her white nightgown, framed by the doorway, looking out over the trees that surrounded their cabin. At night, they tied one of the smaller dogs down at the garden, to keep the deer away, and that morning, Sammy—lightly freckled, with a slight overbite, his hair shaved to a quarter of an inch—walked down there and brought it back up. It was about 8 a.m., and the family was getting on with the business of a normal day. They took turns walking to the outhouse and then grabbed breakfast when they were ready. That morning, it was potatoes and fried eggs.

Vicki's long black hair fell to the middle of her back as she crouched down with Elisheba, who was learning to walk. She would make it five or ten feet before tumbling cheerfully over. She was teething, too, and she fussed a little as Vicki rocked and nursed her. The older children began their daily unscheduled routine of chores and playing. The weather had been in the nineties for more than a week now, but the nights were cooling off, and Vicki hoped the long, sweltering summer was about to break. The herbs and vegetables couldn't take much more sun.

The Weaver kids dressed and took their turns in the outhouse. Kevin had stayed with the Weavers for a couple of weeks, but he was about ready to leave for a farming job in Washington State. He'd been to Spokane earlier in the month, and his mom had been worried that Kevin would be in danger with the militant Weavers.

"Mom, don't worry," Kevin had said. "Nobody's going to shoot anybody."

He woke up that morning on the porch, rolled up his sleeping bag, and talked with Sammy about working on the cabin they were building. He shared a smoke outside with Randy, their rifles resting on their shoulders or under their arms as they talked. Randy came out first in a flannel shirt but went inside and changed to camouflage, a holstered pistol on his waist, a shotgun in his arms. The dogs had been yapping all morning, and Vicki yelled at them to shut up.

A little before 11:00, Sammy and Kevin walked out of the cabin with their rifles and began strolling down the driveway, Striker running in front of them. Sara walked a few steps behind them, and Randy ran out of the house to catch up to the kids, his feet slapping on the packed-dirt driveway. Rachel came last, skipping, a rifle over each shoulder. Near the base of the driveway, Striker alerted on something, a cold bark that meant he'd caught a whiff, just enough to send him nosing off into the woods, half-interested in whatever he'd found. Randy, Sam, and Kevin walked quickly after the dog, as they always did. This time, Randy said later, he hoped the dog was chasing a deer or an elk. It would be valuable meat for the long winter ahead.

With the shotgun under his short, sinewy arms, Randy ran along the dirt road that traced the top of the steep, forested meadow. "You cut down," he called to Kevin and Sammy. "I'll take the logging road." Kevin cradled his bulky 30.06 hunting rifle as he jogged through the field grass, down the hill. Sammy, not even 5 feet tall and eighty pounds, ran with his lightweight .223 assault-style rifle, a .357-caliber handgun on his waist, jumping rocks and fallen branches as if he was playing war.

From the rock outcropping near the cabin, Vicki watched Sara and Rachel walk back up the driveway and saw Randy and the boys chase the dog until they were out of sight. She listened for a second—always concerned when one of the dogs started in—but then she turned and walked back to the cabin, bent over and picked up a rock, and casually kicked at the summer dust.

D ave Hunt had finished putting his camera equipment away and had gone for a little walk to see where Mark Jurgensen should build his undercover cabin. He came back and was watching the Weaver cabin over the shoulders of Frank Norris and Joe Thomas—the three marshals standing in a line—when he saw Kevin, Sam, and Sara walk off the knob and down the hill, probably to work on Kevin's cabin. Hunt saw Randy run after them, followed by little Rachel. Vicki walked back toward the cabin.

"There's a vehicle!" Thomas said into the radio.

Hunt listened. He didn't hear anything. Then he heard the dog bark, and the family began running down the hill. God, no. "They are responding," he said into the radio. "They are responding."

"Give me a body count!" Roderick called back on the radio. Thomas answered him. Only Vicki was walking back to the cabin; all the others could be coming toward the marshals.

Dave Hunt had done enough 'coon and rabbit hunting to know when a dog got a hot trail. When the barks picked up in intensity, he began to get nervous. This couldn't happen on his case. "Get the hell out of there," he muttered to himself. It was probably no more than five minutes, but to Dave Hunt, it seemed to take hours, Roderick calling his position in to them, the dog's bark getting farther from Hunt and—it figured—closer to Roderick, Degan, and Cooper.

• • •

"**D**og's coming! Pull back!" At first, Roderick thought they could take cover, and he slid behind a tree and looked back up the hill, where he saw Striker, with Kevin Harris running behind the big yellow lab, break over the top of the hill, one hundred yards away, aiming straight for them. He realized it could all be coming down right then, but he also thought they could still get away without a gunfight. Torn with adrenaline and fear, the deputies ran alongside the logging road, from one stand of trees to another, twigs and branches crackling under their footsteps and the woods full of confusion. They stopped and turned several times, covered each other, and hoped the dog would turn back.

Instead, the dog was gaining on them. The Weavers usually stayed at the rock outcropping; why were they coming so far down the hill?

"We've got to take this dog out," Roderick said. "He's leading everybody to us."

Over his shoulder, Cooper saw flashes of yellow between the trees, shadows and movement behind the dog that he took for people. Ahead, he realized they'd have to run through a clearing before they reached the next stand of trees. They'd be wide open for fire from above. He realized that he and his buddies might not get away.

"This is bullshit," Cooper said into his radio headset. "We're going to run down the trail and get shot in the back. We need to get into the woods." The dog barked and bayed and still chased them. The marshals fanned out and stopped in the woods, breathing heavily and listening for the rustling of brush and timber. They hopscotched down the hill, shuffling sideways and taking turns as the last in line, covering the retreat of the others.

Cooper told the others to go ahead a little bit and he would take care of the dog if it got too close. He had the 9-mm machine gun with the silencer, and he hoped the Weavers wouldn't hear the metallic clank when he took the dog out. Roderick and Degan made it to the canopied tree line while Cooper continued running sidestep, keeping an eye on the trees where he could hear the dog barking and the crackle of men running behind it. Then it all seemed to happen at once, Cooper seeing someone on the higher trail, the one above them, and the realization that they had fallen into an ambush. He yelled at the man on the upper trail, "Back off! U.S. marshal!" Roderick saw him, too, and yelled at him. Cooper heard the dog bark, turned, and saw it growling at him. He pointed his rifle at the dog, but it ran right past him toward Roderick. Cooper didn't shoot the dog, and when he looked back at the trail, Randy Weaver was running away.

In his peripheral vision, Cooper saw Degan duck into the woods and so he ran behind him. Thirty or forty feet inside the tree line, Degan jumped behind a big stump. Near him, Cooper spied a hole protected by a rock, and he dove into it. The dog was still barking, and Kevin—dressed all in black—and Sammy—in jeans and a flannel shirt— were walking along the trail, coming closer, almost to them. From his stump, Degan saw Kevin and Sam—who was the same age as his youngest boy—walk right past them. When the boys were past, Cooper relaxed a little, thinking they might be safe.

And then several things happened in rapid, foggy succession—the dog moved toward Roderick, Degan rose on his knee to identify himself, and in a thicket of who-shot-first stories, both sides agreed that everything just went to hell.

*I*t hit Randy as he ran back toward the cabin: They had run smack into a Zionist Occupied Government ambush. He had been at the fork in the logging road when a man covered head to toe in camouflage clothing had stepped out from behind a tree and yelled something at him.

"Fuck you!" Randy Weaver had yelled back. He had run about eighty yards back up the road, toward the cabin when he heard the first shots—sharp cracks echoing through the timber. Sammy and Kevin were down there!

"Sam! Kevin! Get home!" He fired a round in the air

from his 12-gauge, double-barrel shotgun. He loaded another shell, but he was too eager and he pushed it too far in and jammed the shotgun. He drew his 9-mm handgun and squeezed off three more rounds. "Sam! Kevin!"

He heard his boy's voice. "I'm comin', Dad!" There were more shots down the hill, someone yelling, a burst from all directions, like the air was being torn in half.

K evin Harris wheeled and fired his 30.06, hitting Billy Degan square in the chest.

Larry Cooper saw his friend knocked backward and saw his arm fly up like a kid asking a question in school. He laid a line of fire right back at Kevin Harris, who fell like a sack of potatoes.

"Coop, Coop, I need you."

"I'll be there, Billy, as soon as I get 'em off our ass. Hang with me." He squeezed the switch on his hand and called for Roderick in his radio. "Get up here, Artie! Billy's been hit!"

But Roderick had his own problems just down the trail. The dog had run up to him, and Roderick had shot it in the back so it wouldn't lead the family to them. Sammy had appeared in front of him, saw that Striker had been shot, and yelled, "You son of a bitch!" Sammy had fired at Roderick. Another round of fire seemed to come from the woods and Roderick dove and bounced, feeling something graze his stomach, just as a bullet tore through his shirt and came within a breath of hitting his chest. The shots seemed to come at him from all directions.

Up the hill, Cooper fired another barrage, darted over the brush, and found Degan a few feet behind the stump he'd chosen for cover, lying on his side, his left arm still in the sling of his machine gun, his right hand up in the air. Cooper cradled his friend, who looked up at him with misting eyes and made a couple of chewing motions and gurgling sounds.

"Come on, Billy, help me and we can get behind the

rocks and we can take care of this." Cooper kicked his pack off and threw it into the brush. He had to stop the bleeding somehow. He felt with his left hand for the entry wound but couldn't find it, so he started dragging his friend back to safety. Degan wouldn't budge. It was quiet and still for a moment, just the last traces of smoke, the awful echo of gunfire.

Billy pointed to his mouth, which was full of blood, and Cooper remembered that his father had done the same thing right before he died, only a few months before. Coop pulled Billy close. He put two fingers on the side of his neck, found the carotid artery, and felt the last three beats of his best friend's heart.

Vicki Weaver heard the gunfire just as she made it back to the plywood house, the place that Yahweh had shown her might be safe from all this.

There was gunfire in the woods all the time. Those who lived up there didn't even flinch at a few rifle shots on a late summer morning. But these reports sounded different, volleys cracking and popping and echoing along the walls of the ridge, and there were so many, it was clear more than one gun was being fired.

Sara heard the shots and the yelling, grabbed her .223, and ran to the rock outcropping. Vicki and Rachel joined her, watching the woods frantically. A few minutes later, Randy broke through the tree line, panting and afraid.

"What happened?" Vicki yelled.

"We run into an ambush!"

And then Kevin came through the trees and up the driveway, wailing and shaking so much, Randy didn't recognize him.

"That's not Kevin!" Randy yelled.

"Yes it is!" he cried.

"Where's Sam?"

Kevin didn't want to tell them.

"Did you see Sam?"

"Sam's dead."

And then, Randy would remember, the family just went "plumb nutty."

*A*t the observation post, the other three deputies heard screaming, heard someone yell "U.S. marshal," and then a loud pop. A few seconds later, there were a couple more shots, and Norris and Joe Thomas turned to Hunt. "Let's go," Hunt said.

They began running through the woods to the base of the hill. They heard more gunfire down below—six, maybe seven shots—and then Roderick came over Hunt's radio. "Dave, get Frank down here! Billy's been hit! He's hurt bad!" Hunt worked with his radio as he ran, trying to get a forest service channel. "Mayday!" he called. But he couldn't raise anyone, and the other two deputies passed him while he fiddled with the radio.

Norris, the medic, didn't know the hill very well, and he turned back to Hunt. "Dave, you gotta get me down there."

"I know." They made it to the fern field below the Weaver cabin, puffing and gasping. Hunt crashed through the deep brush, protecting his face with his arm. And then a barrage of rifle fire cracked around them, Hunt saw the other two deputies go down and figured they'd been hit. The only one without a machine gun, Hunt dropped to his knee with his 9-mm pistol and looked for a puff of smoke to show him where the shots had come from. But he didn't see anything. He rose, saw the other two were okay, motioned them ahead, and followed along the trail into the heavier brush and trees.

They strode toward the Y in the road, tense, their weapons pointed in front of them. Hunt was in the lead again, moving too quickly, his pistol braced at eye level, like a TV cop. Near a stand of trees, the camouflaged Larry Cooper rose up on one knee and motioned the others off the trail and into the woods.

Inside the trees, Frank Norris immediately saw Degan on his back, facing the trail, his head rolled back, eyes slightly open and glazed over. Norris dropped down next to

Degan, pulled a twig from his open mouth, inserted an airway in his throat to keep the passage open, and gave him two quick breaths of CPR. Degan's blood-filled lungs gurgled with the bursts of air, and Norris tried to find a pulse. Nothing. The medic ripped open Degan's shirt and saw a nickel-sized hole between his chest and his collarbone. There was no blood on the outside of the wound. He rolled Degan over and saw that his back was covered in blood. The shot had gone right through him.

"Dave, get Billy and drag him over here," Cooper said.

"Give Frank a few more seconds," Hunt said.

Norris shook his head and keyed his radio. "Billy's gone." He picked Degan's wire-frame glasses off the ground and found a marshal's badge hanging on a low tree limb.

Norris and Hunt dragged Degan over behind a rock and then kept their guns trained on the tree line in front of them. They circled around Degan's body, protecting it. Later, in court, Cooper and Roderick would testify about what had happened: Degan stepped out and said, "Freeze, U.S. marshal" and identified himself with his badge. Then Kevin Harris just turned and shot him in the chest, without provocation. Cooper fired back, dropping Harris. Further down the hill, Roderick said that after the shooting started, he turned and shot the dog to keep anyone else from finding their position. And then both sides had fired back and forth.

Now, lying with Degan's body in the woods just off the trail, they agreed someone needed to go for help. "I'm staying with Billy," Cooper said.

"I'm not going anywhere," Roderick said.

"Dave, you've got to get some help."

Hunt didn't want to go, but Cooper insisted. "Nobody knows how to get out of here except you. Take Frank with you."

Hunt said Norris should stay, in case they needed a medic. "I'll take Joe with me."

Just then, a woman screamed on the hill above them like a siren: "Yahweh! Yahweh!" A man yelled: "You son of a bitch!" and what sounded like a child's voice: "You tried to kill my daddy!" Cooper figured they'd found Kevin's body, right where he'd dropped him. "I guess Kevin bit the bullet," he said to Roderick. There was another round of gunfire. And then more eerie screams, most of them indecipherable, except: "Stay the fuck off our land!"

*T*he family wailed and fired their guns in the air. Randy just kept reloading and firing, until Vicki stopped him and then they just cried and yelled. The bastards had killed their only son.

Kevin's recollection of events was completely different from the marshals'. Kevin, back at the cabin, explained that he and Sammy had been chasing the dog, when a man camouflaged from head to toe stepped out of the woods and shot the dog in the back. And then Sammy lost it. "You killed my dog, you son of a bitch!" he yelled. Sammy fired at the marshals, one of whom opened up on him, hitting his right arm and practically tearing it off at the elbow. "Oh shit!" Sammy had screamed and turned to run away, but the marshals kept firing at him, so Kevin had wheeled and shot one of the marshals to protect Sammy. Another marshal had shot at Kevin, just missing him. But as Sammy ran away, one of the marshals shot at him and a bullet tore through his back and dropped him face first on the trail. Kevin scrambled to his feet, turned, and ran back to Sammy's body. He couldn't find a pulse so he ran back to the house.

Randy and Vicki sobbed as they walked down the trail and found Sammy's body. They called for Kevin to help them. Randy picked the boy up. He was so light. They carried him into the birthing shed, stripped his clothes off, cleansed his body, and covered it with a sheet. They prayed over Sam, cried, prayed to Yahweh, and tried to contain their anger. They crouched with their rifles on the rocks for a while, waiting for the other marshals to come finish them off.

•   •   •

Dave Hunt barreled through heavy brush and timber, over fallen trees and scarred ground. He didn't want to run on the trail because he was afraid Randy was above it, waiting to pick them off as they ran away. Finally, after running forty-five minutes through heavy forest, Hunt and Joe Thomas broke through the woods into Homicide Meadow, near the Raus' cabin. Ruth Rau was standing on the porch, and from the road, Dave Hunt yelled at her: "Call 911! Get the sheriff!" She disappeared in the house and Hunt climbed the steps to her log cabin, exhausted. Thomas took up a position on the road, in case Weaver's friends came barreling up to help him or in case the family tried to get away.

Inside the cabin, Hunt tried to catch his breath as Ruth Rau thrust the telephone into his hands. "Get your kids and get out of here," he said to Ruth. He quickly told the sheriff's dispatcher what had happened: "I got one officer dead. I got more pinned down. I need help quick. I want the state police. I want all the help that I can get! I gotta go back in for more officers that are trapped."

Then Hunt dialed the marshals headquarters in Washington, D.C. He tried to outline the problem as professionally as possible. "This is Operation Northern Exposure. We got one dead, others stuck on mountain." The marshals dispatcher was confused, and Hunt nearly lost it. "I need to talk to somebody who cares, right away!"

Tony Perez got on the phone, and Hunt was never so relieved. Perez was the one they all emulated, the steady and quiet leader of the enforcement division, a man with all the qualities that made a good deputy marshal, the qualities Billy Degan had. But even Perez sounded flustered as Hunt explained what had happened. There was a firefight, he explained. The marshals had been ambushed, and the other guys were up there with Degan's body still. But, Hunt said, he hadn't heard any gunfire for quite a while.

"Dave, I want you to stay on the line," Perez said, while he and Duke Smith, the deputy director of operations, fired up the marshals service crisis center and began notifying

Justice Department officials. When federal agents passed on what had happened, they said the marshals were "pinned down" and "receiving fire." That version of events would continue to make the rounds at the Justice Department and would have tragic consequences.

In Boise, meanwhile, Hunt's marshals colleagues took a harried report from him, misunderstood some of it, and filed an affidavit for arrest warrants on all the adults in the Weaver cabin, saying they'd fired from a pickup truck at the deputy marshals.

For hours, the breathless Hunt talked on the phone, repeating what had happened and what was new. "Local sheriff has SWAT team on the way to the scene, which is no longer taking fire," Hunt said. "Team was trying to pull out when Weaver's dog alerted. Team drew multiple volleys of fire from the house. Degan was struck in the chest. Return fire killed one of Weaver's dogs. The rest of the team is still located in the mountains, but not under fire, unable to withdraw without exposing themselves to hostile fire."

He urged the brass to set up a plan to get the other guys out of the woods. The marshals officials wanted to move slowly, cautiously, and they hinted that perhaps the surviving deputies should pull back, leaving Degan's body for the time being.

But Hunt was a Marine, and Degan was a Marine, and Marines don't leave their dead behind. He also knew that he could never get Roderick and Cooper to leave the body. That won't work, Hunt said. By late afternoon, state and federal officers began showing up in the Raus' meadow, and Hunt was ready to go up and rescue his colleagues. A sheriff's deputy and a couple of border patrol agents wanted to fill a four-wheel-drive Jeep with tires, for protection, drive up the hill and bring the deputies down. But at headquarters, Tony Perez and his boss, Duke Smith, wanted to move slowly, didn't want any more casualties. Up the hill, Roderick concurred. For all they knew, the family could have them completely surrounded and might be waiting to gun down whoever came up there.

At the Raus' cabin, Dave Hunt tried to stay calm. In between phone calls, he paced and smoked on Ruth Rau's porch, talked to Roderick, Cooper, and Norris on their fading radios, and stared off into the woods, wishing he could do something. There was gunfire in the distance, across the valley, and Hunt figured they didn't have much time before Weaver supporters stormed the mountain. In fact, one of Sammy's skinhead friends tried to get through the meadow with his mom, and Hunt sent them back. Hours passed, and Hunt was going crazy.

Finally, with the sun going down, the ten-man Idaho State Police Crisis Response Team showed up, ready to begin the extraction. Joe Thomas begged to lead the team up there and said they needed to bring someone with them who knew the way. Finally, the SWAT commander agreed. Thomas—who had been up there only once himself—turned to Hunt and asked, "When I get up there, which way do I go?" After gathering the right equipment and a short briefing by Hunt and Thomas, the CRT left the meadow. It was almost 9:30 p.m.

Hunt ran in, got on the phone, and told headquarters that the SWAT team was on its way. The official on the other end of the phone said they'd reconsidered and wanted them to hold off on the rescue mission. Still assuming the deputies were under fire, they said they wanted to wait until they had an armored personnel carrier to get their men out.

Hunt walked outside to see if he could still call the CRT team back. It was dark now, and there were dozens of officers just waiting around. Hunt told a couple of marshals that headquarters wanted the CRT to wait. "You tell 'em to go to hell, Dave," one of the retired marshals said.

*O*n the mountain, the afternoon turned cold, the clouds moved in, the temperature plummeted, and the rains started. The batteries for the team's radios were going dead, and Cooper, Norris, and Roderick had trouble picking up Dave Hunt. They crouched in the heavy brush, buffeted by icy rain, their muscles strained and cramping,

exhausted from being constantly on guard. Cooper didn't
know how much longer they could make it.

"Dave," he rasped into his radio. "If you don't have
someone up here by four thirty, we're picking Billy up and
moving out."

They didn't take any more fire, but that afternoon, an
airplane flew over and the deputies thought they heard
gunfire from the top of the hill. There was still no help at
4:30, so the three marshals tried to carry Degan's body
down through the forest. But he was wet and slippery, the
brush was matted and thick, and weighed down by wet,
heavy clothes, Degan's 200-plus pounds wouldn't budge.
They took his belt off, tied it around his chest, and tried
to drag him to the trail. That didn't work either. Drenched
with rain and sweat, they put plastic cuffs on Degan's
wrists to keep his arms from dragging the ground and
slowing them down. Cooper tried to reach under Degan's
shoulder to pick him up, but he felt the hole in his
buddy's back, and his hand came back covered with
blood. He was afraid that if they kept trying to carry Billy
to the trail, they were going to pull his arm right off. They
pushed and pulled his body closer to the trail, hid him in
some brush, and collapsed on top of him in the rain,
waiting for help.

"*I*t's crummy weather," Vicki said between sobs.
"We better get inside." In the cabin, Randy, Vicki,
Kevin, Sara, and Rachel cried, prayed, and talked
about what had happened. Even though the marshals had
shot Sammy, Kevin said, he was sorry he had had to kill
one of them. But for the most part, the family was angry.
The ZOG bastards had ambushed them! They kept an eye
out for more troops, and Sara expected, any moment, to see
an army come up the driveway and begin blasting at them.
They could see trucks and cars in the valley and could hear
the rumble of engines everywhere, pinballing off the
canyons and foothills, which distorted the noise so that it
sounded as if the rigs were right on top of them.

Most of the day, they were too grief-stricken to do anything, but finally the family started gathering blankets, quilts, and sleeping bags and laying them out on the floor. They started bringing food and fresh water into the house and preparing for the attack. But what were they going to do without Sammy? There was no talk of surrender. The agents from the shadow government had started this war and—even if they let the family live—now they would frame Kevin and the Weavers for murder, just like they'd framed Randy on the gun charge. More likely, the government would just gun them all down, the way they'd killed Sammy, with a bullet in his back as he ran away.

Mixed in with the anger, Sara Weaver was afraid. Every noise hit her like a shock of electricity, and every creak of the flimsy cabin made her think of the end. And, just when she got the anger under control and could face the fear, she'd think of Sammy and start crying again.

After wailing and yelling in the afternoon, Vicki was quiet and ashen-faced, distant. She and Sammy had been especially close, and Sara watched her mom climb the narrow staircase up to the sleeping loft with Elisheba. That night, Randy, Kevin, and the girls checked their guns, watched out the windows, grieved over Sam's death, and prepared to be surrounded. Vicki—whose visions had led them to this place and who had only ever wanted to protect her family—crawled into bed and stayed there all night, holding her baby and her Bible.

*I*t's the way seasons changed in North Idaho, not gradually, but like someone slamming a door. A day that started August-sunny had become wet, windy, and cold, and Cooper and Roderick were already miserable by the time the rain began to freeze and turn to a wet snow that covered the ground with a white, watery sheen. At 9:26 P.M., almost eleven hours after the first shots were fired, the Idaho State Critical Response Team finally left the Raus' meadow to bring the deputies in from the forest. In cloud-covered darkness, with Joe Thomas in the lead, each member of the state team put his hand on the shoulder of the guy in front of him and moved slowly through the wet, black woods. Only two of the ten Critical Response Team members had night-vision goggles, and they tried to guide the others up the mountain.

It took them two hours to travel the same distance Hunt and Thomas had run in forty-five minutes. When they reached the exhausted deputies, they were huddled together in a triangle, lying across Degan's body, pointed away from each other and keeping their eyes on the woods around them. The deputy marshals held out a small infrared penlight, visible only to someone wearing night vision goggles.

"There they are!" said one of the CRT members.

Roderick told one of the Idaho state police officers what had happened. "I shot the dog," he said. The CRT posted sentries on the trail while they worked on getting the deputies out.

The state police officers tried to push Degan into the canvas bag—called a jungle stretcher—and Cooper stepped in to help, even though he was tired and cramping. Degan was bigger than any of the men trying to move his body. He was lying downhill on a steep piece of ground and rigor mortis had left his arm cocked up by his head. The men couldn't get a good enough hold on him to get him in the bag. Finally, they loaded him. When the state police tried to carry the jungle stretcher, Cooper stepped in and grabbed the straps on one end. Degan was going down the mountain the same way he came up, with his best friend.

But several times, the wet, overloaded canvas stretcher slipped, and Degan slid out onto the ground. Finally, they put him in another body bag and finished carrying him down the trail. It was slow and torturous, but the drained Cooper refused to allow anyone else to carry the front of the body bag. They reached the meadow forty-five minutes after midnight.

"I need an ambulance!" Roderick screamed. The meadow was filled now, with spotlights, cars, trucks, and dozens of state and local cops, federal agents, and officials. They loaded Billy on a gurney and slid it into a van.

Roderick and Cooper tried to climb in the van with the body, but the FBI agents said no, they couldn't go with him.

Instead, the five deputy marshals were handed over to Mark Jurgensen, the deputy whose undercover operation was supposed to lead to Randy's arrest. Jurgensen rode with them in a prisoner van to the Bonners Ferry hospital emergency room, where a doctor took their blood pressures and temperatures and gave them all some Tylenol capsules and something to settle their stomachs. Between them, the three deputies on the hillside had had nothing to eat except a granola bar that they'd split. They hadn't eaten since dinner the night before.

The deputies kept their guns because of rumors that Weaver's supporters would try to storm the hospital. They called their wives and told them they were okay. After Cooper had finished talking to his own wife, Jurgensen approached him.

"Mrs. Degan doesn't want to put more on you than you can handle, but she wants to know if you'll call her."

Of course he would. Cooper called Karen Degan and explained what had happened, how Billy had died trying to keep anyone from being hurt.

Finally, twenty-four hours after they'd started, the marshals drove back to their condominium at the base of Schweitzer. They talked about the vehicle that Thomas had heard and wondered where the noise had come from. They reminisced about Degan—Cooper and Roderick telling stories about him that brought half-smiles and pained laughter. They cried some. Mark Jurgensen joined the deputies at the condo, took their machine guns, emptied them, and counted the number of rounds they'd fired. Hunt, Thomas, and Norris hadn't fired at all. Roderick had fired once, and Cooper had fired two three-round bursts. Then he examined Degan's gun. Cooper said he was certain that Degan had never fired a shot.

There were seven rounds missing.

twelve

*T*wo hours after the gun battle, the Justice Department was scrambling. In the crisis center on the eleventh floor of the marshals headquarters in Arlington, Virginia, computers, telephones, fax machines, teletypes, and videocassette recorders hummed, spreading the horrible news that one of their best had been killed. Tony Perez and G. Wayne "Duke" Smith—the number three man in the service—worked the phones, gathering intelligence about the shootout from Dave Hunt, who tried to remain calm as he told them the marshals weren't taking any more fire, but that they were in danger and needed to be rescued. A map quickly went up on the wall and a telephone call went out to Mike Johnson, the U.S. marshal for Idaho.

"Where the hell is this place?" someone from headquarters asked Johnson.

"It's at the top of Idaho," Johnson said, "right by Canada."

In Washington, the director of the U.S. Marshals Service, Henry Hudson, met with FBI brass, told them one of his deputies had been killed, and repeated that two others were "pinned down." After the meeting, FBI officials briefed director William Sessions and the head of the bureau's criminal division, Larry Potts, who decided to call in the Hostage Rescue Team, from Quantico, Virginia. The head of that team, Richard Rogers, gathered his top three aides, loaded a helicopter and other supplies in the FBI's private jet, and left for Idaho to set up the details of the mission before his team arrived.

The marshals service's Duke Smith needed a ride west, and so he met Rogers at the FBI airstrip. On the jet, they sat down together to figure out what to do.

By that time, the overstatement of danger had reached high levels in the Justice Department, coursing through offices, faxes, and telephones like a virus. Hours after Hunt made it clear that the deputy marshals were no longer taking fire, one of his bosses told FBI officials they were "still pinned down by gunfire." The same bad information was relayed through a series of top-level meetings, working the Justice Department into a bureaucratic frenzy over William Degan's death.

The version of events spreading through the FBI and U.S. Marshals Service had the Weaver's dog discovering the deputy marshals and the family chasing them through the woods and gunning down Degan, who stepped out to demand their surrender and was killed without firing a shot. Some officials believed Weaver and Harris had fired from a truck and that the deputies took a round of automatic weapon fire from the cabin. Most critically, officials thought the surviving marshals were caught in an ongoing firefight and were pinned down by the Weaver family. Randy Weaver, they believed, was a highly trained Green Beret and Aryan Nations member who might have booby-trapped his mountain with bombs and grenades.

In reality, the situation was hazy at best, and much of the information being spread around Washington was simply wrong. There was no truck, no automatic weapons. Even if Degan was killed before Roderick shot the dog, he had fired seven shots himself, possibly before he was hit. There was every indication the Weavers didn't know what they were chasing (Vicki strolling back to the house, Kevin and Sammy walking along the trail). There was no evidence that Randy Weaver had any more than a glancing affiliation with the Aryan Nations or that he had shot at anyone or had booby-trapped anything. The federal officials flying to Idaho knew nothing about Sam Weaver being killed or the dog being shot. But the biggest mistake they made was to overestimate the "ongoing firefight," since the actual gun battle had lasted only minutes, and Hunt said clearly there had been no gunfire for hours. The misperception that the Weavers had the marshals "pinned down" and were firing automatic weapons at them from the house would color everything that happened for the next two days, as the government's fear of the Weavers began to gather its own momentum, a landslide of blunders, bad information, cold decision-making, and eventually, cover-up.

At FBI headquarters, Potts, the bureau's assistant director in charge of criminal investigations, met with deputy assistant director Danny Coulson. The Weavers possessed every tactical advantage, Potts said. They had supporters in the woods who might come to their aid, turning the mountain into a war zone. There was such a high risk of casualties, FBI agents who went into this situation had to be given the opportunity to defend themselves. Potts figured that Bill Degan had died because he underestimated how dangerous the Weavers were and had stepped out and demanded their surrender. Coulson and Potts concurred: Because of the rugged terrain, the antigovernment sentiment of North Idaho, the Weavers' extreme beliefs, and the fact that they had gunned down a deputy marshal without provocation, this might be the most dangerous situation the Hostage Rescue Team had ever faced. Before Rogers left, he and Potts talked about revising the FBI's rules of engagement—which stipulated that agents could fire only if someone's life was in danger—to allow snipers to shoot at the Weavers without provocation. Potts approved changing the rules, Rogers said later.

Another FBI official, E. Michael Kahoe, talked to a
legal adviser about changing the rules. The adviser said the
FBI could change them if the situation was that dangerous
and if it was the only way to control the situation and
protect people from being hurt. He told Kahoe the final
decision about how dangerous the situation was needed to
be made in Idaho. And, no matter what happened, before
any shots were fired, they had to demand the Weavers'
surrender.

*T*he flight left Washington, D.C., about 6:30 p.m.
eastern time, five hours after the gunfight. Aboard
the FBI's Saberliner jet, Duke Smith briefed Rogers
on everything the marshals service had done the last
eighteen months, how stubborn, committed, and dangerous
the Weaver family was, and how even the children were
well-armed extremists. The marshals service had
commissioned a psychological study of Vicki and Randy
Weaver, done through information gathered by deputy
marshals, Smith said. It showed that she was as zealous in
her beliefs—and maybe more so—than was Randy Weaver.
She wanted so badly to keep the family together, there was
even some fear that she would kill her own children, Smith
said.

They went over aerial photos of a cabin atop a rocky,
defensible knob—a fortress, practically—with deep forest
all around it. Smith told Rogers about Operation Northern
Exposure and gave him all the information he'd gotten
from Hunt and from the case file, including
misconceptions about the ongoing firefight and the
weapons the Weavers might have stored away. Rogers had
the impression of a mountain rigged with grenades and
explosives and a highly-trained family hiding in bunkers,
waiting to shoot anyone who came up. Flying across the
country, they agreed the case called for drastic measures.
Clearly, the family knew authorities were up there; they'd
already proven they would kill federal agents. Rogers and
Smith reasoned that to send HRT and SOG members up

there with the normal rules of engagement would be tying their hands in a very dangerous situation. If it was indeed an ongoing firefight, then FBI agents were in danger the minute they stepped on the mountain. Rogers drafted new rules:

"If *any adult* is seen with a weapon in the vicinity of where this firefight took place, of the Weaver cabin, then this individual *could be the subject of deadly* force.... Any child is going to come under standard FBI rules, meaning that if an FBI agent is threatened with death or some other innocent is threatened with death by a child, then clearly that agent could use a weapon to shoot a child."

Rogers called Potts on the jet's telephone and went over the new rules of engagement. Potts gave preliminary approval to the rules—said they sounded good—but Rogers knew he would have to send the detailed rules in writing before enacting them. Duke Smith called his boss, Henry Hudson, and advised him of the rules as well.

The U.S. marshal for Idaho, Mike Johnson, met the FBI jet at the airport and told them the state police were in the process of bringing the deputies out.

"How far is this place?" Duke Smith asked.

"We've got a good two, two-and-a-half-hour drive," Johnson said. They rented cars and hit the freeway going east at about seventy-five miles per hour. It was forty freeway miles to Coeur d'Alene, just across the Idaho state line, then another seventy miles north, through the woods, to Naples. Near Ruby Ridge, they turned on the wrong mountain road, and—rather than end up in the middle of the night at the Weavers' back door—the officials decided to go to the sheriff's office in Bonners Ferry and find someone to lead them to the federal base camp. It was nearing dawn by the time they finally arrived at the meadow, where fifty federal and state agents were now setting up tents and supplies and guarding the woods against Weaver supporters.

A trailer had been set up as a command post near one of Wayne and Ruth Rau's outbuildings, and Rogers, Smith, and Johnson found Special Agent-in-Charge Gene Glenn inside. Glenn, the top FBI agent from Salt Lake City, briefed them on what had happened so far; they had gotten the deputy marshals down and were trying to secure the woods and the roads leading up the mountain.

The deputies who had been in the firefight were back at their condo by the time Rogers and Duke Smith arrived. And so the officials sat in the command post with Gene Glenn, setting up plans and beginning to revise the rules of engagement without even talking to the deputies who'd actually been in the gun battle. Glenn was told they needed some time to compose themselves.

One problem, Glenn acknowledged, was that they had no one at the top of the knoll yet, "no eyes on the cabin." Rogers said that was okay, the HRT would be there soon and then his sniper teams would move up there.

T he hostage rescue team had trouble getting a flight out to North Idaho from Washington, D.C. They finally arrived on a transport plane Saturday morning, drove to Bonners Ferry, and set up for a 9:00 a.m. briefing at the armory there.

There were hundreds of SWAT teams, operations groups, and special tactics units used by the law enforcement agencies around the country, but the FBI's HRT was the elite, trained to battle terrorism and to handle tactical missions involving hostages or barricaded criminals. Unlike the marshals' SOG team, the HRT was a full-time outfit and its members did nothing but train and go on missions. HRT members had to be experienced street agents, had to endure a rigorous two-week tryout, and then had to pass tough physical and mental training.

The HRT consisted of two sections that were mirror images of each other—the Blue and the Gold—each divided into two separate kinds of members—assaulters and sniper/observers. The snipers would crawl into place on the perimeter of a crisis site, keeping their eyes and guns on the situation, and then the assault team would move into place on the ground, bust down all the doors that needed to be busted down, wrestle the bad guys to the ground and—in the sanitized vernacular of federal law enforcement—"stabilize the situation."

A bright, disciplined graduate of West Point, Lon Horiuchi had joined the HRT after only two years as an FBI agent. There was a friendly, proud sort of competition between the assaulters and the snipers—each chiding the other about who was more important to the team. That was why it was so strange when Horiuchi jumped ship after four years and moved from the assault team to the sniper team. But he knew that to advance in the bureau he needed as many different kinds of experiences as he could get, and with his intelligence, his steady, businesslike demeanor, his crack shooting on the range and his eyesight—a couple feet better than 20/20—he soon became one of the FBI's best snipers. By 1992, Horiuchi, a compact, muscular Asian-American, was in charge of the six agents on one of the Blue sniper/observer teams. In truth, the job consisted of a hell of a lot more observing than sniping and, when Horiuchi and the other members of the HRT flew out to North Idaho, it had been three years since an HRT sniper had even fired a shot on a mission.

The team members knew right away this was a big case. For the first time anyone could remember, both teams—the Blue and the Gold, all fifty HRT agents—were sent on a single mission.

Inside the Bonners Ferry Armory, the camouflaged HRT members sat on their packs or on folding chairs while Richard Rogers briefed them about the mission. Rogers had stayed up all night, working out the details of the operation and the rules of engagement with Gene Glenn and Duke Smith. Now the twenty-year FBI agent stood up and told the HRT members that they were going into a situation that was a continuation of the firefight that had started the day before. He mistakenly said the marshals were still pinned down by fire from the cabin. Rogers gave descriptions and intelligence information about each member of the Weaver family and Kevin Harris. Vicki Weaver, he said, was the most zealous member of the family. He gave them general descriptions of the rugged terrain, the rock outcroppings, and the wooded field beneath the cabin.

There will be "no long siege," Rogers said. And then he gave the HRT members the modified rules of engagement.

"If Randall Weaver, Vicki Weaver, Kevin Harris are observed with a weapon and fail to respond to a command to surrender," Rogers said, "deadly force can be used to neutralize them."

Outside the armory, Duke Smith told his marshals that the FBI was going to "go up there and take care of business." He said that Rogers had assured him the standoff "was not going to last long, that it was going to be taken down hard and fast."

In the armory, the HRT's hostage negotiator, a heavy-set, former street agent named Fred Lanceley, wasn't sure he'd heard right. He'd been involved in about 300 hostage situations, and he'd never heard anything like these new rules of engagement. Clearly, with rules like these, there would be no need for a negotiator.

The rules had been drafted with bad information and with little investigation of the circumstances of the initial shootout, no interviews with the deputy marshals who'd been up there. Then the rules were revised several times, going through a slight evolution of verbs—from "deadly force *could* be used" to "deadly force *can* be used." That evolution would continue and would create problems the FBI had never faced before.

Morning left little doubt how the locals—at least some of them—felt about the shootout and the ever-growing federal army. The old highway curved gradually past the roadblock, where two dozen state and federal officers now stood, wound through picket-fenced pastureland and underneath a railroad underpass where someone overnight had painted the words "Entering Dead Cop Zone."

At first, Bill Grider wasn't too choked up about what

had happened to his old friend Randy Weaver. He'd seen Randy once in the last year, when he'd gone riding up to the cabin on a horse and come across Sam Weaver near the Weavers' driveway. Bill climbed off the horse, and Sam held the dog on a leash toward him.

"Get back on your horse, Bill," Sam said. "Striker is mean."

And then Randy came out.

"Got a cup of coffee for an old friend?"

"Sorry, Bill," Randy said. "But I don't know who is a federal agent, there are so many feds out there."

So when he heard that Randy had finally gotten into it with the deputy marshals—"Randy went on a tear!" one of the sawmill employees told him—Grider figured Randy was a grown man and could take care of himself. I did everything I could for him, Bill decided. Half an hour later, though, Judy Grider came through the doors of the sawmill, sobbing, and so Bill decided to go down to the roadblock with her. The Griders began to burn as they watched cars and trucks full of cops stream past.

"The man ain't done nothin'," Grider said. "He isn't hurting anybody up there, and he has never hurt anyone."

On Friday, police had evacuated nine families whose houses were strung out on the logging roads that etched up Ruby Ridge, and they stood around the roadblock with some of Weaver's friends and a dozen or so reporters and photographers.

Lorenz Caduff was thirty-seven years old, a chef from Switzerland who'd escaped a crowded resort town just six weeks earlier to buy the Deep Creek Inn, a bar, restaurant, and motel on thirty acres just off the old Naples Highway, a quarter mile from the bridge where state police were setting up the roadblock.

Lorenz saw an ambulance sitting in his parking lot and walked outside to ask what was happening. He didn't know anyone, not the Weavers or the Aryan Nations, but he began to worry as police cars raced past.

"What's going on?" he asked in halting English.

The ambulance driver wouldn't tell him.

"This is my place! I want to know what's going on!"

Finally, he told Caduff that a criminal had gotten in a shootout and killed a federal officer. "This is shocking," Caduff said. "We know these things, but they are from the TV, from what you call Wild West."

When armored personnel carriers—tank-like vehicles known as APCs—began rolling on their metal tracks past his restaurant, Caduff did a double take. It was like World War II. It was unbelievable that the American government would do this to its own people. "Do you think we are in danger?" he asked his wife, Wasiliki.

Lorenz pitched a tent in his front yard for some of the people evacuated from Ruby Ridge, and when the temperatures dipped into the thirties Friday night, he allowed them to sleep on the floor in his family's apartment above the restaurant. He couldn't believe it when a Red Cross truck drove right past the evacuated people and turned up the hill to provide food to the federal officers. The Red Cross was founded in his native Switzerland as a neutral aid organization. They weren't supposed to take sides!

Red Cross officials said when they tried to help the people at the roadblock, they were chased off with clubs and sticks.

Two of Randy and Vicki's friends, Jackie and Tony Brown, tried to cross the bridge and were turned back by state police, who manned the line with assault rifles. Jackie stood on the bridge, crying and begging that they at least let her go up and try to bring the children down. "I'm just worried these kids are going to be killed."

More people gathered on Saturday, newspaper and television reporters and as many as twenty-four friends, neighbors, and people who shared the Weavers' beliefs. Their cars stretched for a couple hundred yards on either side of the turnoff to Ruby Ridge. Bill Grider went into the North Woods Tavern and rallied the off-duty loggers and farmers and brought them back to the roadblock. They formed a football huddle and talked about their strategy for protesting. "Tell the Truth" they wrote on cardboard and

"Go Home Feds" on the bottom of an empty half rack of Ranier Beer. By afternoon, they stretched banners along the banks of the old highway—"Freedom of Religion" and "Stop the Violence."

When the APCs and military trucks began rolling up the hill, a few protesters became outraged, screaming and pointing and running up to the vehicles.

"Baby killer!" yelled Kevin Harris's foster brother, Mike Gray, as he jabbed his finger at the window of a Humvee, a short, squat jeeplike vehicle. "Which one of you is going to shoot the baby?"

Kevin Harris's mom, Barb Pierce, sat huddled under a blanket with Kevin's girlfriend and his two-and-a-half-year-old son, Jade, while his stepfather, a Spokane paralegal named Brian Pierce, paced and tried to get some answers from the stone-faced law officers on the other side of the police tape.

Barb had been finishing a customer's nails when she heard about the shooting on the evening news the night before. Now, they didn't know if Kevin was dead or alive, and the authorities wouldn't let Brian go up there to try to talk Kevin down. "I don't share Randy's beliefs," the bearded, constantly frowning Brian Pierce told anyone who listened. "I just want to get Kevin down from there before he gets killed."

Finally, by late afternoon, Brian Pierce couldn't take it anymore, and he ducked under the police tape and began walking up the road. He only made it a few steps before officers grabbed him, handcuffed him, and dragged him off to jail. At least seven more people would be arrested or turned around in the coming days, trying to get past the smothering line of federal officers, either to talk Randy down or help him fight.

A few minutes of fitful sleep didn't keep the shots from coming again or keep the woods from racing past or keep the life from draining out of Billy Degan's face, and in that way, the night faded into morning, and Dave

Hunt and the other deputy marshals woke up, if you could call it that.

Joe Thomas made them a big breakfast—bacon and eggs—and they showered and dressed in the condominium on Schweitzer Basin. They drove to Bonners Ferry, to be debriefed by their own people and to be interviewed by the FBI and by U.S. attorney Ron Howen, who had been assigned the case. After all they'd been through, it bugged Hunt to have to sit down for two hours and be quizzed by FBI agents who had no idea how complicated this case was and how long he'd been working it. The agents themselves were okay and he knew they had a job to do, but he didn't appreciate their cocky, coming-in-at-the-eleventh-hour-to-save-your-butt attitudes. He'd always felt the FBI didn't deserve its bulletproof reputation. It was the agency that investigated Congress and judges, and as a result, Hunt believed, neither was tough enough on the FBI.

But what really got him was the FBI's arcane method of interviewing: the agents took notes, wrote them up, sent them back to the subject for review, and then revised them. Why didn't they just tape-record interviews, so there was no doubt what someone said? It was idiotic, a throwback to the J. Edgar Hoover FBI.

After the interviews, the deputy marshals drove to the armory, talked with Duke Smith and some other marshals officials, and watched the Hostage Rescue Team get ready for its mission. Then they drove up to the meadow, past the protesters, the graffiti—"Entering Dead Cop Zone"—and the signs—"Leave Them Alone!"

They checked in at the meadow, expecting to brief the Hostage Rescue Team about what they would find up at the cabin and what kind of people the Weavers were. Instead, the FBI agents ignored them as they went about the business of preparing for their mission. It was cold and rainy, and the marshals felt out of place, so they drove back to their condominium late in the afternoon. There, they talked about Billy Degan's death and made flight plans to be in Boston for his funeral on Wednesday.

They sat out on the condo's balcony and stared out over

the cool mountain lake they'd talked about fishing once the mission was all finished. This country was the most beautiful Dave Hunt had ever seen, the kind of place where black bears wandered out of the forest and sunned themselves in parking lots, where dense, unspoiled forests lapped up against ski hills and resort towns, so that from their condo balcony, seen through exhaustion and sadness, the Selkirk Mountains seemed like some tempting borderland between civilization and all that was dark and wild.

*H*omicide meadow was a cold, drizzling, muddy mess on Saturday afternoon. Federal agents and state police propped up green army tents in the middle of the field while a cold mist soaked them like grocery store produce. Cars and trucks carved the wet field into deep ruts amid the only traffic jam ever in these parts—police cars, moving trucks, Jeeps, motor homes, telephone company trucks, and every other kind of vehicle lined up in the meadow, unloading men and supplies. At the far end of the pastureland, the Hostage Rescue Team set up its tactical headquarters, a sealed, solid-walled, modern white tent with some sort of generator for heat and air-conditioning that made the other federal agents, in their damp, drafty army tents, stare at it longingly.

In the trailer that served as the overall command center, the top FBI agent, Gene Glenn, was wrestling with logistics that were as muddy as the terrain. By mid-afternoon, they'd established a perimeter all the way around the knoll, but—twenty-eight hours after the gunfight—they still had no one close enough to watch the cabin. Glenn had the armored personnel carriers he needed, and the hostage negotiator was ready to go up there and demand the Weavers surrender, but the state officers wouldn't turn the keys over without authorization from their commander, who was out of town. And Glenn wasn't going to have civilian guardsmen driving the APCs up the mountain to face the Weavers and whatever white supremacists might have joined them, with whatever arsenal Randy Weaver had assembled.

Across the trailer, the marshal for Idaho, Mike Johnson, was working the telephones when Glenn pointed at him.

"Hey, you know the governor?"

"Sure," Johnson said.

"Can you help us out?"

Johnson called the governor's office and, once he'd reached him, handed the phone to the state officer who had the keys to the APCs.

"Yes, sir, Governor. Thank you, sir." They had the APCs. But there was no communications system between the two gunless tanks, and so they still couldn't move out.

In the meantime, Gene Glenn, Richard Rogers, and the other FBI agents were trying to finalize the rules of engagement so they could send the snipers onto the hills around the Weaver cabin. The interviews with Roderick, Cooper, and the other deputies had presented them with new information—specifically, Vicki Weaver apparently hadn't been involved in the original gunfight. There would be no arrest warrant for her, and so, once again, the rules were revised. They wrote the new rules on a large pad in the command post.

One of Rogers's assistants showed him the rules and he told the assistant to scratch out what he'd written and to include "and should" after the verb "can." It was the final, critical evolution of verbs: from *could* to *can and should*.

"If any adult in the compound is observed with weapons after the surrender announcement is made," Rogers read to his assistant, who would brief the HRT, "deadly force can and should be used to neutralize this individual. If any adult male is observed with a weapon prior to the announcement, deadly force can and should be employed if a shot can be taken without endangering the children. If compromised by any dog, the dog can be taken out." For the children, he wrote, "Any subjects other than Randy, Vicki and Kevin, presenting threat of death or grievous bodily harm, FBI rules of deadly force apply."

Since Rogers had arrived in Idaho with Potts's approval

to revise the rules of engagement, there had been little discussion of whether they were appropriate. But it was still up to Gene Glenn to finalize and enact the rules. About 12:30 p.m. Saturday, Glenn got off the telephone with Potts and said that the FBI official had approved the modified rules of engagement. He still had to fax the total operations plan to FBI headquarters and to the U.S. Marshals Service. But FBI agents and marshals officials disagreed on the plan, which a marshal said would "get our men killed." The FBI agents remarked that they didn't hear any better suggestions coming from the marshals. Finally, they came up with an operational plan: surround the cabin with snipers, who would be able to shoot Randy Weaver and Kevin Harris if they came out of the cabin armed. Bring in armored vehicles and demand the family's surrender. If they didn't surrender by Sunday, destroy the outbuildings by ramming them with the armored vehicles. They also considered ramming the house and firing tear gas inside but decided the tear gas might be lethal to Elisheba.

At 2:40 p.m., Glenn faxed the operational plan to headquarters, including the rules of engagement. After thirty-some hours on duty, Potts said later, he had gone home, and the deputy assistant director, Danny Coulson, read over the first page of the draft plan. He realized right away there was no option for negotiations. He said he stopped reading before he read the section about the rules of engagement. Instead, he told Glenn to include a better option for negotiations. Fred Lanceley worked out some plans for negotiations, and Glenn faxed back the negotiations addendum and Coulson approved it. Later, no one at FBI headquarters would admit seeing the modified rules of engagement, even though they'd been faxed to the office as part of the total operations plan. When Justice Department investigators tried to find out who approved the rules, they found no record of Potts's or Coulson's discussion of the rules, a "lack of documentation" that was "significant and serious." Meanwhile, in Idaho, the FBI believed it had permission to shoot any adult who came out of the Weaver cabin with a gun. Problem was, most of the time nobody left that cabin without a gun.

*I*n Fort Dodge, Iowa, David and Jeane Jordison frantically dialed the phone numbers for Dave Hunt, the sheriff, and the other law enforcement officers they'd met and talked with over the last eighteen months. They couldn't get any people, let alone any information. All they knew was the bizarre military talk they heard from officials on the television news: that the Weavers lived in "a compound" or "a fortress"; that the children were armed and dangerous; that the dead marshal was "a hero" who had been ambushed by the family; that federal officers had surrounded the cabin and might be planning to raid it.

None of it made sense to David Jordison. It was no fortress. He'd built better barns than that old plywood mess. And the kids? Dangerous? They were just kids. It almost sounded like they were trying to find enough reasons to go up there with their guns blazing and raid the house, to get some retribution for the death of the marshal. That's why Vicki's dad especially wanted to find Dave Hunt, to see if Hunt could arrange for him to somehow talk to Randy and Vicki. David Jordison had just talked to Hunt on the telephone about a week before. "Don't worry," Hunt had said. The marshals were committed to arresting Randy peacefully, Hunt had insisted, and they wouldn't do anything drastic. "We won't do anything before your visit."

That's what made it so unbearable. It was only a few days before David and Jeane's annual trek to the mountains. They were so excited to finally see little Elisheba. Lanny was even coming this time, planning to see his sister for the first time since she left Iowa nine years earlier. The truck was half loaded with the things from Vicki's winter list, along with five or six boxes of clothes, a wonderful batch of tomatoes, some flour, sugar, and soybeans. Best of all, David had found another gas-powered washing machine to replace Vicki's old one—a 1920s roller washer with a kick start and an engine that he had painstakingly restored. Growing up in the Depression on a farm without electricity and other conveniences himself, David knew how much Vicki would appreciate the new washer, and he was looking so forward to seeing the look on her face when he surprised her with it.

Vicki's sister, Julie Brown, had always known the guns were going to cause Vicki problems. She had resigned herself to her sister's beliefs, but she didn't think there was any reason to give guns to children and to spend your life drilling for war. She'd worried Friday night and Saturday with the rest of the family, but by Saturday afternoon, her husband, Keith, talked her into going to a Ringo Starr concert at the Des Moines fairgrounds to get her mind off what was happening in Idaho.

"There's nothing we can do here, anyway," Keith said.

But as they were walking into the concert, Julie burst into tears.

"What is it?"

"I'm afraid I'll never see Vicki again."

"Oh, Julie," Keith held her. "It's probably all over by now."

But by nightfall on Saturday, David and Jeane Jordison still hadn't heard anything. Jeane held Vicki's most recent letter—dated August 16—which had just arrived a day before. There was no hint of trouble, just small talk about the weather ("We've all been miserable with the heat") and Elisheba ("She's got a tooth that's coming through") and not one word of antigovernment stuff.

> Hope you get this before you leave. We'll be expecting you anytime the last week of August. Be prepared for dust. . . . Take care and don't work too hard.
>
> > Love,
> > Vicki & all

Finally, they just couldn't take it anymore. Just before 8:00 p.m. central time—about the same time the snipers were moving into place around Vicki's home 1,000 miles away—David, Jeane, and Lanny Jordison climbed in Lanny's pickup truck and left for Idaho.

*I*n ten years as an FBI agent, Lon Horiuchi had never heard rules of engagement like these. Back at headquarters in Quantico, Virginia, in one of the classrooms used for the HRT training, the FBI's standard rules of engagement were framed and hung on a wall and members were expected to know the rules as well as their own phone numbers:

> *Agents are not to use deadly force against any person except as necessary in self-defense or the defense of another when they have reason to believe they or another are in danger of death or grievous bodily harm. Whenever feasible, verbal warning should be given before deadly force is applied.*

Those rules were clear. An agent fired only in self-defense or in defense of another person. More important to a sniper, those rules allowed him to make his own decisions about how dangerous a situation was, Horiuchi thought. This time, it appeared that decision had already been made for him. A marshal had been shot and killed, and the Weaver family and Kevin Harris had shown they would shoot indiscriminately at federal officers.

Horiuchi listened to the new rules: Any armed adult male *can and should* be neutralized. And once the family had been given the chance to surrender, that changed to any adult—male or female. That meant Kevin Harris and Randy Weaver could be shot immediately if they were armed, and later, Vicki Weaver could also be shot.

Posted on the walls of the HRT tent were surveillance photographs of everyone in the family and each of the team members stared at the pictures and familiarized themselves with them.

Later that weekend, other FBI agents from around the West who had been called to the scene were given the modified rules of engagement. Some were shocked. One Denver agent turned to another and muttered, "You've got to be kidding." Several felt the rules were inappropriate and planned to ignore them. Another agent later told federal investigators the rules amounted to: "If you see 'em, shoot 'em."

But it wasn't Lon Horiuchi's job to determine the morality of the orders he'd been given; it was his job to react by using his own judgment and the rules of engagement. He had waited all afternoon for vehicles to drive the snipers up to the ridge top and, by 5:00 p.m., when the APCs still weren't ready to go, the snipers decided to go on foot.

Horiuchi talked it over with the leader of the other sniper team. If Randy Weaver and Kevin Harris came out of the cabin with weapons, they would wait until both men were outside, and then they would shoot them. If they saw one target outside and fired, they'd never get the other one out of the house. The team leaders agreed.

By 5:30 p.m., eleven sniper/observers—some of the FBI's top marksmen—were ready to move into position around the cabin. They followed one of the deputy U.S. marshals who had been on Art Roderick's early teams during the first two stages of Northern Exposure. The snipers crept to the adjacent ridge north of the cabin, near where Dave Hunt and the other two marshals from the observation post had photographed the Weaver family the morning before. Because of the steep climb, Horiuchi carried a minimum of equipment—binoculars, a radio, and two guns: the sniper rifle and an M-14 assault rifle. He needed two guns because the FBI was afraid other white separatists might come up the mountain and help the Weavers. But Horiuchi knew the guns were for extreme situations. Mainly, their job was to watch the cabin, gather intelligence, and pass it on to the assault team; let them figure out how to get the people out of the cabin.

The weather was cool and rainy, the clouds slung low over the Selkirk Mountains. The ground was slippery because of the rain, and they spent much energy breaking through brush. When the lowest clouds drifted away, Horiuchi could see snow at higher elevations on the surrounding peaks. He crawled the last 200 yards into position.

The camouflaged snipers wedged themselves between trees and rocks, crouched in bunches of grass, and lay in brush along the hillside, two rising and falling football fields from the Weaver cabin. Lon Horiuchi had a very good spot, across a dip from the plywood home. He watched the cabin through his binoculars. He didn't have to wait long. A low grumble from one of the armored personnel carriers in the meadow drifted up the hill, and once again, one of the Weavers' dogs began to bark.

thirteen

All that overcast and cold Saturday morning, the Weaver family and Kevin Harris cursed and mourned and prayed to Yahweh. They debated what the ZOG devils might do next and guessed the reason they hadn't attacked already was because they realized how badly they'd screwed up by killing Sammy. The family figured they would surround the cabin and contact them with a bullhorn, apologize for Sammy's death, and try to get Randy and Kevin to come out. But there was no way they were going to turn Kevin over and have him framed for the shooting of the marshal. Kevin could never get a fair trial, and since he'd killed the marshal in self-defense, there was no reason for him to turn himself in.

Randy cried because he felt so bad that he hadn't been by Sammy's side when the shooting broke out. Standing inside the beamed living room, Vicki sobbed and cursed in turn, repeating that they would not separate the family. Sara tried to be strong like her mother and remember what they had trained for, but it was difficult. Helicopters and airplanes flew overhead, while the rumble of far-off trucks echoed off the walls of the ridge. There had been little sleep the night before. The family figured they had a day at the most before the agents moved in to surround them.

Sara and Randy walked all around the cabin that morning, shutting in the chickens and feeding the dogs. They saw no one on the misty hillsides around the cabin. Sara and Vicki ran to the root cellar near the cabin, pulled back the door, and descended into the narrow gash in the ground where they kept their food. They grabbed jars of apricots and peaches, canned tuna fish and sardines, and walked back to the cabin. Afraid the feds would try to shut off their water supply, the family filled empty plastic milk jugs with water from the spring and stocked them in the house. They pulled the navy blue denim curtains and prayed that Yahweh would give them the strength to hold off the enemy. The guns were loaded: Sara's and Rachel's Mini-14 semiautomatic rifles, Vicki's 9-mm semiautomatic pistol, which she wore on a holster; Kevin's 30.06—the hunting rifle that had killed Bill Degan—and the Mini-14 of Sammy's, which Randy now carried, along with his own holstered 9-mm. Tins of cheap Czechoslovakian ammunition filled corners of the cabin floor. They would just sit tight in the cabin—as a family—for as long as they could hold out.

They would run across one of Sammy's books, or his bicycle and someone would start crying again. All day, they prayed for help and salvation. For Sara, the day was a blur of frightened preparations until, just before 6 p.m., the dogs began barking again.

Like always, Sara ran out of the house to check on the dogs before allowing her dad to come out. She didn't see anything, and so she returned and got her dad and Kevin, who followed Sara out the door with rifles, into an afternoon weighted with low clouds. Randy said they wanted to check the north perimeter, and Kevin wanted to grab some flashlight batteries he'd left outside. They ran along the hard, rock-strewn ground, worn flat by nine years of foot traffic, bicycles, and dogs. As soon as they got to the rock outcropping, the dog stopped barking.

Randy didn't see anything. But on the way back to the cabin, he stopped and stared full at the birthing shed. "I just gotta see Sammy one more time," he said. He walked over rocks and brush, along the edge of the knob that faced the north ridge. Randy Weaver stood at the door of the birthing shed for just a breath, and then, as he reached for the handle with his right hand, there was a crack. Splinters leaped off the side of the shed.

The bullet ripped through the fleshy part of Randy's upper arm and came out his armpit.

When she heard the shot, Sara spun around and ran toward her father, branches and weeds crackling beneath her feet. Damn! If she had been shielding him, the way she was supposed to, this never would have happened. She darted around the birthing shed and saw her father holding his armpit.

"What happened?"

"I been hit!"

"Get to the house!" she yelled. "Get to the house!"

Randy was running in front, and Sara ran behind him, as close as she could get, trying to shield him, expecting any second to be dropped from behind as Sammy had been. If the bastards were going to shoot someone, she was going to make sure they killed another kid. She winced in terror and ran as fast as her dad could move ahead of her, but the cabin—a few yards off—seemed miles away.

Kevin Harris was bringing up the rear, running and ducking, his rifle dangling in his hand. Vicki had heard the shot as well, and she ran out of the house, holding Elisheba under her right arm, her pistol holstered on her hip. Oh, blessed Yahweh, it was happening again. A few feet from the door, she saw Randy running toward her. "What happened?"

"Mama, I been shot!" Randy yelped as he rounded toward the house.

As she moved back toward the door, Vicki yelled at the hill where the shot had come from: "Bastards! Murderers!" She threw open the heavy door—which had a curtained window at eye level and swung out—in the direction of the hill where the shot had come from. Vicki stood behind the door. Randy and Sara were bursting through the doorway onto the wood floor, and Kevin was gathering himself to dive into the cabin, at the same time trying to push Randy and Sara out of the way.

"You bastards!" Vicki yelled. And then there was another shot.

*L*on Horiuchi felt the recoil from the second shot and had to put the 10-power scope back up to his eye to see if he had killed his target this time. Nothing. The man was gone. He spoke in a low tone into his headset radio that he had fired two shots and that he may have gotten a hit. Two hundred yards from the cabin, he watched the door, but there was no movement. Someone started screaming, a girl or a woman, he couldn't tell. And he couldn't make out any words. The screaming lasted about half a minute. And then it was quiet again.

In the coming year, Horiuchi would testify several times about what happened. "It was overcast, but the visibility was excellent," he said. With the scope, Horiuchi could see both the back and the front of the house, where the old washing machine sat. He could see the blue water tanks and the birthing shed. He could see everything on the north side of the house and anyone who came out either door on the ends of the cabin. He could even see under the stilted back porch that overlooked the steep meadow and the valley beyond.

Horiuchi heard the rumble of an armored personnel carrier at the bottom of the hill, and just as they'd been briefed, the dogs started barking. And just as they'd been briefed, Randy Weaver sent one of his children out to check on the noise. With his scope, Horiuchi picked up an unarmed young girl with a ponytail, running down toward one of the rocky outcroppings and then back to the cabin. Then she ran out again, this time with two men, who ran in and out of Horiuchi's sight, toward the water tanks, then back to the birthing shed. They seemed to be searching the area for something. One of them took a stick and scratched at the ground. Horiuchi heard the sound of a helicopter nearby, he testified later. The chopper "was not in front of me . . . I'm assuming it was somewhere behind me, either to my right or left."

Horiuchi saw a man in his sights whom he assumed was Kevin Harris (and was in fact Randy Weaver), possibly reaching up on the birthing shed and grabbing it to swing around behind it. "He seemed to be looking for the helicopter. He seemed to be moving, trying to get back on the other side of the [birthing shed]. By being behind the [shed], he could take a shot. . . . I perceived that he may be getting ready to take a shot at the individuals in the helicopter." At least two other snipers said later that Weaver and Harris were running as if they were planning to do something aggressive. The snipers also said the helicopter was in danger although neither agent fired his weapon.

Horiuchi testified later that he was too far away to call out "Freeze!" or "You're under arrest!" He figured that as soon as he began firing, the other snipers would shoot, too. With his shouldered .308-caliber Remington model 700 resting on a tree branch in front of him, he took aim at the man's spine just below his neck. Like all shots an FBI sniper is trained to take, this one was intended to kill. But in the flash between the mind's command to pull the trigger and the tiny muscle reflex in his index finger, Lon Horiuchi saw a shudder in his scope and, after the crack of the gun, realized the target had made a sudden move. The shot missed Randy Weaver's spine, and Horiuchi saw it splinter the wall of the birthing shed.

Horiuchi—trained to hit a quarter-inch target every time from this distance with this single-shot, bolt-action rifle— had missed. Still, he figured he must have grazed the man in the back. Horiuchi brought the scope back up to his eye and picked up the three people running for the cabin. He took aim again, this time correctly identifying Harris, but mistaking him for the man he'd shot at the first time. With his big 30.06 in his right hand, Kevin ran six feet behind

Sara and Randy toward the door. Horiuchi decided to finish what he'd attempted at the birthing shed and to neutralize the male and his rifle.

Horiuchi didn't want him back in the house, where he would be protected and would be able to hide behind the children, he testified later. From the cabin, Horiuchi would testify, he believed that the man could shoot at the snipers or the people in the helicopter. Better to drop him outside.

In an instant, Horiuchi estimated Kevin's speed and the distance between them. The target had paused on the porch and looked to be pushing the two crouching people in front of him. He swung his rifle around, gave the target a little lead—one notch, or mildot, from the crosshairs on his scope—and just as Kevin reached the door, Lon Horiuchi squeezed the trigger again. The firing pin slammed forward and detonated the primer, causing an explosion that launched a bullet whose grain and weight were carefully measured for the utmost accuracy, a .308-caliber bullet, somewhat like a hunter might use to drop a 1,000-pound elk, a soft-jacketed, boat-tail bullet designed to travel 2,600 feet per second, to cover that 200 yards in a quarter of a second, to hit, expand and tumble, so that, by the time it made its way out of the target, it had taken with it several times its size in flesh and bone.

In that instant there was a flinch—possibly a hit—the diving man disappeared from Horiuchi's sight, and the recoil from his rifle kicked at Horiuchi again.

*F*or just a second, everyone inside the cabin was quiet. The crack of the last shot echoed through the house, and they all turned back to the door. Vicki was on her knees. Randy looked back and saw her lying there, her head down on the floor, right next to the kitchen door, as if she were praying. Randy figured she was hit, but he was afraid to look, because she wasn't moving at all.

Rachel screamed. She had been standing near her mother, facing the door, when Vicki was shot. Blood hit the ten-year-old, flecks of red in her long black hair. Sara,

Randy, and Kevin had all squeezed through the door about the same time, bottlenecking and then bursting through. Sara had felt something whiz by and then blood splattered on her cheek as she hurled through the doorway and tripped over Kevin, who had been pushed in front of her. Now Sara looked back over her shoulder at her mother, slumped in the doorway. What struck Sara was that there were no reflexes, no movement at all. Her mom—their strength, their warmth, their prophet—was gone. Sara cried out, "No-o-o-o!" They screamed and tore at their hair, until Sara couldn't make any more noise and she realized that they should be quiet, that her mother would warn them all to be quiet.

Vicki Weaver was folded over like a tent, resting on her knees, arms and head, just inside the door. Her hands were still cradling Elisheba, so tightly they had to pull Vicki back and pry the baby out of her grasp.

Randy, wincing from his own wound, turned his wife over. It was as if her skull had exploded. She was almost unrecognizable. Her jaw was blown half off, and blood seemed to be pumping out from everywhere on her head. Elisheba was covered in blood, so Randy and Sara checked all over the crying baby, but found no wounds. She was okay.

Kevin was lying there, too, quiet, a silver-dollar-sized hole in his left upper arm. Sara assumed he was dead as well. The blood pearled and pulsed from his wound and ran down his arm. He was already turning pale. The bullet that had torn through Vicki's head was lodged in Kevin's upper arm, near his shoulder. His chest and arm were pockmarked with bits of bullet fragments and the bones from Vicki's face.

They pulled Vicki's body all the way into the cabin, yanked the door shut, and cried for hours, as quietly as they could. "I don't want to live without Mama," Rachel said.

They waited for the raid that would end them all, probably in the coals of a great government bonfire. They figured earlier that the feds would at least offer them a chance to surrender, but apparently they weren't going to. Clearly, Sara thought, they wanted them all inside the cabin, until they were ready to drop them and then, one by one, the government agents would make them watch each other die. It was the most torturous thing she could imagine. Her eyes raw from crying, Sara picked up her rifle, cradled it, and sank to the floor.

For the first time in two days, Mike Johnson felt as though everything was under control.

His deputies were off the mountain and the experts, the Hostage Rescue Team snipers, were up there. The negotiators would be on their way up soon, and Johnson figured this thing was about to end, one way or another. He stood outside the command post trailer and watched state and federal agents scurrying in and out of the meadow, moving into positions around the ridge.

And then Johnson heard two shots in quick succession. Bang. A pause. Bang. Johnson leaped back into the command post and looked around wildly. On this sound-bending ridge, with a canyon on one side and the long valley on the other, he still guessed the shots had come from on top of the mountain. Johnson had heard them, clear as if they had been fired right in front of him. People were running in and out of the command post, phones were ringing, faxes were faxing, until, after a few minutes, an FBI agent finally found time to brief Johnson on what had happened.

A sniper had seen someone outside the cabin and had fired to protect a helicopter that was surveilling the mountain, the agent said. The sniper thought he'd killed Kevin Harris but wasn't sure. "He didn't see anybody drop because Kevin dove behind the door," the agent told Mike Johnson.

The helicopter landed in the meadow, and Richard Rogers jumped out into the drizzle. He ducked the whirring blades and ran across the field to the

Vicki Jordison, about age twelve, with her younger sister, Julie, and her brother, Lanny. Vicki was very capable and mature for her age.

Vicki Jordison on the night of her senior prom. She had rarely dated in high school.

Randy "Pete" Weaver and Vicki Jordison's wedding photo. They were married in November 1971 at the First Congregationalist Church in Fort Dodge, Iowa. Vicki wanted to be a housewife. Randy wanted to be an FBI agent.

Sisters Julie Brown (left) and Vicki Weaver (right) watch Sara and Samuel teeter-totter in Fort Dodge, Iowa, about 1981. Brown and the rest of her family tried to talk the Weavers out of running away to Idaho.

(Left to right) Samuel, Sara, and Rachel Weaver in 1982, a year before the family left for the mountains. This was one of the last photos Randy and Vicki allowed before they began believing that photography was against God's wishes.

David Jordison with his grandchildren (left to right) Sara, Rachel, and Sammy. Each summer, Vicki's parents visited Ruby Ridge, bringing supplies and helping set up the cabin.

The Weaver cabin in the mountains of North Idaho in the summer of 1984, when it was completed. The cabin was built by Vicki and Randy with two-by-fours, plywood, and scraps from a nearby sawmill.

Vicki Weaver sweeps the rock steps just off the porch of the cabin while her daughter Rachel watches, photographed in the mid-1980s.

The Weaver family in May 1989. Later that year, federal agents taped Randy selling a sawed-off shotgun to a federal informant. The lettering on Randy's shirt reads: "Just say 'NO' to ZOG"— the Zionist Occupied Government.

Sara, Samuel, and Rachel photographed in May 1989, outside the house their family rented. Friends and relatives thought the family's fear and distrust of the government lessened when they lived at the bottom of the mountain.

Booking photo of Randy Weaver from initial gun charges, January 17, 1991. After his release, he returned to the cabin and refused to show up for trial, waiting for the government to come for him.

The remote cabin atop Ruby Ridge, a forested knob in North Idaho, near the Canadian border. Photo taken in March 1992, six months before the standoff began. *Spokesman-Review*/Shawn Jacobson

The Weaver cabin and compound, March 1992. *Spokesman-Review/ Shawn Jacobson*

U.S. Marshals Service surveillance photo on August 21, 1992, the day of the initial shoot-out, shows Samuel Weaver (left) pointing a gun in the air, Kevin Harris (center) with the dog Striker at his feet, and Sara Weaver (right). COURT FILE, *U.S. v. WEAVER*

Kevin Harris on the morning of the shooting, with Striker.
Court File, *U.S. v. Weaver*

Vicki Weaver paces in the yard while waiting her turn in the outhouse at the Weavers' cabin, August 21, 1992, about an hour before the shoot-out. Court File, *U.S. v. Weaver*

After the shoot-out began, as many as 300 state and federal agents quickly moved into a meadow about a mile from the Weavers' cabin, where they set up a command post and tent city nicknamed "Federal Way." *Spokesman-Review*/Colin Mulvany

William Degan, one of the most highly decorated members of the U.S. Marshals Service, flew to Idaho in August 1992 as part of a team that was supposed to eventually apprehend Randy Weaver. WBZ-TV, Boston

FBI Special Agent-in-Charge Gene Glenn (left) and U.S. Marshal for Idaho Mike Johnson meet with reporters early in the siege. Later, Glenn complained that he had been made the fall guy for federal mistakes at Ruby Ridge. His letter to the U.S. Justice Department sparked a new probe into allegations of a cover-up that reached the number-two official in the FBI. *Spokesman-Review*/Jesse Tinsley

Protesters yell "Baby killer!" at federal agents who drive past the roadblock up toward the Weaver cabin. By the end of the week, as many as one hundred people—neighbors, skinheads, angry constitutionalists, and neo-Nazis had gathered at the bridge over Ruby Creek. *Spokesman-Review*/Jesse Tinsley

Five neo-Nazi skinheads are arrested August 25, 1992, the fifth day of the standoff, in a Jeep filled with guns on a back road near the Weaver cabin. Agents with the Bureau of Alcohol, Tobacco, and Firearms received a tip that the men were going to help Weaver. The charges against them were eventually dropped.

Spokesman-Review/Colin Mulvany

Bill Grider, a onetime friend of the Weavers, restrains his wife on the sixth day of the siege as the couple yells at neighbors who have cooperated with federal authorities.

Spokesman-Review/Jesse Tinsley

Idaho state police and federal agents stand at the roadblock at the bridge over Ruby Creek and videotape protesters.
Spokesman-Review/Colin Mulvany

Third-party presidential candidate and retired Green Beret Lt. Col. James "Bo" Gritz talks to a gathering of reporters and protesters about the negotiations in the cabin. Gritz and retired Phoenix police officer Jack McLamb negotiated a peaceful settlement to the stand-off after eleven days.
Spokesman-Review/Anne C. Williams

The inside of the cabin, following the standoff and investigation by federal agents. *Spokesman-Review*/Blair Kooistra

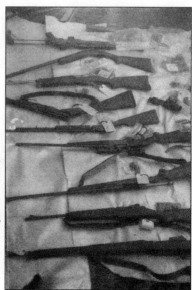

Some of the guns taken out of the Weaver cabin by federal agents. Later, six rifles, two shotguns, six pistols, and thousands of rounds of ammunition were admitted into evidence in the trial of Weaver and Kevin Harris. Jess Walter

Randy Weaver's defense attorneys, Gerry Spence, Kent Spence, and Chuck Peterson, answer questions about the case outside the federal building in Boise. *Spokesman-Review*/Blair Kooistra

Kevin Harris's defense attorneys, David Nevin and Ellison Matthews, enter court during the trial, which began in April 1993. *Spokesman-Review*/Blair Kooistra

Randy Weaver's attorney, Gerry Spence, questions deputy U.S. Marshal Dave Hunt, while Judge Edward Lodge looks on. Hunt spent almost eighteen months trying to get Weaver down from his cabin and was on the team that got into a gun battle with Weaver and his family. Zella Strickland

Lon Horiuchi, the FBI sniper. Behind him is the door to the Weavers' cabin. Zella Strickland

Ken Fadeley, the ATF informant. Zella Strickland

Sara and Rachel Weaver leave the courthouse with Vicki Weaver's brother-in-law, Keith Brown, and sister, Julie Brown, on June 15, 1993. The defense rested its case without calling Sara Weaver or any other witnesses. *Spokesman-Review*/Blair Kooistra

Kevin Harris talks to reporters after being set free by the jury that found him not guilty of murder after a nineteen-day deliberation, the longest in Idaho history.
Spokesman-Review/Blair Kooistra

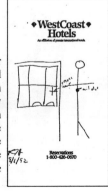

The day after the standoff ended, sniper Lon Horiuchi sketched on a hotel notepad what he saw just before he fired at Kevin Harris and killed Vicki Weaver. It clearly shows two heads in the window, even though he claimed that he couldn't see through the window. The judge fined the FBI for not turning this over to the defense until late in the trial. COURT FILE, *U.S. V. WEAVER*

The sign at the base of the Weavers' property. The words are taken from the King James Version of the Old Testament, the Book of the Prophet Isaiah, chapter 45, verse 23: "I have sworn by myself, the word is gone out of my mouth in righteousness, and shall not return, That unto me every knee shall bow, every tongue shall swear." The Weavers believed Jesus should be called by his Hebrew name, Yashua, and God by his, Yahweh. *Spokesman-Review*/Blair Kooistra

Gerry Spence autographs a book while his son, Kent, looks on, the day Randy Weaver and Kevin Harris are acquitted of murder and conspiracy. *Spokesman-Review*/Blair Kooistra

On December 17, 1993, Randy Weaver was released from jail after serving sixteen months for failing to appear in court. He moved back to Iowa with his three daughters.

Spokesman-Review/Blair Kooistra

Hostage Rescue Team's command post, the off-white, hard-walled tent set up at the edge of the meadow. Inside, he found the sniper coordinator, Les Hazen. Rogers had been flying over the ridge to get a look at what they were up against when he'd heard radio traffic that two shots had been fired.

"What happened?"

Hazen told him Lon Horiuchi had fired the two shots.

"Was anybody hit?"

Hazen said he didn't know. Rogers turned and ran to the overall command post and found Special Agent-in-Charge Gene Glenn. The two men agreed that they needed to get up there as soon as possible. The family knew now they were surrounded, and the FBI had to get the armored personnel carriers up to the cabin, identify themselves, and demand the Weavers surrender.

It was 6 p.m., and the agents still hadn't rigged up a radio system that could communicate between the two APCs. Forget it, Rogers said. He climbed in the lead APC with a handful of other agents and with Fred Lanceley, the hostage negotiator, and they started out.

The low-riding tank grumbled through the meadow and more slowly up the old logging road, a vehicle so wide it crunched the brush on either side. Near the cabin, Randy's red truck blocked the road, and so the APC went around it, but higher up, the driveway was blocked again, this time by a cable gate stretched between a steel I beam and an old trailer. They tried to push the I beam away with the APC, but it wouldn't budge. Finally, the hatch opened and several camouflaged agents jumped out, checked the gate for booby traps and explosives, moved the cable, and climbed back in the APC. The vehicle rumbled up to the house, until it was twenty or thirty feet away. Again, the hatch opened, and this time a telephone with a long cord connected to the APC was lowered to the ground. The hatch closed and at 6:45 p.m., the engine on the APC was shut off, and with the light fading in a low, overcast sky, an easy, authoritative voice intoned over a bullhorn.

"Mr. Weaver? This is Fred Lanceley of the FBI. You

should understand that we have warrants for the arrest of yourself and Mr. Harris. I would like you to accept a telephone so that we can talk and work out how you will come out of the house without further violence. I would like you or one of your children to come out of the house, unarmed, pick up the telephone, and return to the house."

An hour after firing at the family, the FBI was finally making the surrender announcement. There was no answer from the cabin.

The APC sat in the driveway for another twenty minutes or so and the voice called out again that they should surrender and go through the court system. The voice said they were leaving a telephone near the cabin, and they needed to set up some communication to end the standoff peacefully. The voice said he would personally be on the other end of the telephone. "Randall," the voice said. "We need to communicate with you to end this thing peacefully."

"Vicki," the voice said, "maybe one of the children could run out and grab the telephone." But there was still nothing from the cabin.

*T*hey must think he's crazy. After what they did to Sam and Vicki, did the ZOG devils really think Randy was going to send one of his daughters skipping out the door to pick up a telephone? So they could shoot a little girl, too? Did the One World Government snipers need more target practice?

Perhaps they would set fire to the cabin now. Maybe it would be a ground assault. Or tanks. Helicopters, perhaps. Whatever, it was only a matter of time before the cabin was surrounded and they were all murdered, as Gordon Kahl had been murdered, and now, as Sammy and Vicki had been murdered. Sara knew she was going to die soon. There was no way out now; they were going to be shot dead one at a time.

The family could hear the tank moving around the house, like a cat toying with an injured bird. Rachel and Sara held tightly to their rifles and sobbed as quietly as they could. Kevin moaned and grimaced with pain, and Sara sneaked across the room to take a look at his wound. Elisheba cried and squirmed in the arms of whoever held her, trying to find her mother, who lay in a heap on the kitchen floor, her heart having pumped most of her blood into a pool around her. Vicki's legs, covered by the long denim skirt, protruded from underneath the table, and so they had covered her with an old, green army blanket.

*T*he APC backed all the way down the hill, trailing the telephone line from the cabin. The going was slow and it took an hour and fifteen minutes to play out a mile of cable from the cabin. They reached the meadow about 9:45 p.m. and hooked up the telephone. There was no one on the other end.

The snipers were soaking wet and exhausted, and the temperature was knifing toward freezing. Dark and overcast, there was nothing more for them to see, and the HRT officials worried about them getting hypothermia. Besides, Rogers wanted to talk to them about the shots that had been fired. So, about 8 p.m., they had called the snipers on their radios and ordered them back to the meadow.

The snipers came into the HRT command post, changed their wet clothes, and gave their reports to the team commanders. Rogers grabbed Horiuchi, pulled him outside the tent, and asked what had happened.

According to Rogers's testimony, Horiuchi told him about the people running outside the cabin and about the helicopter he'd heard over his shoulder. He said the two men were moving into position to fire at the helicopter and so he shot at one of the men, whose movement caused him to miss. He got another clear shot, fired, and saw someone flinch, but he didn't actually know if he hit anyone. There were ten other snipers up there; he'd expected at least one of them to start shooting as soon as he fired his weapon, but no one else had fired.

That's not what struck Richard Rogers first. He couldn't

believe that one of his snipers had missed from 200 yards.
Maybe even twice.

K evin was going to die next. That was obvious to Sara,
who watched him in the last glow of dusky sunlight
that seeped through the heavy curtains. He was in the
worst shape, and Sara didn't figure he'd last the night.
Randy noticed it as well. Kevin was pale as a sheet.
Clearly, the fragments had gone into his lungs; he was
coughing up pools of deep, crimson blood, and the pain
was almost more than he could bear. Usually strong and
quietly jovial, Kevin grimaced and drifted in and out of
consciousness while Sara poured a bottle of hydrogen
peroxide over the blackening wound and it boiled over
with infection. She dressed and cleaned the wounds, but
Kevin was losing a lot of blood and was in agonizing pain.

Finally, about 9 p.m., three hours after he was hit, Kevin
asked Randy to shoot him again, this time to put him out of
his misery.

"Through the head." He winced. "Don't tell me when
it's comin'. I want you to finish me off."

Sara and Rachel began crying again. Randy, too.
"Kevin, I can't do that."

"Please," Kevin asked more quietly. No, Randy said.

Sara watched Kevin through hot, blurred eyes. He was
the bravest boy she knew. She prayed and read the Bible
but wished her mother was there to help them understand
Yahweh's will. It was up to Sara now. She knew her mom
would want her to help take care of the family.

They whispered a little about what they would do, and
especially about how they might get their story out. It all
seemed useless now. Curled up on the wood floor, with
rain slapping against the sheet metal roof and agents of
Babylon crawling all over the hill, there seemed no point.

"They're gonna kill us all and cover it up," Sara said. It
was like she was in a trance, unable to sleep but not quite
awake, alert but dreamy. It was dark as a cave inside the
cabin, and the only noises were the constant weeping and

Kevin's gravelly moans. Sara watched the doors and windows and waited for the Apocalypse that she'd known was coming for as long as she could remember. It was worse than her parents had ever told her, worse than any nightmares. Sara would look across the floor at her mother and the fear would quickly give way to anger, and she vowed that when they finally came, she was going to take one of the bastards with her if she could.

Every thirty minutes, the telephone rang in their yard, a stupid, jangling trick to get them outside for more sniper practice. They would tense up and watch the door where Vicki'd been shot until it stopped ringing.

Randy's wound was bad Sunday night, too. He was afraid that he was going to die before morning and that he wouldn't even be there to save his daughters. He knew damn well they were all gonna die. Likely, the government agents were waiting for morning to storm the house. They heard noises beneath the cabin—agents setting bombs? listening devices? gas canisters?—and Randy pounded on the floor and yelled until the noises stopped.

Nobody slept except Kevin, who drifted in and out of consciousness when he wasn't coughing up dark blood. The baby woke up crying in the middle of the night, looking for Vicki's milk. "Mama," she said in between wails. "Mama."

That started everyone crying again. Randy patted Elisheba on the back and gently cried himself. "I know, baby. I know, baby. Your mama's gone."

fourteen

"**S**hoot me!" the man yelled at federal agents. Wild-haired and leaning, a can of Budweiser in his thick, laborer hands, the mechanic walked from the old highway toward the Ruby Creek bridge, glaring across the police tape at federal officers with machine guns. "If you don't shoot me tonight, I'm going to come back and shoot you tomorrow! I'm a Vietnam vet and I'm on Randy's side, just like all these other people out here."

It was 2:00 a.m. Sunday morning, the bars were closed, and cars trolled up and back on the old Naples highway, along the wide curve, their headlights tracing the line of trees and the faces of tired federal officers, who stood on the bridge, behind a yellow police tape that suddenly didn't seem quite as authoritative. The Vietnam vet wandered away from the tape and toward Bill Morlin, the reporter who'd broken the Weaver story five months earlier. Morlin had never covered a story quite like this. The government kept them three miles away from the cabin and—going on the third day of the standoff—hadn't even held a press conference yet. The reporters and photographers slept in their cars, at the roadblock, expecting any minute for shooting to break out between the angry mountain people and the flak-jacketed federal officers. This was as tense as he'd seen the roadblock, and Morlin carefully approached the drunk veteran, who said the standoff was the fault of the government, which sent Weaver to Vietnam and then ignored him when he came back. The Idaho woods were full of Vietnam vets like Randy, he said, guys who couldn't cope with the real world.

"You put a strand of ivy on the wall, then come back twenty years later and it's covering the whole wall," he said. "So you don't have to ask why there are men like Randy up there."

The mechanic said he'd met secretly with a group of other veterans, and they would retaliate if the government killed Randy and Vicki. First, they would destroy microwave and telephone communication facilities on Blacktail Mountain, near Sandpoint and at a place called "the pit" near Naples. Then the federal agents would be caught on that mountain with no way of getting help from the outside, and they'd be easy targets for the mechanic and his army of one hundred vets, who weren't going to let a brother from Vietnam hang like that. The protesters listened to his little press conference and nodded in agreement; they had no idea that Randy Weaver had never been to Vietnam.

*F*orty-six hours after the shoot-out with deputy marshals, a secret microphone built into the telephone picked up the whining of a dog. The tiny mike also transmitted muffled voices and footsteps inside the cabin that Sunday morning. But when FBI negotiators at the base camp in Homicide Meadow rang the telephone in Randy Weaver's yard, they got no response and presumed the family still wouldn't leave the cabin to pick up the phone.

They had to presume because, strangely, the cabin was left unattended by the FBI from 8:00 p.m. Saturday to mid-morning on Sunday. Finally, at 10:00 a.m., Lon Horiuchi and the other snipers got back into position on Ruby Ridge, nestling behind rocks and trees on the hillside overlooking the cabin again. The APCs rumbled back up the ridge, too, stopped on the driveway, below the cabin, popped open their hatch doors, and released fourteen camouflaged agents, who crept with machine guns along the hillside just below the cabin, checking for mines or other booby traps. They secured the springhouse and the other distant outbuildings and found cover in case shots came from the house or the woods behind them. While Horiuchi and the other snipers kept a close eye on the cabin from 200 yards away, the assault teams closed the perimeter and surrounded the point from forty or so yards away, on downslopes below the cabin's line of sight. They also kept watch on the woods behind them, in case any of Weaver's supporters made it past the other lines of law officers on lower points of the ridge.

When the assault teams were in place, the lead APC groaned up the rest of the driveway, around islands of boulders and card-house outbuildings, finally parking thirty feet from the front door of the cabin, on the side that opened into the kitchen, the side without the spectacular view, the side where Vicki Weaver had been standing when she was killed. The hatch opened again, and this time, the HRT Commander Richard Rogers spoke through the loudspeaker.

"This is the FBI," Rogers said. "We have a warrant for the arrest of Kevin Harris and Randy Weaver. Come out with your hands up, unarmed. You will not be hurt. We want to take you into custody and put you into the legal processes of our system."

No answer.

Rogers said they needed to set up some sort of communication with the FBI, to find a way to end the standoff.

Every fifteen minutes, the FBI agents rang the telephone. "Pick up the phone, Randy," said the hostage negotiator, Fred Lanceley, who took over the megaphone from Rogers. "We aren't going to hurt you. It's just a telephone." Lanceley knew from his briefings that when Randy heard the dogs bark, he sent his children out to see what it was. He hoped Randy might do the same when the phone rang. "You can send out one of the children. It's safe, Randy."

The assault agents could hear muffled voices and a baby crying, but there was no answer to the negotiator. Lanceley tried Vicki Weaver again.

They rang the telephone in the yard thirty-four times, about every fifteen minutes, from 8:30 a.m. to 4:00 p.m. The curtains didn't even rustle. Finally, in the afternoon, a van drove up to the base of the bumpy dirt driveway and stopped well below the cabin, its back doors thrown open so a ramp could be lowered to the ground. Inside the van, a technician punched commands into the remote control computer and then the robot walked out.

It didn't walk so much as roll, on rubber tracks like those of a bulldozer. It was a modified bomb disposal robot, like a big, rolling trash can or R2-D2 from the movie *Star Wars*—a silver barrel with mechanical arms coming out of the top and sides and loaded with every manner of hardware: cameras, floodlights, microphone, amplifier, and receiver.

One arm ended in a single-barrel shotgun.

*D*ave Hunt hated every minute of this waiting. The rain had stopped, the high clouds were in the process of burning off, and it was incredible how quickly the mud turned to dust. Hunt's chest and head throbbed with a cold that had fed off his exhaustion and two days of drizzle. Hunt watched camouflaged FBI agents with their M-16 rifles getting ready to go into the woods and saw FBI officials walk right past himself, Roderick, and Cooper as if they were invisible, as if they'd screwed up and were of no help now that the FBI was here to bail them out. Finally, the three deputy marshals walked over to the Hostage Rescue Team command post, stuck their heads in the tent, and listened as the agents were briefed by HRT officials. It was tough for Hunt to listen as they went over the information he'd spent eighteen months gathering— "Vicki Weaver is the spiritual strength"—and some intelligence—"there might be booby traps and explosives"—that he'd long ago dismissed. They treated Randy Weaver as if he'd had some special guerrilla training in the army, like some stereotypical Green Beret who built bombs out of nothing and had flashbacks of Vietnam.

For more than a year, Hunt had been telling people that Randy had never served in Vietnam. He had some demolition training, but he was really just an equipment operator, a 'dozer driver, not some kind of killing machine. Hunt knew Randy was dangerous all right—not because of his ability to build booby traps and bombs—but because of his unbending beliefs.

Hunt's surveillance photos hung on the walls of the off-white HRT tent, so the FBI agents could familiarize themselves with the terrain, the cabin, and the targets. The pictures sparked something else in Hunt, a feeling of incompleteness. He was, above all, an enforcement deputy, *the guy* charged with bringing in *the fugitive*. And his fugitive was still up there. For eighteen months, he'd imagined the conversation he'd have with Randy Weaver— "We need to get this thing straightened out before it gets out of control . . ."—and that Randy would walk down the hill with him peacefully. Now, one of the marshals was dead and Hunt was stuck here, watching FBI agents with machine guns getting ready to crawl into the woods to bring in his fugitive.

Outside the tent, the flow of men and materials continued into Homicide Meadow. Two Huey helicopters landed in the meadow, which was close to bursting with green tents and military vehicles. It might seem like overkill to someone on the outside, but Hunt knew how difficult it would be to cordon off this entire mountain, especially against armed people who knew where every logging road connected with every cattle trail, people who might find a way through the brush and timber to the Weaver cabin.

They paced outside the command post trailer for a while, watching all the activity. Hunt just wanted off the mountain. He wanted to go home to his wife and get some sleep, instead of pacing around this hot, dusty field, kept in the dark about what was going on. Finally, sick of the wait, Hunt, Roderick, and Cooper left Homicide Meadow to catch a flight to Boston for Billy Degan's funeral.

*F*ifty-four hours. Fred Lanceley wrote out the things he wanted to say to the Weavers and ran them past the other two negotiators. At the van where they controlled the robot, Lanceley and the head of the HRT, Richard Rogers, talked about where to proceed next. They spent most of Sunday negotiating through the robot's loudspeaker ("Don't worry about the robot. Randall? Why won't you talk to us? Why won't you let us talk to Vicki?"), and there hadn't been so much as a peep from the cabin.

"Yeah, I think we can wait this guy out," Lanceley told Richard Rogers. "But I think it's going to take a long time."

Sunday evening, Rogers decided to create some space for the armored personnel carriers to move and to clear the line of sight for the snipers and assault team members. He told the agents to start by moving the birthing shed, and so they drove an APC over to the small, barnlike building to push it out of the way. But just before the APC rolled over it, Rogers stopped the agents and sent a handful of assault team agents to make sure the shed was empty. It had dawned on him that, the night before, one of the family members could have hidden in the shed and planned to ambush the FBI agents around the cabin. Camouflaged HRT assault team members crawled and dashed serpentine-style over to the little building, slammed up against the sides, burst through the door, and poked their machine guns inside, red-laser targets knifing through the darkness.

Flashlights lit the eight-by-fourteen shed, which was like the attic in an old house that had been turned into a bedroom—a small, self-contained living quarters lined with dark insulation. Just inside the door were cupboards on the right side, bumping into an elevated bed and a nightstand with a plant in a wide-based jar. Bags of clothes and old tennis shoes were piled against the other wall. On the bed was something big, wrapped in a clean white sheet. After checking for booby traps, the agents got closer, unraveled the sheet, and found a boy, stripped naked, his body cleaned and wrapped in a green sheet. He was a little boy, not even 5 feet tall, seventy or eighty pounds, with a T-shirt tan. He was uncircumcised. There were bruises on his knees and shins, like someone who runs through the woods a lot.

"I have found the body of a young white male," an agent said over his radio.

In the van, Fred Lanceley heard the radio transmission and was stunned. My God, he thought, Weaver is killing his own kids.

"Randy, we found the body of a young man," Lanceley said over the robot's amplifier. He asked what arrangements Randy wanted to make with the body. "I understand you have strong religious convictions," Lanceley said, "and I don't want to violate them. Please communicate your wishes to me by just speaking up."

Nothing.

They wrapped the body back up, put it in the APC, and drove it to another shed, farther away from the cabin, where a deputy coroner looked at the boy and determined he'd been shot in the arm and the back. And then they took Samuel Weaver away.

Sunday night, the spotlights came on with a "thunk," like the closing of a door, and lit the front of the house eerily, like a Christmas Nativity scene in front of a church. They were the only electric lights for almost a mile in any direction.

*   *   *

"*Y*ou killed my fucking wife!" Randy yelled at the door. Glaring white light painted the cabin walls, seeped through seams in the curtains, holes in the walls, and cracks in the doors, like flashlights probing the dim cabin.

On the floor, Sara couldn't stomach the idea of *them* having Sammy. The whole family had cried when the voice talked about moving Sammy's body. They were going to take his body away, cover up the fact that he'd been shot in the back by the marshals, and probably try to pin Sammy's death on Randy or Kevin.

Sara felt the spirit of Yahshua, the one the pagans call Jesus, the Messiah of Saxon Israel, moving in her and giving her strength. Huddled on the floor with what was left of her family, she hoped the cowards would try to storm the house, so she could at least get a shot at one of them. More likely, they would use tear gas to lure the whole family outside and then gun them down. But she was prepared for that, too. When it happened, she would put Elisheba someplace safe, check her gun, then burst out the door and start firing at anything that moved.

"Mama, Mama." Elisheba sobbed. She wanted to be nursed again, and she fussed over the milk, water, and apricots that Sara and Rachel tried to feed her.

Sara was busy—the way Vicki would have been. She got almost all the food out of the kitchen, changed Kevin's bandages, cleaned his pus-filled chest and arm, and dumped the last of the hydrogen peroxide on the wound, which fizzed again with infection. The rags putrefied quickly, and the stench of his arm and chest was getting to all of them, a horrible reminder that he probably wouldn't live, and a glimpse of what was in store for the rest of them. Sara remembered her mother's herbs, and when the hydrogen peroxide was gone, she used the herb goldenseal to help get rid of the infection.

They were in the middle of their third night without any real sleep, jigsawed around the living room in a pile of blankets, quilts, and sleeping bags, behind couches and chairs. They listened to a radio and talked in low whispers about what the feds would do next; they guessed a raid or tear gas or a firebomb. The thing they didn't expect was the constant psychological warfare, which was—in many ways—worse than anything the feds could've done physically.

"Good morning, Mrs. Weaver," the negotiator called out. "We had pancakes this morning. And what did you have for breakfast? Why don't you send the children out for some pancakes, Mrs. Weaver?" That had started the whole family sobbing.

The spotlights were another cruel stroke, blurring the time between day and night and keeping the whole family from sleeping, until Sara's head bobbed forward, and she snapped awake and wondered whether a few minutes had passed or a few hours, whether she'd heard something outside or something in a dream, whether she was alive or whether Yahshua had finally taken her away. Everything ran together: the phone ringing every fifteen minutes, urging them to come out and die; the tanks tearing up the ground around their cabin and circling them like sharks; the constant, unnerving drone of Fred, the negotiator. Maybe they were trying to make the family snap so they'd come out of the cabin firing their guns and the ZOG agents could claim they were justified in killing them all. Or maybe it was just torture, making sure they suffered plenty before shooting them all in the back and then claiming that Randy had killed his own family.

And now their latest trick was this robot they were always talking about. Don't be afraid of the robot. The robot is just so we can talk to you better. Randy knew exactly what it was for—to punch a hole in the house and shoot gas in, to kill the family or drive them into the open where they could be plugged by the fucking snipers. Randy held his rifle up and said he was going to start shooting if the robot came any closer.

"Back off!" Randy yelled. "You'd fucking better back off or it's all over!"

•   •   •

*S*eventy hours. Fred Lanceley arrived at the command post at 7:47 a.m. on Monday and found out that—just twenty minutes earlier—Randy Weaver had yelled something. He was cussing and screaming, the other negotiator said, and possibly even preaching. Much of it was unintelligible, but one thing was clear. He wanted the robot moved.

Damn. Lanceley had missed his chance.

"Randy, I'm sorry I missed what you had to say," Lanceley said. "Please, tell me what you wanted to talk about?" The negotiator pleaded with Weaver to talk, to say anything, but he got no response. "We just don't understand, Randall. We've done everything you asked us to do. We've cooperated the best we know how. We backed the robot off when you said back off."

Frustrated, Lanceley resumed his usual speech about how the Weavers should pick up the telephone. He warned them when helicopters and APCs moved around the cabin and just tried to keep the family calm. Besides being a negotiator, Lanceley believed he understood human nature. He had a bachelor's degree in psychology, master's degrees in business and criminal justice and was almost done with his psychology doctorate. The marshals' intelligence reports had included a psychiatrist's evaluation of Vicki that indicated she was the backbone of the family and would do anything—perhaps even murder her own children and then kill herself—to keep the family from being broken up. So Lanceley tried Vicki again, spending several hours trying to reach her.

"Vicki, how's the baby?" Lanceley mispronounced Elisheba's name. "I share your concern about Elizabeth. I need to know if there's anything that can be done for the baby. Milk? Diapers? Food? If you need anything, all you have to do is call out, and I'll see what I can do."

Nothing.

This was becoming one of the most frustrating cases he'd ever worked. That night, he broke into some deep conversations about death and about having the courage to face adversity, hoping to make Randy see how serious this had become and to sober him up and make him take some responsibility for his kids' safety.

Nothing.

Wayne Manis tooled along the old Naples highway and looked for the turnoff to Ruby Ridge. The friendly FBI agent from Coeur d'Alene had been off the white supremacist beat ever since he'd helped break The Order in the mid-1980s. But like every other federal agent in the West (and many from the East), he'd been called to help cordon off Ruby Ridge. Manis would've come sooner, but he was on another remote mountain in north-central Idaho, twelve hours away, trying to build a horse corral at a hunting camp in a driving snowstorm. He and another agent had driven back to Coeur d'Alene on Sunday, got the message about the Weaver case, and immediately set out for Boundary County. Now, Monday morning, Manis eased through Naples and into the foothills of the Selkirk Mountains. In the back of Manis's red Jeep was a 9-mm machine gun—the same gun he'd used in the siege of Order leader Robert Mathews's home on Whidbey Island.

Manis rounded the corner near Ruby Creek and couldn't believe what he saw. There were cars and pickup trucks everywhere—perhaps eighty vehicles, lined up along this little country road for nearly a mile. And the scene at the roadblock itself was even more stupefying.

Along the left bank of the road, a few feet before the Ruby Creek bridge, huge, white satellite news trucks hummed, their big dish antennae pointed away from the mountain. There were two dozen reporters and photographers working the roadblock, with their own portable outhouse. On the other bank of the dirt road, fifty protesters held signs—"Government Lies/Patriot Dies" and "Death to ZOG"—and yelled at the trucks and cars that were allowed through the roadblock. To Manis, the people looked dirty, like homeless stragglers just shouting and raising hell. He'd never seen such a concentration of angry racists.

They yelled at him as he drove through the roadblock, "Leave them alone!" Manis shook his head as he passed dozens of federal and state agents in bulletproof Kevlar vests and jungle hats.

As he rattled up the hill toward the command post, Manis expected to find a sedate mountain operation, with perhaps a half-dozen officers standing around in a field. Instead, he found a military camp. The road broke through the woods and into a meadow, where a Red Cross van was serving chow and hundreds of cars and trucks were parked on one end, covering a section of the meadow as big as a football field. Huge military-style tents, stuffed full of army cots, were set up everywhere. A barn had been converted into a staff office, a place for meetings and briefings and even a steno pool with secretaries, typists, and computers. Across an old horse trail from the barn was the command post, a forty-foot travel trailer with an awning coming out of it.

Manis stared all around the meadow. This was about the same size as the contingent of federal officers that had battled The Order on Whidbey Island.

"*I*t appears as if Samuel Weaver was killed during the initial exchange of gunfire, but that can't be definitely stated until the autopsy is completed," said Gene Glenn, the special agent in charge of the Weaver operation. In a field on the federal side of the Ruby Creek bridge on Monday, newspaper, television, and magazine reporters from around the country fanned out in front of Glenn and Mike Johnson, who conducted the first press conference since the shoot-out began. Glenn said they'd just discovered the body the night before. He said nothing about Lon Horiuchi's two shots.

"Samuel's death is a tragedy, as is the death of Deputy Degan. I emphasize we are taking and will take every reasonable precaution to avoid further loss of life or injury," said Glenn, his gray, earnest face cocked sideways and his hound dog eyes turned down at the corners.

"However, it must be understood that Harris and Weaver have been charged with serious crimes and they pose an immediate threat, not only to law enforcement officers, but to the community as well."

Johnson would go one step further. "What bullet killed Samuel Weaver is still under investigation. It's a possibility shots came from Harris's weapon."

Vicki's parents, David and Jeane, and her brother, Lanny, had made it to North Idaho by Sunday night at 8:00 p.m., 1,500 miles in twenty-four straight hours of driving. The next morning, they'd driven to Bonners Ferry, where they met with the county prosecutor, Randall Day, who told them to go back to Sandpoint to wait for instructions from the FBI.

The Jordisons found a motel across from the parking lot where Randy had last met with Kenneth Fadeley. They checked into a second-floor room and waited for the FBI to show up. In the afternoon, an agent finally did show up, a friendly middle-aged man with a deep Texas accent who asked them to make an audiotape for the agents to play on top of the mountain to coax the family down. On the tape, they all pleaded with Randy and Vicki to at least protect the children, and they said the Weavers wouldn't get their story out unless they surrendered.

The FBI agents visited their motel room several times, asking questions about the cabin. Once, an agent wanted David Jordison to draw him a map showing where all the furniture was located. Julie Brown, who flew out to be with her family in Idaho, couldn't believe what they were asking. "What if you use this map to raid them and something happens to someone in the family? How will my dad live with that?"

They needed the map, the FBI agent explained, to figure out where the booby traps were.

"Aw, you fools," David said. "They have a baby. Are they going to booby trap their house with a baby running around inside?" Another time, FBI agents warned the

Jordison family that they believed Vicki had shaved her own head and had become suicidal.

After the agent left on Monday, the Jordisons turned on the television news. Once again, the Weaver story led the local news, as a somber anchor said fourteen-year-old Samuel Weaver had been killed.

David Jordison sat down on the bed. He couldn't believe it. He thought about the mountain streams he and Sammy used to fish, the four-inch trout he'd find, the trails they'd go on, and the birch walking sticks Sammy was always bringing him. My God, he thought, they killed a little boy. David had never thought too much about government before, but now he was baffled. Maybe Vicki was onto something. After all, the FBI had just been in their room. Why hadn't they told him Sammy was dead?

In Jefferson, Iowa, Randy Weaver's mother, Wilma, answered the telephone and a reporter asked for her reaction to Samuel's death. She set the phone down and told Randy's father, Clarence, that, apparently, Samuel had been killed. Slowed by old age, Randy's parents had never visited Idaho, and they hadn't seen Samuel in nine years. Wilma returned to the line and asked why no one from the government had called to tell them. Of their sixteen grandchildren, said the elderly woman, Samuel was the only one who would've had their last name.

"I was just writing him a letter to tell him that and tell him how proud I was of him," said Wilma Weaver. "I guess I won't mail it."

*T*hey handed out extra plastic handcuffs, and the camouflaged agents at the roadblock strapped them to their belt loops and turned to face the crowd. Half-a-dozen agents bolstered the presence at the highway and officers who had leaned casually against the aluminum railings of the Ruby Creek bridge now stood at attention, watching the crowd with darting eyes.

It took only a few minutes for the word to spread that Sammy Weaver was dead.

A pretty, college-aged woman with red hair stepped forward from a mess of yelling men in cowboy boots and baseball caps. She held up a sign that read "FBI—Rot in Hell." The shouts and taunts started in again, and the crowd pressed toward the roadblock, leaning across the tape and screaming phrases as familiar as the chorus to a song: "Baby killer! Baby killer!"

When a reporter asked about the dead marshal, Vicki's friend Jackie Brown spit through her tears, "I hope they get a dozen more." They had murdered Sammy. But at least, Jackie said, Vicki and the girls were okay.

Construction crews waiting to get past the roadblock were taunted by protesters who didn't think they should be helping the feds. "Shame on you! Shame on you!" yelled a woman in a T-shirt that read "Leave ME Alone."

One man with a goatee held a "Death to ZOG" sign high over his head and preached at the driver of a semitruck waiting to haul railroad ties up the hill: "How much do you have to be paid to compromise your people? . . . How many dead patriots will you bury? How many will you put dirt in their faces and watch them and their families cry as you sat by and you drove the truck? You drove the truck! Walk away! Wipe the blood off your hands now! Stand up and be a man! Don't fight for a system of tyrants!"

The driver, unlike the television crews, ignored the man.

By Tuesday, the protesters had covered their campsite with a blue tarp and were cooking cheese sandwiches and bacon over a barrel stove. Foot traffic flowed from the Swiss innkeeper Lorenz Caduff's nearby Deep Creek Inn, which quickly became home base for protesters and reporters alike.

"Sir," a skinhead asked, "are we allowed to come in here?"

Lorenz said anyone was welcome in his restaurant. "I have a Korean housekeeper, and there's a black reporter here. If you are polite and nice, you can come in."

The Swiss chef served up scrambled eggs and roast beef sandwiches and listened to the conversation, shocked when he heard that government agents had shot a little boy in the back. He just couldn't understand it. He sent his own wife and kids to Sandpoint, to keep them out of trouble. "They have killed a boy? Why is this?"

At night, the protesters gathered around the television crews filming live shots and warmed themselves in the bright lights, listening to the tired TV reporters to make sure the "Jews Media" reported the story correctly. People continued to show up at the roadblock: white separatists joined by Randy's neighbors, bikers, woodsmen, grandparents, and Vietnam vets like the guy in a Jack Daniel's T-shirt that very nearly stretched all the way over his substantial beer gut.

Television lights glinted off the shaved domes of about a dozen skinheads—two separate troupes from Las Vegas and Portland who had called their various sponsors and been advised to report for battle in the great race war in North Idaho. Fresh off their success in a race riot in Denver, the Vegas skinheads—resplendent in their classic storm trooper jackets and slick-shiny heads—jumped right into the fray, launching glowers and derisive cheers of "Baby killer" at federal agents. One of the Vegas skinheads claimed to have hidden their guns in the woods.

The Portland skinheads kept more to themselves until Tuesday afternoon, when five of them piled into a Jeep Cherokee and drove back toward Naples, then turned down a back road that led to the canyon behind Ruby Ridge. As the Jeep moved toward the back side of Randy's cabin, a helicopter appeared over the tree line and hovered above the skinheads. They turned down another dirt road and state police and ATF cars raced up behind them, lights flashing. The Jeep stopped and Lance Hart—the ATF agent who'd arrested Randy Weaver—warned the young men to step out of the Jeep one at a time.

"Get on the ground!" The agents pushed the young men

to the ground, handcuffed them, and searched them. They
found a carload of guns and ammunition, and a sign that
read, "Whites Must Arm." One of the skinheads wet his
pants while they were being stopped. Federal agents had
been watching them since the day before, when they'd
gone to a gun shop, bought rifles and ammunition and
asked how to get to Randy Weaver's cabin. They were
taken into custody and charged with possession of a
concealed weapon, a charge that was later dropped.

At the roadblock, some of the Vegas skinheads were
defensive and pointed out that they had weapons, too, and
they would use them if it became necessary.

"We are here to let people know that those people
behind the yellow line are our enemies," said Johnny
Bangerter, a twenty-three-year-old skinhead who looked
like an angry Curly from *The Three Stooges* and who was
the second cousin of the governor of Utah. "Every federal
institution and government is our enemy because of its
action of killing a child in cold blood. We are ready to
fight, and it could be a bloody one. This is going to be a
second revolution in America."

Ninety-six hours. Given enough time, people will
find a rhythm for everything. The APCs moved up
and down the ridge, running over the gunshot body
of Striker—which no one had bothered to move—twenty-
seven times. There were reports that a federal officer had
put up a sign that read: "Camp Vicki." By Tuesday, Lon
Horiuchi and the FBI snipers and assaulters—working in
twelve-hour shifts—crouched more comfortably on the
hillsides around the cabin, the tension gradually displaced
by boredom. They were still alert but were no longer
running only on adrenaline. Chickens milled around the
yard, while the family's remaining dogs, which had long
since given up barking at the military vehicles, wandered
down to assault team agents who stood next to wilting
flower gardens and who scratched the dogs' ears and
petted them. In the coming days, agents closer to the

cabin began noticing a rancid smell coming from the house. Some of the more experienced agents knew what it was. There was someone dead inside the cabin.

Fred Lanceley showed up for work that morning and decided to lighten it up a bit.

"Randall. This is Fred. Good morning. I thought you might like to know that we are taking care of your dog, the one with the mismatched eyes. The last time I saw him, he was eating a big plate of spaghetti. We are calling him J.R., because he looks like one of the guys that works with me. Let me know if Elizabeth needs anything. Over the next few days, I hope to demonstrate to you, to Mrs. Weaver, to Kevin, that despite all that's happened, everything is being done to insure that this situation ends without further violence."

The weather had cleared for good, and the FBI agents snuck glances at the incredible view, the glacial valley played out before them in soft greens and browns, roads and houses small-scaled by all that was undeveloped. It was a startling view, a reminder that in some places, civilization is still subject to the flow of wilderness.

Down below, FBI agents had begun investigating the scene at the Y and were surprised to discover seven shell casings from Degan's gun spread along twenty-two feet, meaning not only did he fire his weapon, but he may have been moving when he did it. Along with Samuel Weaver's death, it was another indication to FBI investigators that the marshals' initial version of the shoot-out was not the whole truth. It also appeared the family wasn't as dangerous as they'd first believed. They still hadn't fired out of the cabin. The rules of engagement—which, until this point, would've allowed the snipers to kill Kevin and Randy again if they saw them—were changed back to the normal rules.

To Lanceley, the clear, warm Tuesday morning—five days since the shoot-out—seemed like a good time to talk about religion. Their intelligence showed that Randy fancied himself quite a preacher, and so Lanceley probed that for a while, like a dentist looking for a cavity.

"Randall, I'm from Virginia, and until a few days ago, I had never heard of your religious beliefs and convictions, and even today all I know is the government's version of what they say are your religious beliefs. I would like to hear from you, you know, what's going on here. What's happening? I just don't understand. Randall, these people aren't going away. It may take a few days to demonstrate that to you, but tell me what's happening. Let's discuss the problem and see if you and I and Vicki and Kevin can work it out."

Nothing.

Every time the ZOG agents mentioned Vicki's name, it was clear to the family what they were trying to do. They knew damn well they'd killed Vicki, and they were just showing the rest of the family how the standoff was going to end and making it clear that no one was getting out alive.

"Did you sleep well last night, Vicki?"

Always that coy, smart-ass tone. And every time they said her name, every eye in the cabin went to the blanketed body beneath the kitchen table, lying graceless in her own blood on the wooden floor.

Randy had screamed at the door on Sunday that Vicki was dead, and ever since then, the feds had turned that against the family, using their own grief to try to destroy them. Randy wasn't going to give them any more information. He was done communicating.

The Beast's agents had crawled underneath the cabin during the night again, fastening their fucking listening devices to the floor. Randy pounded with his feet and the noises stopped, but all night Randy and Sara sat up with their weapons, waiting for the devils to try again.

Kevin seemed to be more alert, but his wound was still infected and oozing, and he was in pain. Sara crawled past her mother's body into the kitchen and brought back several bowls of cold green beans. When she gave a bowl to Kevin, he tried it, then asked if she could warm it up. At least he was feeling better.

They listened to the news on the radio, took turns going to the bathroom in a portable toilet they had inside the house, kept themselves clean, and talked about how to get their story out, how to keep the government from covering up the murders of Sammy and Vicki. It seemed hopeless to Sara.

Another sleepless night moved hazily toward dawn, and Sara resumed her prayers to Yahweh, wishing she'd listened better to her mother's explanation of how to discern His will. She would give anything to have her mother's guidance.

And then—sometime before sunrise—the negotiators started in again. At least once every hour, day or night, they tried. It didn't matter to them.

"Is there anything we can do? Please let me know."

Clearly, Sara thought, the feds wanted them to get no sleep at all. She didn't know how much longer she could hold out. Her eyes played tricks with her in the closed-up cabin, and her exhaustion was cut with rage until she wasn't even sure what she would do next.

The voice said, "Behind every strong man there is a good woman. Mrs. Weaver, please support your husband by coming out. I want to resolve this situation peacefully. I'm sure you want the same. Are the children all right? Can we get some milk for Elizabeth? Let me know if I can be of any help just by calling out."

Sara glared at the door through teary eyes and prayed that Yahweh just get it over with and allow the ZOG devils to firebomb the cabin.

*F*red Lanceley listened to the report from the graveyard shift negotiator on Wednesday, who—at 4:30 a.m.—had appealed to Vicki again but had gotten no response. They were getting nowhere. Meanwhile, Rogers and the others still working on assault plans, faxing them back to headquarters for approval. The pressure to assault the cabin was building, just as it had when it was Dave Hunt's case. Lanceley figured that if he didn't get some response from the Weavers soon, the HRT would have to go tactical, firing tear gas into the cabin or raiding it with agents. If he only knew what was going on inside the cabin. Agents had only gotten two microphones tentatively attached to the side and bottom of the cabin, because whenever they tried someone stomped on the floor. The listening devices didn't seem to pick very much up—mostly muffled voices and the chirps and whistles of a couple of pet parakeets.

The robot seemed like the best way to get close enough to the house to hear what Randy had to say. As he always did, Lanceley told the Weavers about the next move they were making, so they didn't panic when the activity began again. "Randall, I can understand your concern about the telephone, and I can also understand why you might not want to step out on the front porch. So here is what they are going to do. The robot will go to the window on the right side of the porch. The grippers will try to punch the phone through the window and the glass will break. After the telephone is delivered, the robot will back off. You have nothing to worry about."

"Get the fuck out of here!"

Lanceley sat up and smiled. In fourteen years as a negotiator, he'd never heard profanity that sounded so beautiful. "Randall, this is a good opportunity for a dialogue. Let's you and I start talking?"

Randy said something the negotiator couldn't understand.

"I can't hear you, Randy."

The rest was like bad radio, breaking up on Lanceley. "Get out of here . . . this kike son of a bitch . . . you fucking pig . . . lying motherfuckers."

"Randall, I still can't hear you."

"Get the fuck out of here."

At least it was an opening. One hundred and twenty-four hours into the standoff, Fred Lanceley leaned forward and spoke clearly and slowly. "Randall, I can understand your anger, but you and I have got to try to resolve this. Let's you and I—Let's you and I see if we can resolve this. Let's you and I—Let's you and I see if we can start something anew right here."

"Get the fuck out of here."

fifteen

Karen Degan gripped tightly the hands of her two sons and walked into the slate-roofed Sacred Heart Church, where nineteen years earlier she'd married a strong, quiet football player. Now, on Wednesday, August 26, 1992, she was going to his funeral.

The streets were filled with mourners—3,000 cops, federal agents and judges, Marines, football players and old friends, escorted by dozens of police cars and motorcycles. Billy Degan's funeral gripped Quincy, a shipyard town just south of Boston that, for a couple hundred years, produced more than its share of big boats and Irish cops. Somber officers walked past brick businesses and beer-and-shot taverns until they reached the church, where 200 deputy U.S. marshals stood ten deep in dark suits and sunglasses, their badges striped in black. Six Marines carried the casket into the church while a customs agent played "The Marine Corps Hymn" on bagpipes on the church steps.

Dave Hunt was fogged in by a painful cold he'd gotten on the mountain, and he watched Degan's wife through bleary, tired eyes, amazed by her poise and strength. The boys were slices of Billy, one eighteen, the other fourteen, both tall and earnest in their dark blazers and cop-kid haircuts. Hunt had never seen such a funeral. Degan was a lifelong Marine and deputy U.S. marshal—two outfits that knew how to send someone off with full honors. The Marines had to bury heroes all the time, thousands of them. The U.S. Marshals Service had lost its share too, three hundred men in two hundred years, but rarely one with Billy Degan's reputation and abilities. They shut down the Boston federal building, and among those at the funeral were the mayor of Boston and the governor of Massachusetts. In a cop town like Quincy, in the belly of three centuries of civilization, it was hard for Hunt to imagine the place they'd just come from, that damned twilight zone of a mountain, where people were threatening federal officers, cheering on neo-Nazis, and holding signs that celebrated Degan's murder. It was such a shame. Hunt watched the pain in Larry Cooper's face as he listened to a parade of people talk about Billy.

"Bill Degan died in the company of his best friends," said Henry Hudson, "performing the duties he enjoyed best in the job he loved most."

The mayor of Boston, Ray Flynn, said, "It is a sad day. This is a personal tragedy for all the people in the state. We lost a dedicated law enforcement officer and somebody who had devoted his whole life to defending and protecting others."

Outside, Quincy police patrolled the church in bulletproof vests, in case Weaver supporters tried to strike at the funeral. They jumped when a nearby transformer blew, popping like gunfire, then relaxed when they realized what it was.

Inside the church, Degan rested in a steel-gray casket while the Reverend Cornelius James Heery, pastor of Sacred Heart, praised him as a family man and a good marshal. The pastor said it was okay to be troubled by the events in North Idaho and to wonder, "Why did Bill have to die on a hillside so far from home?"

*   *   *

*B*unkered in behind their furniture, the Weavers girded themselves for what was coming; they must not be afraid to fight, and they must not be afraid to die. But first, they had to get their story out. Much of the time during the last five days, the family turned on the radio and tuned it to the end of the dial, where one of the Spokane television stations was simulcast on radio. The ZOG agents seemed to know when they were listening to the radio, and they spoke through the loudspeaker during those times, but the family was able to pick up some news. They were glad to hear that followers of Yahweh and those who saw through the shadow government had gathered to protest. But they couldn't believe the misinformation that was being spread by the Jewish-controlled media. The government was calling the family white supremacists and members of the Aryan Nations, when in fact, they were separatists or racialists who belonged to no group. They didn't want supremacy over other races, just separation. Also, the marshals claimed Kevin shot Degan first and didn't even mention that the dog had been killed. They even went so far as to say Kevin might have shot Sammy. And there was no mention at all of Vicki being killed or of Randy and Kevin being wounded. It proved the government had no intention of letting anyone in the family go and that they would kill everyone to cover up what had really happened.

"We have to figure out a way to get our story out," Randy said. Sara was taking on her mother's role, and she didn't hesitate to grab her mom's yellow legal pad and take over the job of writing as Kevin and Randy dictated.

Wednesday August 26, 1992

Approximately 11:30 Friday morning, August 21, 1992, the dogs started barking like they always do. . . .

And then Sara laid out Kevin's version of the shoot-out with the marshals:

> The men were still shooting at Sam, so I shot one of the sons of bitches. . . .

and described the shots that killed Vicki and wounded Kevin and Randy.

"Needless to say, we understand snipers are everywhere," she wrote.

> They killed Sam, wounded Randy, killed Vicki, and wounded Kevin. The feds totally covered up the murder of Gordon Kahl . . . amongst numerous other cover ups [sic], especially against white rascialists [sic] and/or Tax Protesters. If they think we are going to trust them, (We didn't trust them before they shot us) they're crazy! Yahweh is starting to heal Kevin. We constantly pray he'll be okay. Since Vicki, Kevin and I (R.) have been shot we haven't left the house, and do not plan to unless we are starved out. Then we will most certainly take the offensive. It appears as though the feds are attempting to draw fire from the house as an excuse to finish us all off. If they even so much as crack a window pane on this house with a robot, telephone, gas, grenades, ect . . . ect . . . [sic] it's all over with.
>
>   Our heart felt thanks goes out to all our sympathizers. Our faith is in our creator Yahweh Yahshua the Messiah. We do not fear the One World Beast government. They can only take our lives. Only Yahweh can destroy our souls. Samuel Hanson Weaver and Vicki Jean Weaver are martyrs for Yah-Yahshua and the White Race. Even if the rest of us die, we win. Hallelu-Yah!

Keep the faith. To all our families and brethren,
We love you.
Hallelu-Yah!

And then they all signed it, yet another pronouncement
of defiance from a family that was being bled to death.
Sara wrote Elisheba's name in capital letters and then tried
to figure out where to hide a six-page letter so it might
survive a firebomb.

*O*ne hundred twenty-six hours. The FBI agents
continued working on their plans for a tactical raid,
while the negotiator, Fred Lanceley, tried to build on
the momentum from Randy's profanity-strewn ravings
about the robot. There had to be some way they could get a
telephone to Randy that wouldn't frighten him, Lanceley
said.

"I ain't taking no goddam telephone," Randy yelled. He
said something else, but it tailed off in a stream of words
Lanceley couldn't pick up, until the last one—"sister."

Lanceley pounced. "I heard sister, was there something
else?" He said that some FBI agents had reached Weaver's
dad.

"Is my father here?" Randy called.

"I don't know if he is here." Lanceley looked around the
upper command post and an FBI agent shook his head.
"No, none of your folks are here. I just checked."

"I want to talk to my sister!" Randy yelled. His sister,
Marnis Joy, had visited him that summer and now he called
out her name and where she lived.

All afternoon, Lanceley told Weaver about their
attempts to find Marnis, until, a little after 6:00 p.m. on
Wednesday, the negotiator said they'd found her. "Your
sister will be here in the morning."

But Lanceley wasn't sure about bringing Marnis in.
"There is some concern out here that if you speak to your
sister, you will tell your story to her and then commit

suicide. Randall, is that your intent? Are you going to commit suicide? You are going to have to promise me that you will not harm your family or harm yourself."

"I promise you no one will be hurt! I promise you no one will be hurt! I want my sister at the back door!"

"Randall, I don't know if we can do that. I'm not sure we can bring her just walking up to the house." Lanceley didn't know Weaver's sister at all. What if she was as paranoid and zealous as Randy? How could Lanceley be sure she wouldn't yell out to the house, "Brother, why don't you just kill your family and kill yourself like we talked about."

Finally, on Thursday, Marnis and her boyfriend were driven up the driveway to the base of the knob, one hundred forty-four hours after the standoff began.

"Page two," Paul Harvey said. And then he talked about the standoff in North Idaho, and he pleaded with Randy Weaver to give up. "You can negotiate an end to this standoff right now, and believe me, Randy, you'll have a much better chance with a jury of understanding home folks than you could ever have in any kind of shoot-out with two hundred frustrated lawmen."

Vicki's family had remembered that Randy listened to Paul Harvey every day on the radio, and so they sent word through a television reporter, who contacted her radio affiliate, who called Paul Harvey's producer and passed along the message. For most of the three-minute plea on his national show, Harvey criticized the federal government. "I wonder, too, with all the crass criminals we have running around in this country, this focus on you certainly constitutes grotesque overkill and frankly, from an objective distance, it looks pretty silly."

Randy, Kevin, and the kids heard the show, but people on the outside just didn't understand. If they gave up, they'd be gunned down, one by one. Nobody below the mountain knew that Vicki was dead and that the snipers had tried to kill Randy and Kevin, too, without any warning. No, the family was already martyred. It was only a matter of time before the feds finished them. What was left of the Weaver family was never leaving that cabin.

But that's why it was so important that they get their story out. That's why Randy wanted to talk to Marnis. But he worried the feds were going to wait until Marnis got to the back door and then shoot her and blame it on Randy. The negotiator had talked about suicide, which Randy took to be a suggestion that he kill himself and his family. Clearly that was their plan, to try to talk him into it, or maybe to paint him as some lunatic killer, and then gun everyone down and claim Randy did it. When the negotiator asked Randy if he was going to kill his family, no doubt they were taping the exchange to somehow justify their murderous attack.

The negotiator made a big production of the fact that they'd gotten Randy's sister for him, and then when he asked that she come to the back door, the one with the porch overlooking the valley below, he played a game like they didn't know which door was the back door.

"Randall, is it the one with the robot or without the robot?"

Finally, the family heard Marnis call out Randy's name over the loudspeaker.

"Marnis!" Randy yelled from inside the cabin. "Vicki is dead!"

But Marnis droned on about Randy surrendering, and it was apparent to the family that she hadn't heard, that the feds were blocking Randy's voice somehow.

"What happened?" Marnis yelled. "Is everything all right?"

As loud as he could, Randy yelled. "No, it's not! Can you hear me? They are playing games! Don't believe a goddam thing they are saying. They are afraid to let the truth out."

"Brother, I just can't hear you," Marnis said. "In our visit last summer, I didn't tell you that I've got a hearing loss. I've got a hearing loss from the factory where I work and I just can't hear you."

The girls couldn't stand it anymore, and they called out, "Aunt Marnis! Aunt Marnis!"

But Randy was pissed. If they weren't going to let her hear that Vicki was dead, then Randy was done talking.

*L*anceley could feel the trust disintegrate after Marnis said she couldn't hear very well. Randy, he figured, probably believed that his sister had gone over to the FBI's side. Marnis had talked nonstop since then, with no answer from the cabin.

Lanceley got back on the loudspeaker. "Randall, Marnis and I are going to come forward with the APC so we can get closer to the house so both of us can hear what you are saying. This is not working, and we have to establish some communication."

No answer.

"Randall, since you are not answering me, what I'm going to do here is I'm going to take it as an affirmative answer from you. Marnis and I are going to come forward in the APC unless you tell me I can't.

"Back off, Fred!" Weaver yelled. "Back off!"

Lanceley looked over at Marnis, a pleasant, middle-aged woman, weeping and drained from hours of negotiations. They were finished for the day.

The next morning, Thursday, one hundred sixty-eight hours into the standoff, Marnis went to another hillside with an FBI agent who brought a parabolic microphone—the kind of big dish used to pick up the sound of tackles and snap counts at football games.

Marnis made impassioned, emotional pleas for Randy to give up and let the courts handle his case, and for the children to come out of the cabin.

No answer.

Lanceley could hear the parakeets chirping inside the cabin. Across the hillside, Marnis was sobbing, and Lanceley lost his temper.

"Randall, you asked me to bring your sister here. I brought your sister to you, and then you treat her like this.

You can hear in her voice how brokenhearted she is. You owe her an explanation for what you are doing here. I think you owe her an apology."

Nothing.

Marnis tried again but it was clear Randy wasn't talking anymore.

"Randall," Lanceley said, "you know, the word around town . . . is that you are a family man, that you see the family as a unit. Isn't your sister part of your unit, Randall? Don't you owe your sister an explanation for what is going on here? I mean, you asked her to come out here. You have broken her heart here, and you won't even tell her what's happening."

There was no answer from the cabin.

Marnis was exhausted again, and Lanceley let her go home. It seemed as if this case could only end in a raid. The FBI played tapes of Vicki's parents, her sister, and her brother but got no response from the cabin. Later, the negotiator said they had more tapes to play, from Vicki's father and from a right-wing presidential candidate named Bo Gritz.

Randy jumped at that one. "I want to talk to Bo Gritz! I want to talk to Bo Gritz in person!"

*T*he people parted, and Bo Gritz strode through the protester's side of the roadblock. They joined his procession as he passed, chin high and weathered blue eyes calmly taking in the police tape and the armed officers on the other side. Two men slipped in alongside Gritz and one of them poked Bo's thick shoulder and introduced his friend. "Bo, he was in Vietnam, too."

Gritz's curly gray head stopped suddenly, and he looked the man over, slapped him on the back, and said, "Welcome home, son."

The veteran was a little confused as the Gritz entourage moved away. "Thanks," he said. "I've been home twenty years."

In his autobiography, Gen. William Westmoreland

called retired Lt. Col. Bo Gritz "The American Soldier."
He was rumored to be the model for Rambo, a crafty
Green Beret commando who possessed that leadership
quality possible only during war—he seemed invincible.
He was cited sixty-two times for valor, more than any other
soldier during the Vietnam War, and later became
commander of Special Forces in Latin America.

Gritz faded away after his retirement, only to resurface
in the early 1980s, when he led several trips back to
Indochina, promising to bring back POWs abandoned by
the U.S. government. Financed by industrialist H. Ross
Perot and some distraught widows, Bo skulked through the
streets of Thailand and Pakistan and crept through the
jungles of Laos but never brought home anything but a box
of bones that turned out to be pig remains. Gritz was
criticized for taking thousands of dollars from families to
whom he'd given false hope. But his reputation only grew
among the far right wing and by 1992—seemingly more
eccentric by the day—he was the candidate for president of
the Populist Party, which had last nominated the former Ku
Klux Klan grand dragon David Duke.

Sure, Gritz said, he remembered Randy as an engineer
and munitions expert. Or, no, maybe he'd never met
Randy. "Doesn't matter," he snapped at a reporter, a
member of what he cheerfully called "the faggot press."
When his campaign was contacted by one of Weaver's
supporters, who said Weaver had a poster of Gritz, the
candidate immediately changed his schedule and flew to
northern Idaho. He arrived Wednesday, about 2:00 p.m., in
a caravan of long, fifteen-year-old American cars, marched
directly up to the roadblock, and announced, "I'm Bo
Gritz, and I want to speak to the agent in charge."

Federal agents stared blankly at him and told him to
wait. For the next two days, while his two-person
campaign staff worked the crowd and the reporters—
"Gritz. Rhymes with whites."—Bo repeatedly volunteered
to negotiate Randy's surrender. "It takes Special Forces to
understand Special Forces," he said. He got no answer.

Frustrated, Gritz and Jack McLamb—a retired Phoenix

police officer who published a radical right newsletter—
marched into the Deep Creek Inn and asked Lorenz Caduff
if they could use a table to settle this whole matter. Lorenz
said sure. Gritz and his aides quickly wrote out a citizen's
indictment, charging Idaho Governor Cecil Andrus, FBI
Director William Sessions, U.S. Marshals Director Henry
Hudson, and FBI Special Agent-in-Charge Gene Glenn
with twelve felonies ("10. Felonious abuse and coverup of
abuse in refusal to allow Lt. Col. 'Bo' Gritz to mediate
settlement . . ."). Lorenz watched over their shoulders,
wondering if he was witnessing the beginning of the
overthrow of the American government.

At 10:00 a.m. on Friday, Gritz swaggered back up to the
roadblock in a Thai safari suit specially made to avoid
wrinkles, his navy blue epauletted shirt stretched tight over
a red-meat-and-potatoes stomach. The silver-haired
McLamb stood next to him, and fifty protesters fanned out
on either side. Gritz held the citizen's indictment in
sausage-thick fingers and growled through his bushy gray
mustache, "Is there any agent here that represents Gene
Glenn from the Federal Bureau of Investigation?"

The agents on the other side stared back blankly.

"Knowing that Gene Glenn is present on these
premises, we as citizens of the United States and under the
Constitution of the United States hereby present this
citizen's arrest for Gene Glenn."

Gritz read the indictment and handed it to a protester
who leaned over the yellow police tape, set it down, and
placed a big rock on top of it.

"Consider yourself served!" the protester said, to the
applause of his compatriots and a hug from a nearby woman.
After a week of waiting and feeling powerless, the angry
white people at the roadblock finally had a champion. If only
the people at the roadblock were registered voters, Bo Gritz
could have claimed another seventy or so votes. That
afternoon, the people—gap-toothed and scruffy, skinheads
and long-hairs, camouflaged vets and flannel-shirted
loggers—cheered as federal agents led Gritz through the
roadblock. They ushered him right past the citizen's arrest

that sat where they'd left it, anchored by the people's rock, on which one of the people had written "Justice. Ours or God's. Your choice."

*I*t wasn't the citizen's arrest that had gotten Bo Gritz past the roadblock. Gene Glenn just couldn't think of any alternative. He and the other federal agents had debated whether or not to allow Gritz up to the cabin. Richard Rogers, for one, was against it. But in the end, it had begun to seem like the last hope for avoiding a raid. They drove Bo up the hill, where he met with Glenn, got the ground rules, and talked about what they hoped to accomplish.

Then they drove Bo past the lines of marshals and FBI snipers and up to Ruby Ridge in an APC, Lanceley right next to him. Lanceley quickly briefed Gritz about how to proceed, and when they reached the cabin, a little before 7:00 p.m., the hatch was opened and Gritz was given a megaphone.

"Randall. This is Bo Gritz."

"Bo, is that you?" Weaver seemed distrustful and asked for some sort of proof.

Bo had the same problem as the other negotiators, and he wriggled his head out of the hatch to try to hear better. "Is everyone all right in there?"

Randy's voice strained through the plywood walls. "No."

Lanceley poked Gritz in the leg. "Follow up on that."

"I've been shot," Randy yelled. "Vicki has been shot through the head, and Kevin has been shot through the arm and into the chest."

Lanceley said later he was devastated. For much of the last week, he and the other negotiators had been asking for Vicki, negotiating directly to her, asking her about the baby, inviting her outside. He thought about the pain he must've put them through, especially the children. All these gadgets—listening devices and closed-circuit televisions—and they spend a week negotiating without knowing Vicki had been shot. Lanceley's job was to try to settle this standoff, yet, for all he knew, he may have made it worse.

"So I can hear you better, we're going to pull the APC forward still," Bo said. "I'm going to talk to you from the side of the APC, direct voice." The military vehicle moved toward the rock outcropping, and Gritz just hopped out. Lanceley and Richard Rogers stood behind the APC, nervous as Gritz edged closer and closer to the cabin. Bo moved away from the rocks, out into the open, and walked closer to the cabin, until he was right next to it. All around the rocky point alert FBI snipers watched through their scopes as Gritz approached the cabin, put his hands on it, and talked to Randy. Behind the APC, Lanceley tried to hear what they were saying but couldn't make it out.

"Bo," Randy called through the wall, "they shot Vicki, and they won't tell anybody."

D avid and Jeane Jordison, their son, Lanny, and their daughter Julie, were led past the roadblock to a clearing where Gene Glenn was waiting for them. They'd met the warm FBI agent-in-charge the day before, and now he told them to hold hands; he had something to tell them. They formed a small circle with Bo Gritz, Kevin Harris's parents, and with Glenn, who said he had good news and bad news. The good news was that the girls were all okay, and although Randy and Kevin were injured, they were okay, and the standoff was going to end soon. When she didn't hear Vicki's name, Julie held her breath.

"The bad news—" a tear pearled in Gene Glenn's eyes, and his voice quavered, "Vicki is no longer with us."

It was a blow that hollowed them all out at first, until the thoughts rushed back like water to fill the emptiness: memories, black sadness, and intuitions—hindsight, maybe—the overwhelming feeling each of them had had for days that they would never see Vicki again; that it had to end this way, and that in some way, she had been gone for years. David didn't seem to understand at first, and Lanny was just dumbfounded. Julie's knees buckled, and Jeane felt lost, crying in a clearing in the woods 1,500 miles from her home, holding hands with an FBI agent who explained that they had accidentally shot one of her babies. They sobbed and squeezed each other and tried to muster questions about how it had happened.

"There will be time for that, and believe me, I have as many questions as you do," Glenn answered, and they were glad to have him there. He seemed so understanding and so genuinely sorry, and they were relieved, at least, to have people like him in their government.

Julie looked up the dirt road, through the dark forest. In a whir of grief, anger, and pity for her sister, she knew one thing: They had to get those girls out safely, and she had to take care of them.

*T*he first sign of trouble Friday night was the lights. They were huge spotlights, pointed into the crowd at the roadblock, the last bugs of summer frying in the yellow beams. Flak-jacketed officers stood three deep in a military line on the suddenly claustrophobic bridge, passing out riot gear, plastic handcuffs, and canisters of tear gas. The protesters exchanged worried glances and wondered if they were going to be attacked.

And then the police tape was lowered, and Bo came down the mountain. They gathered around him, and he blew out a deep breath and rubbed his head. The night was cool again, and the mountain folk gathered around, pulled into their worn coats, and nudged past reporters and camera crews.

Bo swung his head around slowly. "I want all of you in the vigil to join your hands. I want you to get close. I want you to get warm. I've got good news, and I've got bad news.

"The good news is that we went right up to the top of the hill and established an immediate dialogue like I knew we could with Randy. Wasn't any problem at all." Then

Gritz explained that the girls were okay, that Kevin had been shot, but was doing well, that Yahweh was taking care of him. He said he expected to resolve the standoff the next day, Saturday. "The communication was tough," Bo said. "Tomorrow I'm gonna throw a bullhorn right through the winda' so we can talk."

A deep breath. "I want your hands joined. That's an order. I've got a reason for it. The bad news—now get a grip on yourself—is that Vicki was killed."

A woman cried, "Oh, no."

Someone else said, "Damn."

Bo talked fast, trying to keep the crowd in control. "I want you to listen to what I gotta say, so you get it right. Randy and Sara seem to be okay and in pretty good spirits. I had a good dialogue with Randy. The only thing that is shameful is that I could have had this conversation three days ago. . . . A wonderful woman, a pioneer woman, has had her life taken," Bo said, "and she's in God's hands as we speak.

"There's a bureaucrat up here that's guilty," he continued, trying to talk through the protesters' anger. "Somebody is going to be brought to the bar of justice. I believe we're gonna find some fat bureaucrat who authorized this to go down."

The protesters formed a wide prayer circle, while uncomfortable reporters backed away. Judy Grider gathered some of the patriot women into a chain, and they turned and faced the roadblock, their arms interlocked, Judy yelling in a singsong growl, her Bible study voice, "Feel the strength of the women of Yah! And believe in your hearts that together we stand. Put arm and arm together. Now!

"And show everybody that we are women of Yah! And you will not let this come to pass again because we are the virtuous women! You are his enemy and you will never receive us in your hands again. From our savior, never will you take another woman down! Never!"

One of the Vegas skinheads screamed, "We're goin' to war!"

The ATF agents and state police watched the crowd closely, their eyes shifting from angry face to angry face.

Tall and intimidating, Bill Grider puffed on a cigar and stood a few feet away from the lines of federal officers, casting long shadows from the spotlights as he tried to stare the enemy down. "You've got a lot of hair on your asses. You're all standing for what you think is right. This isn't right, is it?" He poked a finger at them, jumpy, as if he might leap over the police tape. "You proud to serve these fuckin' morons? They don't care if you die. They put you out here in front! You're fuckin' nothin'!"

They prayed and screamed all night, forming and reforming the circle, the collective rage threatening to send them all up like bone-dry kindling. Reporters took short odds on whether or not the roadblock would erupt that night as a white-bearded man who looked like he'd been shaken from the Old Testament led two dozen protesters in prayer.

"For Kevin and the girls and for little Sam," he prayed. "Father guide us and help us and lead us. Yahweh saves! Yahweh saves! Praise Yahweh." And then he lost it, too. "Arise O Israel! It is time for war!"

*L*orenz Caduff had been most of a week without sleep. Strange people slept on his floor and in tents all over his yard. Reporters from Boston and Los Angeles used his telephone, and people displaced from their homes begged him for help. He ran from table to table, delivering food and catching bits of conversations from the angry people who said the government had murdered that Weaver woman, shot her right in the head even though she had no gun. The government was filming the protesters, they said, and eventually would track them all down and kill them, too. It was all so confusing.

Saturday morning, Lorenz gathered grapes, a gallon of milk, a half-gallon of apple juice, and a baby bottle and put them all in a box. Bo Gritz, Jack McLamb, Jackie Brown, and a local Baptist preacher were sitting at a table, preparing to go up the mountain and end the standoff, and so Lorenz brought the box over and gave it to them as a peace offering to take up to the cabin. On the side of the box, he taped a tiny Swiss flag, the symbol of neutrality.

The four negotiators were taken through the roadblock, up to the meadow, and finally, to the base of the driveway and to the huge rock near the cabin door. Gritz moved immediately to the cabin and began talking, but the family seemed distant again.

Inside the cabin, Sara had no intention of giving up. They had agreed to let Bo Gritz come up to the house the day before just to buy some time and to make sure they got their story out. But Sara didn't completely trust Bo. She knew her dad liked him, but Sara had her mother's intuition about people, and she didn't like the way Bo kept asking to speak to Randy face-to-face.

No, Sara argued, they should sit tight, mourn Vicki's and Sammy's deaths, read the Bible, and pray to Yahweh for guidance. Under no circumstances should they budge from the cabin. Sara knew what would happen if she let her father give up. The feds would tell Randy to send the girls out first, Sara, Rachel, and Elisheba. Chances were, they would kill all of them. But even if the girls lived, they'd kill her dad and Kevin. Ever since Saturday, when they'd shot at the two men without any warning, Sara knew what the One World agents were trying to do. With Kevin and her dad dead, they could blame Vicki's and Sam's deaths on the men. Sara's parents had told her many times what would happen next: The government would say the girls were mentally disturbed and throw them in nuthouses, where they would be drugged and used as guinea pigs for scientific experiments. Sara would die before she'd let that happen.

Gritz wanted to come in the house, but Sara insisted no one except her family was ever coming in the cabin again. Bo used his secret weapon. "Sara, Rachel," he said, "do you want to hear from Jackie?"

"Yes!" Their voices rose. Jackie had visited them almost every week for the last year or so. Doe-eyed and earthy,

Jackie Brown had a strong personality like their mother's, a sense of pioneer competence and warmth. For the past couple of years, she had been almost like an aunt to the kids.

"I'll walk her to the door," Bo suggested.

"No," Randy called. "You stay there."

Jackie was standing behind the Weavers' outdoor shower with Richard Rogers, the head of the HRT. He told her that if she stayed in there a long time, they would assume she had joined the family or was a hostage. Jackie nodded. Rogers asked if she wanted a bulletproof vest, but she shook him off. As she was walking away, she said, "Maybe if I wear it backwards." She walked around to the back of the house—the one that jutted out over the ridge—climbed the rickety steps, and stood on the porch, the valley opening up behind her. She told the Weavers to unbolt the door and step back, and when they did, she slipped inside and locked the door behind herself.

It was dark and clammy inside, and messy, not at all the way Vicki kept the house. When her eyes adjusted, Jackie saw Kevin slumped in a reclining chair in the living room. He was jaundiced and thin, and when she felt his pulse, his heart was racing. The girls were standing next to their father in the corner, wearing their gun belts. When they saw she was alone, they came over, grabbed Jackie, and sobbed. Sleeping bags and blankets were spread over the floor, and Elisheba sat in the middle of the nest, playing with blocks.

Randy told Jackie what had happened so far, how they'd been under attack since the initial shoot-out. "Kevin saw Sam get hit in the right arm, and he just wanted to stop 'em from shootin' at him. What the hell would anybody else do?" Bloody sheets and towels were piled in a corner of the room.

Jackie gave the family the care package, and they were especially grateful for the milk. Kevin slurped the apple juice down and asked for one of Jackie's cigarettes. Jackie and the girls cried and hugged each other, but the family told her there was no chance any of them would get out alive. Rachel wouldn't let go of Jackie's waist. She and Sara told Jackie to ask their grandparents, David and Jeane, to not be mad at them.

Gritz continued his negotiations from outside the cabin, asking if he could take Vicki's body out. Randy said no, not during their Sabbath. Their Sabbath was from 6:00 p.m. Friday to 6:00 p.m. Saturday.

"My daughters aren't ready for that," Randy said.

Jackie didn't see the body that first trip into the cabin. When she got ready to leave the cabin, Sara had something for her. She ran into the other room and came back with a feminine napkin. Sara had folded the six-page, handwritten note into the napkin, and now she pointed at the note, without saying anything, because she knew the agents were taping their conversations. Randy whispered that Jackie should give copies to three reporters whom she trusted and keep the original. Jackie nodded, lifted her skirt, and put the napkin in her underwear. An FBI agent patted Jackie down lightly and then let her go. When she got down the hill, an exhausted Jackie grabbed her husband, pulled him to the side, lifted her skirt, pulled out the napkin, handed it to him, and said, "Here, copy this."

*T*he next step was getting Kevin out of the cabin. "Bo, you tell Kevin and Randall that the best thing for Randall is for Kevin not to die," Lanceley said before they started negotiations Sunday morning, August 30. "If Kevin dies in there, without appropriate medical aid, who knows what the government will try to put on him. Maybe the government will try to get Randall for killing Kevin because he didn't allow Kevin to come out."

"Good," Gritz said. "Kevin coming out is the best thing for Randall." Bo hoped that when it was just him and the girls, Randy would start to realize coming out was best for the kids. That morning, they attached a body wire to Gritz, and an FBI physician stood near the rock outcropping, yelling out questions for Bo to ask about Kevin's wound. Gritz stood outside the cabin walls and relayed the messages back and forth. "The doctor says that sounds pretty serious."

Kevin's condition had improved, but Gritz and Jack McLamb told the family that young guys who get shot often get better for a while and then deteriorate quickly and die. The family didn't know whether to believe them.

"I'm not going to throw Kevin out," Randy said.

"You're not throwing him out," Bo said. "Listen, if I were the jury, and you allowed Kevin Harris to die in that cabin, I would fry you because you can save Kevin's life right now."

"Bo," Randy said, "the girls say they don't want him to go."

"Damn, Randall. You are the head of that family. You are the man. Make those decisions."

A little after noon, Randy opened the door and let Bo and Jack McLamb come inside.

Sara was still the holdout. She shielded her father, watching Bo and Jack suspiciously. She heard Bo say that Randy might be charged if Kevin died, and she also understood them to say that if Kevin left the cabin, no charges would be brought against her dad. Even so, Sara figured that as soon as Kevin walked out that door, they were going to finish him off. It made no sense. Why would they try to kill him in cold blood a week ago and then let him waltz out of the cabin? Maybe they'd wait until he was at the hospital, but they were going to kill him.

But Kevin didn't want Randy charged with another murder, so he gathered his strength and stood up. McLamb—who was a quieter, more thoughtful Bo Gritz—promised to stay by Kevin's side all the way to the hospital. Kevin walked through the back door and onto the porch overlooking the long, green valley. He grew tired and sat down on the steps, then stood and continued walking, McLamb helping him. Sara was impressed. Still keeping an eye on Bo Gritz, she watched Kevin walk out of the cabin and decided he was one of the toughest boys she'd ever seen.

•   •   •

*V*icki was next. Bo started talking about her as soon as Kevin had left the cabin. He told Randy that Vicki's body was evidence and could prove that she was purposely killed by the FBI. Jackie Brown had pointed out that having a dead body in the cabin wasn't healthy and that Vicki wouldn't have stood for it. By late afternoon, Randy and Sara decided Bo and Jackie were right. Gritz left and came back with Jackie and a blue, felt body bag, which he set on the floor. Randy and Bo pulled back the army blanket, and Jackie saw a bloodless, crushed version of her friend, still wearing a striped sweater and denim skirt, a bandana, and knee socks. Bo knelt down and took the holster and pistol off Vicki and set them on the table. Even though FBI agents reported smelling a dead body inside the cabin, Bo and Jackie insisted there was no stench and no rigor mortis. Blood and fluids filled Vicki's chest as Bo lifted her. Randy sobbed as he helped ease his wife into the bag.

Jackie and Bo carried the body to the back door and down the stairs as the girls wailed. "Please," Sara said, "don't let Mama's body touch the ground." Bo staggered down the rickety steps and finally just threw Vicki over his shoulder. They left the body on a cot next to the back porch, and Jackie walked down the hill and asked for some paper towels and water. Two FBI agents filled a couple of buckets from the spring and handed her a roll of paper towels and three white bath towels. Back in the cabin, Sara wanted to help, but Jackie—in her most motherly voice—said no, she couldn't. She dropped to her knees on the kitchen floor and cried as she mopped up her friend's blood. It was the worst moment of her life.

Lanceley couldn't believe it. He was impressed with the negotiating of Bo Gritz and, especially, Jack McLamb, whose quiet charm was a nice balance to Bo's bluster. Lanceley was glad they were making progress, but now things were moving so quickly, the FBI was making mistakes. Bo, Jackie, and Jack McLamb were being allowed to go in and out of the cabin without being searched, and now Jackie was cleaning up the blood, which would be evidence later. He shook his head as she tossed the bloody towels over the high porch onto the ground.

It looked as if the family was ready to come out, and so Gritz and McLamb turned the negotiations toward finally ending the standoff. The back door was left open and natural light filled the cabin for the first time in more than a week. Then, Randy peeked out the front window and saw the robot for the first time.

"There's a shotgun on that robot!"

Bo looked out the window. "Yeah, there sure is."

The family became scared again. For days, the feds had been saying that the robot would bring the phone, that there was nothing wrong with the robot, and yet there it was, right in front of them, a shotgun. They weren't leaving.

Jackie accompanied Vicki's body down the road to the meadow, where FBI agents checked her fingers to see if she'd fired a gun before she was killed. Jackie also brought the girls' pet parakeets. At the roadblock, she collapsed on a friend's shoulder and cried.

Gene Glenn led the reporters back up the hill again for another press conference. "I can't overemphasize how pleased we are," he said. "We're optimistic that there will be further progress."

But, at the cabin, Bo was demoralized. The family wanted to die. He'd seen similar reactions in Vietnam—a kind of survivor's guilt, so strong it made rational people almost eager to be killed. Vicki and Sammy had been shot to death, and Bo figured that Sara, especially, believed she would somehow desecrate their memory by leaving the cabin alive. And once you gird yourself up to die, it can be difficult to back down from that decision. At the base of the driveway, Bo talked to FBI agents who told him that if the standoff wasn't settled soon, they were going to raid the house.

Over Bo's wiretap and the microphone under the cabin, Lanceley had heard Sara talk her father out of surrendering. It occurred to him that she had taken her mother's place as the matriarch of the family. "There isn't much of a man in there," Gritz told him. By Sunday afternoon, Lanceley heard the Hostage Rescue Team preparing for a direct assault, and his morale fell to its lowest point. Ten days. Two hundred forty hours. The FBI wasn't going to wait any longer. The negotiations had failed.

Lanceley tried not to let Gritz see his concern, though. "Let's you and I sit down now and work out our game plan for tomorrow," he said.

But Bo knew. "Negotiations are out," he said. "Tactics are in."

That night, Fred Lanceley drove to Sandpoint to sleep in a bed and take a hot shower. He hoped they could raid the cabin without anyone else getting hurt, but he wondered if that were possible. Whatever happened, he knew that by morning, it would be over.

sixteen

Vicki would've wanted those girls to die up there."
Vicki's sister, Julie Brown, wasn't quite sure she'd
heard right. "I'm sorry?"

One of Vicki's friends explained that the Weaver girls
weren't afraid to give up their lives for the white race, that
Vicki had become a martyr and a hero, and that she would
want the girls to do the same. Julie and her family had
reluctantly come down to the roadblock from their motel
that day, hoping to get some news, but now—surrounded by
angry racists and crazy mystics like this woman—Julie
wanted the hell out of there. These people were nuts.
Earlier, Julie had just about lost it when she saw some of
Vicki's friends selling her letters to reporters. They said they
were raising money for the Randy Weaver defense fund, but
Julie was suspicious.

The crowd had swelled to more than one hundred. By
the second Sunday, sightseers from Washington and
Montana drove slowly past, craning their necks to see all
the protesters and the armed federal officers. Campfire
smoke and dust filled the air as the self-styled patriots and

white separatists shuffled around the roadblock, some happy there had been no more violence, others disappointed that the great race war hadn't started yet.

Wayne Jones, the security chief for the Aryan Nations, who had vouched for Randy's character at his arraignment, didn't think Weaver would give up peacefully. "He made it very clear to me that he was going to take it all the way to the end," Jones told a reporter. Randy was "a man of honor and a man of his word, and I fully expect him to do what he has said he will."

Other people were sorry the standoff was dragging to a close because the protest was going so well. Seemingly, everyone had come. Richard Butler and other Aryan Nations leaders showed up for a while. The wives of Order members Gary Yarbrough and Robert Mathews were there, agreeing that Randy was doing what their husbands would do. Both said they'd met the Weavers at Aryan Nations meetings, and Jeannie Yarbrough said she'd been up to the cabin. The Trochmanns were there, too, the family Randy was supposed to spy on, the men who would later form the Militia of Montana. Carolyn Trochmann, Vicki's midwife, flipped sixteen slices of French toast over the barrel cookstove for the patriots and skinheads who rested underneath the campsite's blue tarp. She cried as she described Elisheba's birth and said she was proud of Vicki and Randy and the stand they'd made. She wished she could have joined them on the mountain.

Skinheads crouched on rocks in the middle of Ruby Creek and shaved their heads with disposable razors, while a photographer took their pictures and snipers watched from the woods. One of the skinheads said he heard a language he couldn't understand, proof that United Nations soldiers had been called in to quell the uprising and institute the New World Order. "I think it was Belgian," he said.

Other rumors surfaced, some more likely than UN soldiers speaking a nonexistent language. The creeks had been poisoned with phosphorus, people said. Farm animals and pets were being slaughtered by federal agents.

Helicopters were planning to dump diesel fuel on the cabin. They didn't just discuss the plots, they raved about them, looking wildly over their own shoulders and talking out of one side of their mouths, until at one point, a man rushed to the roadblock and, quickly losing his calm, began insisting that they disperse immediately because he had intelligence indicating the feds were going to provoke a gun battle to kill Bo Gritz. "Obviously, the feds aren't telling us everything," one camouflaged man said. "We know. And we know they know we know."

White patriots linked arms and sang "Onward Christian Soldiers" as skinheads gave the Nazi salute behind them. A Jewish father's rights advocate from Seattle, who was also an anarchist, tried to convince one of the skinheads that they had much in common in their hatred of government, but the nineteen-year-old skinhead refused to acknowledge the Jew, who was the spawn of Satan. "You don't exist to me," the skinhead said. He talked instead about the band he was trying to form, which would play folk music, "kind of like Joan Baez, except racist." Skinhead groupies flirted with the young men from Las Vegas ("I'm in my third year of German") while Vietnam vets stood in cadres of five or six and waited for orders from Bo Gritz, or anyone for that matter. Moderate people showed up at the roadblock, too, saying their eyes had been opened by the case. Satellite news trucks hummed as reporters from *Inside Edition, USA Today*, *People* magazine, and the *New York Times* covered the carnival.

Gritz was cheered at the roadblock Monday morning, but he knew that even if the FBI hadn't raided the cabin overnight, it was going to be tough getting Randy out. Bo decided to enlist the help of some of the Las Vegas skinheads, whom Randy had met at an Aryan Nations meeting. He knew Randy respected them and thought they'd been unfairly characterized by the TV station he picked up on his radio. Bo agreed they'd shown remarkable restraint and done a good job keeping the crowd from getting violent. He grabbed a few of the skinheads and asked them to sign a note that Jack McLamb had penned, pleading with Randy to take his battle to the courts now. Not everyone agreed that Randy should give up, and Bo could only find two skinheads to sign the note. Then Gritz and McLamb trudged up the hill once more, knowing time was running out.

M onday morning, Fred Lanceley consulted his best source of intelligence to see if the cabin had been raided. In his Sandpoint motel, he turned on CNN, which reported the standoff was still going. He called the command post and Gene Glenn told him to get up there, they had decided overnight to try another shot at negotiations.

Lanceley met Bo at the command post. As they'd done the previous two days, the agents placed small transmitters on McLamb and Gritz—Bo's in the pocket of his faux military shirt, behind his sunglasses, so that if Randy patted him down, he would feel only the glasses. The FBI agents made it clear they were running out of time, that this was likely the last chance for negotiations. They set up Operation Alaska, a plan for Gritz and McLamb to subdue the family if the negotiations broke down or they thought they were in trouble. They would wait until the baby took her nap, then McLamb would grab Sara and Rachel and pull them to the floor and Gritz would jump on Randy. He'd say "Alaska" into his transmitter, and the FBI assault teams would burst in and—hopefully—there would be no violence. Of course, there were no guarantees. Gritz thought the plan was a very good one. Some FBI agents were upset by how cooperative Gritz was up here and how critical of the government he was once he got to the roadblock.

At 10:00 A.M., Gritz and McLamb walked toward the house, but before they could reach the door, Randy called out. "Bo, we're not going to talk anymore. There's nothing against you, but . . . we have prayed all night and we have asked Yahweh and we will stay here. They can kill us if they have to."

"That's right," Sara called. "Bo, we're just going to stay here." They would surrender, Randy and Sara said, but not until September 9, which they'd used Bible passages to calculate was the Feast of Trumpets, an important religious holiday for the Weavers. It was the same holiday, nine years earlier, which they'd believed was Yahweh's deadline for finding a cabin.

Gritz told them that federal agents would never wait that long. He pleaded with Randy, got angry with him, and finally, shoved the note signed by the skinheads through the crack in the door. Randy read the note, seemed pleased, and opened the door to talk about it.

Bo had begun to view Sara as the real leader of the family, and he set to work trying to convince her to surrender. She was firm. The lives of Vicki and Sam had to be worth another nine days, she and Randy said. Bo explained that the government needed to get inside the cabin because it wanted to get the evidence. They were spending a million dollars a day on the standoff, Gritz said, and the FBI was becoming impatient. If they didn't surrender that day, the FBI would raid the cabin, and the girls might very well be injured or killed. Randy agreed with Bo that he should give up to keep his kids from being hurt. But Sara still said no. The ZOG agents had already tried to murder the whole family. Why should she believe now that it was safe?

Gritz and McLamb promised to shield the family. Bo said he would handcuff himself to Randy and wouldn't leave his side until he was safely off the mountain. They promised to make sure an investigation was conducted on the case and that Randy got a good attorney. They promised the girls they could return on September 9. Still, Sara said no. So Randy talked to her. Gritz may have believed Sara was the one he had to negotiate with, but she insisted her father was really in charge. He told her that if there was anything he could do to keep anyone else in his family from being hurt, he was going to do it. He told her he was going to surrender to protect them. Finally, about 12:15 P.M., after eleven days, Sara went along with her father's wishes and gave up. If that was the way Yahweh wanted her to die, then so be it.

*T*hey packed a few things into small cloth bags and looked around the cabin once more. Rachel took off her holster with its .38-caliber snub-nose pistol, and Sara unstrapped her 9-mm. Randy took off his gun belt and then remembered something. He removed his Aryan Nations belt buckle and handed it to Gritz. "I don't want them to get this," he said. They changed Elisheba's diaper, took a deep breath, and stepped toward the door.

Randy came out first, cradling Elisheba in his good arm, holding Bo's hand with his other hand. Behind them came McLamb, holding hands with Rachel and Sara. Some of the FBI agents had such a low opinion of Randy, they bet that he was holding the baby as protection.

On the porch, standing in the sunlight for the first time in ten days, Sara waited to be shot dead. It didn't happen. The whole family was crying as they walked down the steps and along the rocky path to the driveway, and that's when some of the fear fled from Sara. She saw snipers and camouflaged agents hiding behind rocks and trees, staring at the family from all over the knoll, and she knew one of them had murdered her mother. She wondered what they would do to her father now.

"All this, for one family," she said as they walked down the driveway, tears streaming down her cheeks, agents peeking out from everywhere. Several armored personnel carriers sat at the base of the driveway, and Sara glared at the agents she saw, hoping they were ashamed for declaring war on a peaceful man and his little family, on people who just wanted to be left alone. There were dozens of people at the base of the driveway and even more in the meadow down below. One of the agents—a tall, older man with gray hair and a potbelly—bent over and asked Sara if there were any booby traps in the house. "Is there anything else up there?" he asked. "Because we don't want anyone else to die."

"You're worried, aren't you?" Sara snapped. She shook her head. "There's nothing else up there."

When the family was gone, FBI agents came down from the hills to look at the cabin. Lon Horiuchi walked all the way around it and stared for a long time at the door where he had fired and at the window where Vicki Weaver apparently had been standing when his bullet tore through her face.

Down the ridge, in Homicide Meadow, Wayne Manis waited outside a medical tent next to the HRT building. Gene Glenn stood near him, somber and tired, glad it was finally coming to an end. He'd asked Manis to be in charge of Randy Weaver once he reached the command post. A caravan of military vehicles came down the road and into the clearing and opened their doors. Randy Weaver stepped out—scrawny and pale, wearing a drab olive T-shirt and faded jeans, his shaven hair grown out into rough stubble. Bo Gritz got out with him, but the FBI led Randy away.

They put Randy on a stretcher and carried him into the medical tent, where Manis ran alcohol swabs over his fingers, then dropped the swabs into an evidence bag and sealed it. Randy was exhausted and quiet.

They loaded him in another rig and drove him to the Raus' barn, which had been turned into the field office, where his daughters now stood, waiting to say good-bye.

With his hands cuffed in front of him, Randy leaned forward in the car, an agent on each side of him, said good-bye to his girls, and told them not to worry. Rachel cried, but Sara tried not to, firing angry looks at the agents, pissed off that Bo had already abandoned her father. She told the FBI agents, "If anything happens to him, you're going to pay."

Nearby, Dave Hunt watched quietly. He'd come back to the mountain after Degan's funeral, and now, finally, it was all over. They brought Randy Weaver within about ten feet of him, and Hunt stared at Weaver's gaunt, blank face. For eighteen months, he tried to get that man off that mountain, did everything he could think of to meet with him face-to-face. Now, here was Randy Weaver, right there in front of him. All this trouble and death because of one little man. It seemed to Hunt this was what Weaver wanted all along, to be some kind of hero or martyr, to be some kind of Aryan legend, to sacrifice everything for some half-baked ideals that he didn't even really comprehend. *I hope you're happy*, he thought, as they led Randy away. *I hope you're satisfied.*

They drove 200 yards east, to a dark marshals service helicopter. They loaded Randy in with Mike Johnson, two armed deputies, and Wayne Manis. Randy was secured by waist chains shackled to his wrists and ankles. The flight to Sandpoint took only fifteen minutes, and then they landed next to a private Learjet, quickly loaded Randy aboard, and were off again after only five minutes on the ground. On the flight to Boise, Manis sat next to Weaver, who was almost catatonic. Mike Johnson had just one question for Weaver. "Did you ever leave the mountain?" Weaver said no. Johnson—ever the politician—leaned back and told Manis there would be photographers in Boise. If he wanted, Manis could stand on one side of Weaver while Johnson stood on the other.

"That's okay," the FBI agent said.

A few years earlier, Manis had been featured in *The Silent Brotherhood*, a book about The Order that had become required reading for some white separatists. When Wayne introduced himself, Weaver said he recognized the name. For the first time since coming down from the mountain, Randy seemed mildly interested in something. "I've known about you for a long time," Randy said. "I've read about you."

"T he glory goes to God Almighty," said Bo Gritz, framed against an impossibly blue sky and an American flag unfurled by his supporters. "And if the media doesn't use that, then you're everything Randy said you were."

Bo said he'd awakened at 2:30 Monday morning with a vision that the surrender would take place at noon. "I was maybe fifteen minutes off, and it may be that my watch needs to be readjusted.

"The government learned something here," Gritz said. "The government learned there are times when common sense pays off. It doesn't have to be in a book of procedure." But he complimented the agents and, especially, Gene Glenn. "Everybody up there did his job," Bo said.

And then Bo thanked the skinheads for their letter, which he said helped resolve the standoff. "By the way," he added. "Randy told me to give you guys a salute." Bo raised his right arm in what looked like a Nazi *Sieg Heil* but what he claimed later was just a wave. "He said you'd know what that meant," Bo said. There was no mistaking the salute he got in return.

One young skinhead sat dejected on a stump. "It's just not what we expected," he said. "I wish he'd taken a few more out."

On the mountain, FBI agents crawled all over the cabin, gathering evidence. They looked for fingerprints and blood samples and measured the bullet hole in the door. They taped off the cabin and brought up a Humvee full of reporters, who were allowed to walk over the rocks outside the cabin but not let inside. The family's arsenal was spread out on a white sheet on the ground—two shotguns, seven rifles, five pistols, and thousands of rounds of ammunition. FBI agents pointed to tins of armor-piercing bullets, which they said might have penetrated the APCs if the family had used them. Still, for an investigation that was based on trafficking in illegal weapons, it was clear that none of these guns was illegal.

A relieved Gene Glenn answered questions from reporters about the standoff. "The key line is, there has been restraint," he said. Glenn said Bo Gritz worked heroically to end the standoff, even though Glenn never got clearance from FBI director William Sessions to use him. "Maybe I should have," he said, smiling, "but I didn't." When someone asked what he would say to the people of northern Idaho, Glenn became serious. "The message that I would like to say is, 'We are very sorry.' . . . There are no winners in a situation with all this sadness."

When he was done, Glenn climbed one of the boulders and stared out over the disappearing hillside to the green-and-brown valley below, spread out before him like a deck of cards in God's hands. He asked no one in particular, "How could there be so much evil in such a beautiful place?"

*T*here was a party that night in the wood-paneled bar at the Deep Creek Inn. Print reporters hit their deadlines and TV guys finished their live stand-ups, and they met at the Deep Creek, where a few protesters and local cops had already gathered to toss a few back. The booze was a nice match for eleven days of fatigue and adrenaline, and thirty people got pretty drunk pretty fast. They told war stories, swapped ZOG jokes, and posed for pictures: the angry Bill Grider with a grin on his face and his arms around the reporters he'd been scowling at all week; one of the Las Vegas skinheads posing with a black reporter from Boston, their faces dissolved in laughter as they Nazi-saluted the camera.

When Lorenz came to check on the rowdy bar, its occupants broke into applause. Everyone—reporters, cops, and protesters—was fond of the Swiss chef. They had listened to his earnest, bent-English questions about what was happening and smiled at his naive beliefs about America and justice. They had eaten his food, used his telephone, and slept on his floor, often for free. Some of the bar and food tabs would never be paid, and when the phone bill came, Lorenz would find dozens of calls he couldn't account for. The innkeeper looked out at a room full of people who had argued and debated for eleven days, now friendly with each other, as if it had all been some sort of insane play.

"Ladies and gentlemen," said the bartender, a young newspaper reporter who had stepped in when the regular bartender was swamped by the fast-drinking crowd.

Everyone raised their glasses. "To Lorenz, a man who really knows how to host a standoff."

Lorenz didn't smile. He looked down, thought for a moment, and then spoke. He said his eyes were opened by the last eleven days and that the news reporters needed to realize how important their job was. They had to find some meaning in what had happened, some explanation. A few of the reporters almost sobered up. "I hope you will be responsible in telling this story," Lorenz said. "And I hope you find the truth."

Rachel and her grandpa walked through the grocery store in Sandpoint, looking for snacks to take back to the motel. The ten-year-old only wanted chocolate doughnuts. After all she'd been through, David Jordison wasn't going to say no. Rachel stayed close to her grandfather as they walked through the store's bakery, past shoppers who paid them no attention. She hadn't been away from the cabin, hadn't been off the mountain, in eighteen months, and now she was in a crowded grocery store, all kinds of faces streaming by, a blur of strangers and strange-looking people who were pushing carts with gallons of milk, loaves of bread, batteries, cookies, and frozen foods. She clung to her grandpa and whispered, "I sure wish I had my gun." That night, as she got ready to go to sleep in the motel, Rachel asked if anyone had a flashlight she could borrow in case she had to get up and go to the bathroom.

Back at the motel, Elisheba toddled around in nothing but a diaper while Sara talked with Jackie and Tony Brown, with her cousin from Colorado, John Reynolds, with a couple of the nicer skinheads, and with her other relatives. Then she agreed to tell a reporter from the *Spokesman-Review* what had happened. In a motel room in Sandpoint, she started slowly, trying to get everything out. She was tight-lipped and tense, her eyes swollen from crying. She broke down a couple of times as she described how her brother had been shot, how her mother had been killed, how she'd crawled around Vicki's dead body to get food and medicine for the people in the cabin. But mostly she was furious. She concentrated, trying to make sure she had the sequence of events right. She scolded herself when she started crying: "Get this out!" After three hours, with the interview ending, a workman banged against an outside wall, and sixteen-year-old Sara jumped, her eyes open wide, her fists clenched.

"Sara Waited in Fear for Feds to Finish Job," the headline read. The subhead was, "'I couldn't watch them pick us off one at a time.'"

Exhausted, Julie Brown flew back to Iowa to be with her own family; then she tried to figure out how to get the girls back there and away from all the trouble. But Sara wouldn't leave. She wanted to stay with Jackie and Tony Brown or with other friends near the mountain, to guard the cabin, so they could go back the next week in time for the Feast of the Trumpets. "I want to stay here," she said, "around people who understand me." And the younger girls couldn't go either, said Sara. "I won't let what's left of this family be broken up." Her grandparents were trying to talk them into coming to Iowa and going to public school. No way. Sara said their hearts were in the right place, but she would never go to school.

Julie Brown knew she had to get them out of Idaho. She thought of Vicki's wild-eyed friend who had told her the girls should have died on the mountain. Although Jackie and Tony Brown weren't like that woman, there was no way Julie was going to let the girls stay in Idaho. Even Vicki wouldn't have allowed that. She called every few hours on September 1, and each time Sara would have some other excuse for not being ready to go. "Maybe tomorrow," she said.

Julie had met a social worker while she was in Idaho, who told her the surprising news that the Weavers had received food stamps for a while, years earlier. It made Julie sad because she knew how much Vicki would've hated to do that. But it also reminded her that Vicki would do whatever was best for her family. That was what Julie had to do now. She called the social worker back and told her what the woman at the roadblock had said. Julie said she considered it a threat against the girls' lives. The social worker agreed that the girls should go to Iowa as soon as possible.

Then, Julie called back to the motel and got on the phone with Sara. She told her there were threats on the girls' lives, and she stretched the truth, saying the social worker had called her and promised to send the girls to a foster home if they stayed in Idaho.

"Right now, you have two choices," Julie said, hoping to scare Sara into leaving Idaho. "You can stay there and be institutionalized or you can come back to Iowa." It sparked Sara's vision of orphanages and foster homes, passed along by her parents: such places would be infected with the AIDS virus, and the children would be brainwashed and fed psychedelic drugs until they cracked and were no longer dangerous to ZOG.

On September 2, Sara relented again, this time only for the good of her sisters. She said good-bye to her friends and to two of the skinheads she'd met at the Aryan summer congress years before, Johnny Bangerter and David Cooper—the handsome and intense-looking skinhead the others called Spider. At the Spokane airport, Sara refused to let her sisters out of her sight and began to change her mind about flying. It dawned on her relatives that the only aircraft these girls had ever seen were the planes and helicopters that buzzed over their cabin, taking photographs for the government or for tabloid television shows. In addition to everything else they had gone through, the girls were scared because they were about to fly for the first time in their lives.

A ll that was left was the cleanup. With Weaver safely in jail, Wayne Manis returned to Homicide Meadow after everyone else had packed up the tents and gone home. He took down all the electrical equipment and got rid of all the classified garbage. After the investigators had gone,

Manis visited the cabin once more. It gave him the creeps: the rock fortifications, the supply of beans and herbs and grains, which could have lasted a year, maybe two, maybe even more. He stood on the porch and realized it could have gone even worse. What would have happened if Sara had come out at some point firing her rifle. What could they have done?

Manis was angry, not just at the family, but at the system, too, which allowed law enforcement agencies to compete over criminals and informants. Not that he thought the Bureau of Alcohol, Tobacco, and Firearms had done a bad job on the case. No, Manis had reviewed the bust and thought it was a clean case, not entrapment from his standpoint. And they'd arrested Weaver using a pretty slick ruse, with no bloodshed. But he didn't understand why the ATF was trying to develop its own informant on the Aryan Nations in the first place. There should be one agency—the FBI—in charge of domestic terrorism. Otherwise, agents and informants were falling over each other, undercutting the other's investigations. Hell, he wished the ATF had just come to him if they'd wanted intelligence on the Aryans or on Weaver. He probably would have told them Weaver wasn't connected enough to worry about. That he wasn't worth the effort. And then, standing on the empty knob, he tried to imagine the whole thing had never happened.

On September 3, three days after the standoff ended, Wayne Manis drove down the dirt road and turned onto the old highway, the last federal officer to leave Ruby Ridge.

A few protesters straggled at the roadblock for a day or so after that, and then even they left. But the flowers and wreaths remained for months, alongside a painting of Jesus standing behind a boy at the wheel of a ship. On the bullet-pocked dead-end sign marking the bridge, someone had duct-taped "Mother & Child" below the word "DEAD."

The Weaver case continued to hang over the Inland Northwest like a still summer storm, building anger—not just among the radical right—but in a broad swipe of people all over the political spectrum. The case continued to draw people to the Northwest. Weeks after the standoff, a group calling itself Citizens for Justice convened a meeting to demand an investigation of the government's actions. One of the speakers at the meeting was Louis Beam, a former KKK leader who immediately announced his plans to move to Idaho, in part because of the support Randy Weaver had received.

A week after the standoff ended, a drifter from Texas walked into the bus depot in nearby Spokane. He had come to North Idaho to protest the siege but had arrived too late. Inside the bus station, he pulled out a handgun and fired three shots at two strangers—a black man and a white woman whom he'd been watching for a few minutes. The twenty-nine-year-old man was hit in the stomach and arm and the nineteen-year-old girl in the side. Both lived but were permanently disabled. When they arrested the drifter a day later, he said he was disappointed the standoff was over and had wandered around aimlessly for a few days. Then, God had told him to shoot the couple because black people weren't supposed to mix with white people.

*L*orenz and Wasiliki Caduff had met on a park bench in Switzerland in 1974, when he was a successful nineteen-year-old antique shop owner and she was a sixteen-year-old sales clerk. They were married a year later. They visited the United States the first time in 1980 and loved the open spaces and the peace of the West. Switzerland, Wasiliki liked to say, was a country of people packed like chickens in a pen. Back in Switzerland, Lorenz got a master's degree in management and opened his own restaurant. But after five years of eighteen-hour days in a drug-rich resort town, Lorenz and Wasiliki decided to take their three children and leave Switzerland for good.

With a partner, they bought the thirty-acre Deep Creek

Inn for $210,000 cash. For six weeks, business was good and they knew nothing about the Aryan Nations or constitutionalists or Randy Weaver. They were in heaven. But then, the ambulance pulled into their parking lot, and the standoff started.

Lorenz had sent Wasiliki away during the trouble, but when she returned, Lorenz was different. He couldn't sleep, couldn't eat. Nervous and shifty-eyed, he made vague references to being in trouble, being under investigation. A few crank telephone calls chided him for supporting the Weavers, and Lorenz began to worry that he was in trouble because of a trust fund he'd agreed to start for the Weaver girls. He'd only volunteered because he had the only safe in the area, he explained to Wasiliki. He didn't condone those beliefs! He just felt sorry for a family, that's all. She tried to convince him there was nothing to worry about. But Lorenz was haunted by images of tanks rolling past his inn and couldn't sleep at night, worrying that he had done something to bring the government after him. Then he stopped talking about it completely. If he talked about it, he feared, the government would kill him.

In October, two months after the standoff, Lorenz was so troubled that Wasiliki took him to visit friends at a nearby lake. And there he just snapped. He ran down the rural road, hiding in the bushes while his wife and friends looked for him. Then, the peaceful chef ran up to a house and dove through a picture window into a stranger's living room, stood up, and began strangling a woman in the house, screaming that they were coming to get him. The woman's husband hit Lorenz on the head with a baseball bat, and he collapsed. When he woke up, he ran away again and hid in a field until the police finally found him, curled up and whimpering.

He was taken to the psychiatric ward of a hospital in Spokane, but when they asked him what was bothering him, Lorenz wouldn't talk. He sobbed and pulled at his hair and said he was frightened because he'd never done anything violent before. He whispered to Wasiliki that if he told the doctors he was troubled by the Weaver standoff, they would kill him. The next time she came to visit him, she almost didn't recognize him. In just two weeks, his hair and beard had turned completely gray. He stared blankly at her.

"Lorenz," she cried. "Don't you know me?"

Then, there was some recognition. "Yeah. You are Wasiliki. Go away, I don't want to hurt you."

"I came to move you to a different place," Wasiliki said.

"This place is hell," he said.

"I know."

He did better at the hospital in Coeur d'Alene, and in November, he came home for a little while. By then, they had more than $6,500 in doctor's bills, and there were still unpaid bills from the standoff. Business was dying at the Deep Creek. Lorenz thought they were being boycotted because he had sent food up to the Weavers and started a trust fund for the girls. He seemed okay for a day or so, and then he began crying and fretting again, saying the government knew that he had provided food and shelter to the protesters. They would surely come and kill him for that. He seemed on the verge of snapping again, so Wasiliki sent him north to be with some friends in Canada. He had another nervous breakdown in Canada and beat his friend with his fists and a stick. After he saw what he'd done, Lorenz ran to a hospital and tried to turn himself in, saying he was worried he was going to hurt someone else. They didn't have a psychiatric ward at the little country hospital, so they called Wasiliki and she bought an airline ticket for Lorenz to fly home.

At the airport, he screamed and cried and wouldn't get on the plane. His friend decided to drive Lorenz back to northern Idaho, but a half-hour from the border, Lorenz begged his friend to stop. He grabbed his Bible and stepped out of the pickup onto the side of the road. A cattle truck was barreling down the highway, and Lorenz walked in front of it and was killed. Authorities couldn't say whether it was suicide or an accident, but Wasiliki insisted her husband would never kill himself. She said he was probably trying to cross the road to look for a telephone to call her.

Even so, she always considered her husband another victim of the standoff. "Deep Creek was Lorenz's dream," she said. "I will never meet such a wonderful man." Wasiliki sold the inn later that year and moved back to Greece, where she was born. Back in the Northwest, folks read about the bus station shooting and about Lorenz's death and wondered when the world would start making sense again, when the violence would end, when the storm would finally pass.

"We don't say that word here," Julie Brown said gently. She realized Rachel hadn't meant anything by the word "nigger." These girls had simply been raised that way. But for lifelong liberals like Keith and Julie Brown, it was heartbreaking to hear such talk from a ten-year-old. Rachel seemed uncomfortable, too, maybe embarrassed, and she wrinkled up her face and turned back to the rodeo in front of them. "Look at that black on that horse?" she tried. Julie smiled at her. The last thing they wanted to do, she said, was change the girls. But Keith and Julie had decided that Sara and Rachel would respect the Brown's rules when they were visiting. "I just want to show them some other choices," Julie said.

At the Dayton Rodeo, on Labor Day weekend—just days after they'd left their cabin—Sara and Rachel watched the horse- and bull-riding and were spun around on some carnival rides, but there was little joy in their faces. They were clearly distracted, especially Sara, squinting into the sun, in a black T-shirt and a baseball cap. She still insisted that they not be photographed—because of their mother's admonition about images—but when they weren't looking, Keith clicked off a few frames.

The girls split their time between Randy's sisters and Vicki's parents, spending most of it with David and Jeane, in the one-story house they'd moved next to the old farmhouse, where Lanny now lived. It was a small, grandmother's house, full of collectible spoons, tiny dolls, and fifty years of photographs. As soon as Rachel and Sara moved in, they stripped the pictures off the walls and the knickknacks off their dressers in their bedrooms. Such images, Sara said, were pagan and disrespectful to Yahweh.

Besides the usual cooking and cleaning and volunteering at the local hospital, now Jeane had to take care of three tough, headstrong girls.

"These are turkey, aren't they?" Rachel asked as they sat down to a dinner of barbecued hot dogs one day. They still refused to eat unclean meat.

"Yes, they're turkey," Jeane said patiently.

"Don't feed them pork," Sara called from the hallway, where she was on the telephone with a supporter.

"I won't, Sara," Jeane said.

Neither girl would go to Vicki and Sammy's memorial service, which was held at the Reorganized Church of Latter Day Saints in Fort Dodge. Churches were pagan, they said. But, finally, one of Randy's nieces talked the girls into at least standing outside the church and greeting well-wishers.

Sara refused to go to school and wouldn't allow Rachel to go either. Jeane's sister brought over some schoolbooks, and they watched a little public television, but Sara didn't want them to be brainwashed. Neighbors donated care packages of clothes and toys for the girls, and some days, Rachel would change in and out of her new clothes two or three times.

Sara told her story over and over again, to relatives and supporters who called or stopped by. Rachel wouldn't talk about what she'd seen. She was more interested in television and Nintendo.

Rachel and Sara watched videotaped news programs that showed the lines of federal officers, and Sara yelled at the television. "You bastards! Which one of you shot my mother?" At night, she stayed up, sometimes until 3:00 A.M., talking on the telephone to supporters, friends, and the Las Vegas skinheads, who played Peter Pan's Lost Boys to her Wendy. Often, they called to get advice. Once, when they were sure federal agents had surrounded their house and were going to come in shooting, Sara calmed them down and told them that wasn't how the federal agents would do it. She especially liked the soulful one named David.

Elisheba was a whirlwind, and Jeane raced around behind her, putting pictures back up, cleaning up spilled Cheerios, and trying to keep the little girl from tearing off her diaper. Elisheba was nervous around strangers, especially men. But when Julie Brown came over, Elisheba took one look at the long black hair and the Jordison family face and attached herself to her aunt's leg. She stared at Julie with huge eyes, as if she recognized her. And then she heard Julie's voice—Vicki's voice—and the ten-month-old baby smiled and looked up, confused. She clung to her aunt's leg as Julie tried to leave.

Strapped with their own grief and questions, David, Jeane, and the rest of the Jordison family struggled with the girls those first weeks. They were drained by their own burgeoning distrust of the government and by the unimaginable sorrow and anger of these two girls. Lanny watched Rachel move around the farm and was amazed. She had watched her mother's face blown off, had been splashed with her blood, and she was stone quiet about it. "I can't believe she doesn't have nightmares," he said.

They had to confront their feelings about Randy. At first, the Jordisons blamed Randy as well as the government for what happened. But now, they began to see troubling questions: What were six marshals doing in the woods that day? Why would they shoot the dog during a gunfight? Why did they shoot Sammy in the back, when he was running away? Why did they need so many agents to settle the standoff?

Vicki's death troubled them the most. Why would they fire at the family before giving them a chance to surrender? They reasoned that an expert sniper couldn't miss twice— one shot grazing Randy, the other accidentally hitting Vicki? So, if it wasn't an accident . . . they couldn't handle the next thought. Government leaks to the press said that

Kevin, Randy, and maybe even Sara had fired at a helicopter, but Sara said that wasn't at all true. The Jordisons began to picture FBI snipers crouched on hillsides, aiming at Vicki's head while she stood in the doorway, and—POW! Sara claimed that Randy had been yelling for days that Vicki was dead, and the family knew the government had listening devices under the cabin. And yet, Gene Glenn had pretended to be choked up when he told them—a week into the standoff—"I have some good news and some bad news." Had they just been conned by some great actor? The Jordisons were a patriotic family, but that was stretched and finally broken by the horrible feeling that their government had murdered little Sam and their beloved Vicki and now was covering it up.

Even Julie and Keith Brown, the liberals in the family, began to see the case differently. They still believed Randy's and Vicki's paranoia brought some of the problems on, but their own faith in the government bled away completely. And their discomfort with Randy's beliefs paled before their anger at federal law enforcement.

And then, there was Sara. She sulked and stared out over the ruffled soybean fields, leaning against the barn her mother had roofed twenty years before. No one in the family could reach her. She refused to see a counselor or a psychologist and wouldn't allow Rachel to see one either. She boiled with anger until Julie worried that Sara would run away and join the skinheads in some battle against the government. With Vicki gone, Julie and Keith began to think that Randy might be a calming influence on Sara and that she might be lost without him. They worried that Sara would explode if her father was convicted of murder. Julie came to the conclusion the only thing that might save Sara from the same fate as Vicki would be for her fear and anger to subside and for their world to return to some kind of normalcy. She needed to trust that there was some fairness and justice out there. And the only way Julie could see that happening was if Randy got off.

In the fall of 1992, there wasn't much chance of that.

seventeen

*T*hose guys are going to get the big needle, David Nevin figured. Lethal injection. Nevin read the August 31, 1992, newspaper accounts of Randy Weaver's and Kevin Harris's ambush of federal officers and their standoff in a fortified cabin on the top of a castlelike point. A lawyer would have to be a masochist to represent a militant neo-Nazi who killed a federal agent and then held 300 government agents at bay for eleven days.

"I'll need a day to think about it," Nevin told the federal judge, Mikel Williams, who had called to ask Nevin to serve as Kevin Harris's court-appointed attorney. The next day, Nevin agreed to take the case.

He flew to Spokane and drove with another lawyer to the hospital where Kevin was staying, made his way through an army of marshals guarding the door, and slid into the hospital room where Harris was spiderwebbed with tubes, recuperating from surgery on his bullet wounds. He was bloated from the intravenous feeding and was so pale, it looked as if he'd already served twenty years.

"I'm David Nevin," he said. He told Kevin that he'd been appointed to represent him, and Harris looked up at him wearily. At first glance, David Nevin was thin and aristocratic, with graying, wavy hair that—it became apparent as you talked to him—had probably been quite long at one time.

"Hi," Kevin whispered. "I have a lawyer. You don't have to worry about it."

But the lawyer who'd promised to represent him wasn't really qualified, and so Harris's parents, Barb and Brian Pierce, auditioned Nevin, asking if he'd ever represented someone charged with killing a law enforcement officer before. Nevin said no, but he assured them he would be up to the task.

It wasn't much of a case, Nevin thought on the short flight back to Boise. It was a losing proposition on every side. As a court-appointed lawyer, Nevin would be working for forty to sixty dollars an hour, less per hour than his overhead. They were going to throw everything at this Harris kid; he was going to have to work eighty hours a week to match the full technical and legal might of the U.S. government. He'd be facing Ron Howen, who collected neo-Nazi convictions like other people collect stamps. He'd be representing an accused cop killer with racist beliefs, a combination that was poison to a jury, even an Idaho jury. There were a million reasons to turn down the case and one very good reason to take it: Gerry Spence.

Two days earlier, newspapers reported that Spence, the famed Wyoming defense attorney, was considering representing Weaver. Bo Gritz had called him during the standoff and Spence had agreed because Gritz told him it might bring Weaver down from the mountain. The Boise legal community buzzed with the chance to see perhaps the greatest trial lawyer in the country—a bombastic, entertaining, plain-spoken storyteller who claimed to have never lost a criminal case. Nevin had read a couple of Spence's books and had even attended one of Spence's psychodrama workshops in Wyoming, where he gave an uncharacteristically flat demonstration of his trial methods.

Nevin loved the idea of working with Spence, and he firmly believed in the defense attorney's responsibility to defend society's worst, but something about the case squirreled with his conscience. He saw the need to challenge government and to keep it from abusing citizens and their rights. But, at the same time, he hadn't gone into law to represent white separatist cop killers.

Whites were certainly separate in the Shreveport, Louisiana, neighborhood where David Nevin grew up in the 1950s and 1960s. With his chin resting on his back fence, the young Nevin could see the "colored quarter," blocks of two-room shanties—tin-roofed, board-frame homes with no paint, no lawn, and no running water. His divorced parents impressed upon Nevin the importance of the civil rights struggle and taught him equally that there was a dominant order in the South—a legacy of abusive power among the white majority—that was evil and immoral. So, alongside a fair hatred of racism, Nevin grew up believing that, often, the establishment is just dead wrong.

He was everydude in the late 1960s and early 1970s—a shaggy-haired, liberal, itinerant college student, checking in at the University of Connecticut, the University of Iowa, and finally, Colorado State University, where he graduated in 1974. He got a job in the woods of Colorado on a construction crew—hard, tiring work that made him realize there had to be a better life out there and that made him decide to go back to school for something better: heavy equipment operator.

Or law school. Either one beat swinging a pick. Nevin did the obligatory aimless trip through Europe and then applied to law schools and, once accepted, flew from Luxembourg to Spokane, and from there hitchhiked to Moscow, home of the University of Idaho. He graduated from law school in 1978, moved to Boise, and eventually took a job in the public defender's office. Like most PDs' offices, Boise's was horribly understaffed; at any given time, 150 cases crossed Nevin's desk, with time to prepare only five of them. It was straight triage, like bringing one ambulance to the scene of a fifty-car accident every day.

Two weeks on the job in 1978, Nevin had his first criminal trial. He faced the toughest prosecutor in Boise—the plug-muscled, steely-eyed perfectionist Ron Howen, who unnerved Nevin with his nonchalant confidence. During jury selection, when Nevin asked a potential juror whether he could believe the defendant could be innocent, Howen snapped up in a voice so authoritative, it sounded like the law itself: "Your honor, I object. The issue before this jury is not innocence. It is guilt or not guilt."

About the same age as Nevin, Howen was a no-nonsense logician who would sooner lose a finger than plea-bargain a case. He swung the law like a club, almost always going for the broadest charge and the stiffest sentence. Boise defense attorneys joked that Howen had the perfect formula for prosecution, four words: "No" to plea bargains, and "What happened next?" to witnesses.

Nevin's client in that first case was a black man named Tommy Fort, who, in an effort to drive to Salt Lake City and see his girlfriend, had been arrested stealing five gallons of gas. Tommy had climbed the fence at a private fueling station, but the pump wouldn't work. So he broke a window, got the keys, flipped on a pump and filled a beat-up, hand-held, five-gallon tank. Howen charged Fort with burglary and grand theft, which Nevin thought was a little stiff for someone trying to pinch a few gallons of gas. It seemed to him like overprosecution.

But Howen told the jury this was more than just a simple theft. He called as a witness the owner of the private fueling station, who also delivered fuel oil to tanks all over the city. When he'd broken into the building to turn on the pump, Tommy had grabbed a ring of keys that would open fuel tanks all over the city. Howen had the witness identify the keys and testify to the fact that they could open thousands of tanks.

In his first closing argument ever, Nevin said, "Mr. Howen would have you believe that Tommy Fort was involved in a conspiracy to control the fuel oil market in the city of Boise—" He showed the jury the battered gas tank, implying he would have to do it five gallons at a time.

Then came Nevin's encore. How could Tommy be guilty of burglary, breaking *into* a building to commit a crime, when his crime had been committed outside?

The jury hung on the serious charges, and Tommy Fort was found guilty only of attempted petty theft. In the real world of lawyers—where innocence is unashamedly measured in degrees—it was a huge victory. And so were the next four cases Nevin tried, bringing cross-eyed looks from the other PDs and jokes that he might never lose.

As soon as he began to wonder himself, he did lose. But he quickly gained a reputation as one of the top defense attorneys in Boise, a natural in front of a jury. After he left the public defender's office, he continued to hold his own against Howen. But their biggest case together went to Howen, the 1986 trial of Elden "Bud" Cutler, security chief of the Aryan Nations, who hired an FBI agent to behead a witness in the Order trial.

Over seven or eight cases, Nevin and Howen developed a grudging respect and a good working relationship. The prosecutor seemed intrigued by the young upstart who managed to smudge some of his intricate prosecutions. Both rose in the Idaho legal community, Nevin as a top criminal lawyer and death penalty specialist, Howen as a top federal prosecutor and the scourge of white supremacists.

Nevin learned a lot about what a prosecutor was trying to accomplish by talking to Howen. "My job is to put on a good enough case that the defendant has to take the stand," Howen told him. "Because then, I've won my case."

Howen taught him another important lesson about being a lawyer, something that came back to Nevin as he decided to take Kevin Harris's case. Howen said, "What you have to do as a lawyer is get away from the fear of losing."

Chuck Peterson wasn't thinking about winning or losing. As soon as he read that Gerry Spence had volunteered to be Weaver's lawyer, Peterson had only one thought: he needed to be on that case. Peterson wasn't in the upper echelon of Boise defense attorneys like Nevin. He was talented and confident, an in-your-face litigator who'd been trained as an army lawyer on a base with 28,000 soldiers and just seven defense attorneys. But after the army, he hadn't fit in at his first big law firm. So he hung his own shingle with a handful of other lawyers and tried to make a name for himself. But Boise—home to federal, state, and county courts, federal agencies, the Idaho legislature, and corporate offices—was up to its lapels in lawyers, 1,333 in the district, 40 percent of the lawyers from an area with 20 percent of the state's population. Chuck Peterson felt like just another suit.

But working on a case with Gerry Spence—who never even wore a suit—could change that. Since Spence was from Wyoming, he would need local counsel, and Peterson figured that was his chance to finally display his talent. "I just gotta figure out a way to get on that case," he said to one of his partners.

Like Nevin, Peterson was a student of Spence's, and he recalled the Wyoming lawyer's advertisement years earlier: "Best Trial Lawyer in America Needs Work. . . ." On August 31, while Bo Gritz was still trying to get Sara Weaver to surrender, Peterson sent Spence a fax, offering his assistance as "Best co-counsel in America . . ."

That afternoon, Peterson—young-looking and intense, with a sharp face and spiked, blond hair—got ready to leave his office and go to his first night class at nearby Albertson College. Burned out on the law, he was thinking about becoming a counselor. As he was getting ready to leave, the secretary said he had a phone call. It was Gerry Spence. "Right," Peterson said. Before sending the fax that morning, he'd told some of his partners about it, and he laughed at their clever joke as he punched the button for the speaker phone.

"This is Chuck Peterson."

There was no mistaking the tugboat baritone. "Chuck. This is Gerry Spence. I got your fax."

"Oh?"

"You know," Spence said, "that's something I would

have done if I was your age. I'm going to get in my airplane and fly to Boise, and you can pick me up at the airport."

"Sure."

"You arrange a way for us to meet with our client," Spence bellowed.

"Okay."

After they hung up, Peterson grabbed one of his partners, Garry Gilman. "You may as well come along. It's going to be a hell of a show, no matter what." Chuck Peterson never made it to those counseling classes.

"*L*ook," Gerry Spence growled at Randy Weaver. "I haven't decided if I'm going to defend you or not, but if I defend you, I'm not going to listen to that bullshit." Spence told Randy that he had black and Jewish relatives. "I'll defend you, maybe, but it isn't going to have anything to do with your beliefs, and I don't want to hear about any of that stuff, and you're not going to say any more about it."

That was about the last time Randy talked to his attorney about the Jewish conspiracy.

Gerry Spence had arrived in Boise the day before as grand and confident as Peterson expected, 6 feet 2 inches tall, 230 pounds—but in a way, even bigger than that. A fringed buckskin coat draped over John Wayne shoulders, gray felt cowboy hat pulled down over his silver pageboy haircut and settled just above his squinting eyes, Gerry Spence seemed for all the world like the gunfighter he fancied himself. He and his lawyer son, Kent, had driven with Peterson to the Ada County jail—a drab, concrete block of a building out by Boise's strip malls and office parks. Randy Weaver sat in front of them, tired and beaten, just hours after his surrender. Peterson sized up his potential client. His head was shaved, he was pale and weak-looking, shackled, wearing an orange jail jumpsuit and orange thongs. Spence had expected a wild-eyed, charismatic kook, and what he was seeing was just a tired, little man. But he also sensed that he was listening to someone who was telling the truth. Peterson and Spence looked at each other: This guy kept all those officers at bay?

That night, they just listened to Weaver's tale: set up by the government on a phony gun charge, his son shot in the back, his wife shot in the head. Every few minutes, he'd start weeping and couldn't finish his story.

Kent Spence didn't seem too enamored of the case and, as Peterson dropped the Spences off at their hotel, he figured he'd never see them again. He guessed they'd fly out the next morning and Peterson—who had told the local judge he'd represent Weaver—would be stuck with this dog of a case.

But they were still in Boise the next morning, when Spence lectured Randy about not talking about his beliefs. That day, at the jail, Spence had pointed questions for Weaver: Who shot first? Where were you? What did you see? Randy answered clearly and consistently, and then he started in on his beliefs, how the Jews were probably behind the whole thing, and that's when Spence stopped him. "Keep that stuff to yourself."

That afternoon, they drove downtown to meet with the judge. "Well, what do you think?" Spence asked the other attorneys on the way.

Peterson didn't think they had much of a case, but he stayed quiet. He was there because of Gerry Spence, not Randy Weaver. "I don't know. What do you think?"

"I don't know," Spence said.

The lawyers turned into the parking lot at the federal courthouse, hours before the arraignment was to begin. There were television and print reporters camped all over the courthouse steps, and Spence turned to his son. "We're in this case."

Kent Spence—the actor they'd have to hire to play his dad twenty-five years younger—grinned, shook his head, and turned to Peterson. "You in?"

"If you're in, I'm in."

"We may never get paid," Gerry said.

"Oh, well," Peterson said. "If you're in, I'm in."

Back at Peterson's office, Spence sat down and wrote two press releases, one with his name on top and the other with Randy's name.

STATEMENT of Gerry Spence:

> *I was told that if I would agree to represent Randy Weaver he would come down off of the mountain and surrender. Hoping that my agreement to represent him would prevent further bloodshed, I have made my appearance on his behalf in federal court today.*
>
> *Mr. Weaver and our co-counsel do not see eye to eye on many issues.*
>
> *We do not believe in white separatism.*
>
> *We do not share Mr. Weaver's religious beliefs.*
>
> *But our personal beliefs and his are not important to this case. In America, all of our religious and political beliefs are protected by our Constitution. . . .*

And then, Spence composed Randy's statement, detailing in it how his wife and son were murdered and explaining he had come down from the mountain only to protect his girls.

"I have never believed that I could get a fair trial in a government court," Spence wrote for his newest client. "I was assured that Mr. Spence, one of the great lawyers of the country who has spent his life fighting for people and the cause of freedom, will see that I get a fair trial. I believe Mr. Spence will see that my rights are protected. If I did not believe that I would still be up there.

"I have authorized Mr. Spence to undertake my defense understanding that he and I see eye to eye on very few political and religious issues. As a matter of fact, we are poles apart in our beliefs. But one thing he and I agree on, and that is people ought not to be murdered by their own government."

Besides single-handedly reviving the buckskin industry, Gerry Spence had another rare quality, one vital among prominent people of the twentieth century, the ability to define *himself*. In a world narrated by mass media, fame cast people arbitrarily as Good Guy or Bad Guy, but Gerry Spence would have none of that. Through a chain of bold and unlikely court victories and a string of books about his exploits and philosophies, he crafted himself as the Lone Ranger of the law, not just a good guy, but something more, a mythical figure, a hero. Big. It was some trick, creating a hero from a profession that mostly inspired mistrust and scorn.

In the courtroom, Spence was his own work, painting himself—broad and imperfect—for the jury just as soon as he was turned loose on them. He hoped to make the jury forget the client and concentrate on the lawyer—who didn't fight the other side's battle of details and evidence, but who crafted a homespun story, a worldview, an ethic that he invited the jury to adopt as its own. "Folks, you and I know this case isn't about a murder. It's about . . ." Every chance in front of the jury he repeated his story and every question was framed through that worldview. Outside the courtroom, he spun another kind of story, revealing and reinventing himself in a string of books that mixed his exploits, his flaws, and his philosophies like a stiff drink that he just kept topping off; he was a man who just couldn't stop writing his autobiography.

Gerry Spence sprang from Colorado ranch and farm families in January of 1929, the winter before the Great Depression. One grandfather was a member of the Zion Religion, an Old Testament, fundamentalist faith that decreed, among other things, that it was a sin to eat pork. Some Protestant fire made it to Gerry's mother, but that was pretty much the end of that line in the Spence family. Gerry's own faith was the Western ideal of individualism.

His father was a chemist at the University of Wyoming, but the Depression sent the family spilling out of Laramie and into California, looking for work. They ended up back in Wyoming—in Sheridan—where Spence's dad worked for a railroad and the family grew up poor. Spence's mother made coats and gloves for her three children from the hides of elk, deer, and antelope, and the family took in tourists to get by. Gerry helped out, washing and ironing sheets, driving horse teams, and selling cinnamon rolls door-to-door. By fifteen, he was a pushy loudmouth who terrorized the rest of the family—perfect attributes, his devout mother realized, for a preacher. Gerry thought he was better equipped to be a lawyer.

He graduated high school at a time when it was still possible to go off to sea for adventure, and Spence had a fine time as a merchant marine. He drank rum, smoked cigars, visited whorehouses. But he refused to pay union dues, and the other merchant marines tossed him overboard and emptied a garbage chute on him. He eventually quit, moved back to Wyoming, was married, and got into law school, but his mom didn't approve of his gambling or his godlessness, and they argued almost as much as Gerry and his wife did. During his first year of law school, Spence's mother committed suicide.

Her death was hard on Spence, and at first he blamed his own boozing and debauchery for her sorrow, and ultimately, her death. He graduated first in his law class at the University of Wyoming, failed the bar, passed the second time, and moved to Riverton, Wyoming, to join a law practice. In 1953, he ran for county attorney as a Republican, knocking on doors and leaving notes for those people he missed. Every registered Republican got a check "to be cashed for honest law enforcement," and at twenty-four, Gerry Spence became the youngest county attorney in the state. He was a buzz saw. He shut down the Little Yellow House brothel in Riverton, revoked liquor licenses, and even prosecuted himself for shooting ducks outside of shooting hours. He won a second term but didn't run for a third, deciding instead to aim for Congress. He got pounded and set up a private practice.

Gerry Spence in the early 1960s was the picture of Republican establishment—four kids, a member of the Kiwanis, the Elks, and the Sheridan Country Club. He represented insurance companies looking to screw those little guys he'd later champion.

There was a midlife crisis: lots of drinking, some sensitivity training, and an affair with a woman named Imaging. He campaigned to be a judge, was rejected by the governor, sold everything, and moved to Mill Valley, California, where he planned to go to art school. That didn't work out, and he and his wife divorced. Spence wandered back to Wyoming, married Imaging, and sobered up. He went back to his insurance law practice and got rich. In his first book, *Gunning for Justice*, Spence wrote that he was in a grocery store when he saw a hobbled accident victim that he'd earlier defeated in court. After years of representing the establishment, Spence had a "crisis of conscience" and vowed to spend the rest of his life on the side of criminal defendants and people suing big corporations.

The cases—he called them "little people trying to get big justice"—made him famous in the West and in the larger legal community. Spence's reputation quickly went national with a string of high-profile cases, beginning with the 1979 Karen Silkwood case, whose heirs sued Kerr-McGee Corp., charging she had been contaminated by Kerr's Oklahoma nuclear plant. Spence won a $10.5 million civil suit, but it was overturned on appeal, and the family eventually settled for $1.4 million. He represented a beauty pageant queen, Miss Wyoming, in a case against *Penthouse* magazine, which published an article, purporting to be fiction, describing a woman who resembled Spence's client but boasted incredible sexual talents. With his commanding pronouncements and disdain for legal technicalities, Spence cut through the "legalese," turned witness's testimony into his own speeches, and strained courtroom decorum so much that the opposing attorney in the Miss Wyoming case moved for mistrial four times—during Spence's opening statements alone. Miss Wyoming won a $26.5 million defamation award in that case—also overturned.

When rich, powerful clients requested his services, he transformed them into little people too, so that in 1990, Imelda Marcos became "a small fragile woman" whose only crime was being "a world-class shopper." Charged with embezzlement and racketeering in a New York trial, the former first lady of the Philippines hired Spence for a reported $5 million. His loud, plainspoken tactics didn't play as well in front of a New York judge, who rode Spence throughout the trial. Even his co-counsel in that case tired of his antics. But Spence won the case, and Marcos was acquitted. He did many cases for free but commanded huge fees when his clients could pay—40 to 50 percent of the settlements at a time when other top attorneys got 33 percent. He bragged that he'd never lost a criminal trial and hadn't lost a civil case since 1969.

Spence's ethics were welded to his emotions, and he used the same zeal to win a murder acquittal for an old ranch hand and sheriff named Eddie Cantrell as he did to convict another accused murderer while serving as a special prosecutor in 1979. In that case, Spence wore a bulletproof vest and placed bodyguards around the courtroom to illustrate the viciousness of the defendant, a man named Mark Jurgenson, who'd killed an old friend of Spence's. At his most stirring and manipulative, Spence asked for and got the death penalty. Later, he would write that it was the worst thing he ever did. In another epiphany, he decided he was against the death penalty. He even tried to get Jurgenson's sentence changed to life in prison.

But in 1991, Jurgenson was electrocuted, the first execution in Wyoming in twenty-six years. In some ways, Jurgenson's death was the most startling example of Spence's power in the courtroom and, also, his ability to reinvent himself—a trait critics called hypocrisy.

And there were plenty of critics: prosecutors and

attorneys who'd faced him, companies whose pockets had been lightened by his rhetoric, and judges who'd spent entire trials trying to keep him in line. But by September 1992, Gerry Spence had something that transcended all of that: celebrity. Gerry Spence came to the Randy Weaver case in September 1992 as probably the best-known attorney in the world, a perfect mix of brilliance and self-promotion.

*I*n Boise, Idaho, it was as close to a dream team as you're likely to get. On Weaver's side of the table were Spence, his son, and the sharpest of the Wyoming contingent, Gerry's assistant, Jeanne Bontadelli—the left side of his brain. ("Jeanne? Have I read that memo?" "Yes, Gerry, you have.") Peterson and Garry Gilman completed Weaver's team.

Nevin was assisted by another court-appointed lawyer, Ellison Matthews—for twenty-four years one of the top criminal lawyers in Boise. "Ellie" was the antithesis of the manic attorneys on the two teams. They would be flying off the walls, arguing circles around minute bits of law, and—on those rare occasions when he said anything— Matthews would interject quietly and forcefully, right to the heart of whatever they were talking about. Spence called him a three-worded laser.

At first, of all the lawyers, only Gerry seemed optimistic. And only he grasped what the case would finally be about. Nevin was edgily preparing for a regular murder case, one that didn't look particularly promising. His guy had admitted shooting the federal marshal. Peterson—assigned by Spence to Weaver's initial gun charge and the failure to appear—worried about the tape recordings of Randy with the informant Ken Fadeley and about the letters Vicki Weaver wrote, pronouncing their hatred for government and their intent to never show up in court. Spence paid some attention to the evidence, but he talked more about freedom of religion, until the other lawyers began to wonder if that wasn't what the case really was about.

They got some clues in September, at the preliminary hearings, the first chance for defense attorneys to really see the case against them. During Kevin Harris's preliminary hearing, the prosecution asked for a delay because Boise police needed to provide security for Vice President Dan Quayle, who was making a campaign stop. Nevin was suspicious. He knew prosecutors would rather skip the preliminary hearing and take the case straight to a grand jury, a group of citizens who met in secret, without the defense present and usually rubber-stamped indictments.

Nevin feared that they were delaying the preliminary hearing to give a grand jury time to indict Weaver and Harris, which would nullify the hearing and keep the defense attorneys from hearing the evidence. U.S. Attorney Ellsworth promised the delay wouldn't keep the two men from having a preliminary hearing, and Nevin said that as long as that was true, he didn't mind the delay.

On September 15, Kevin Harris's preliminary hearing began with ten motions from Ron Howen filed just the night before. The judge told Howen that many of the issues had been settled already. Later, as prosecutors questioned a relatively unimportant FBI witness for hours, Nevin began to get suspicious again. He leaned over to Matthews. "Do you think he's just stalling?"

That afternoon, the grand jury returned an indictment against Harris, and the prosecutors immediately asked that the preliminary hearing be canceled. David Nevin felt like he'd been kicked in the teeth.

Nevin remembered a case years earlier in which a witness could have destroyed his case, and he'd asked Howen whether the prosecutor was going to call that witness. The prosecutor had said no, and even though Nevin opened the door for that witness to testify, Howen stood by his word and didn't call him. Deep down, Nevin believed Howen was an honorable man, but he couldn't shake the suspicion that the delay had been intentional. It made him wonder what incredible pressure Howen must be under.

With Spence onboard, the battle quickly became one of public perception. Two days after Spence's press releases accused the government of murder, an unnamed Justice Department official offered another version of events. "I've heard enough of this talk that we shot a mother with a baby in her arms," the official told the Spokane newspaper. "That's just not true." The official gave an inaccurate account of the August 22 shooting, claiming snipers were given the order to fire only after Vicki and other members of the family shot first at a helicopter.

Gene Glenn's assistant in Salt Lake City, Dave Tubbs, also told reporters there was more to Vicki Weaver's shooting than her husband was letting on. "I can tell you we don't shoot mothers with babies in their arms," he said. Tubbs said the FBI rules of engagement preclude agents from firing unless their lives are in danger. "So, obviously, if a shooting occurred that resulted in the death of Vicki Weaver, something happened to precipitate that shooting. . . . We don't go around killing people unless there's good reason."

But in Boise, that was one of the problems facing the assistant U.S. attorneys assigned to the case—Ron Howen and a terse coilspring of an attorney named Kim Lindquist. They had Degan's gun, which had been fired seven times. They had Sammy Weaver, who had been shot in the back. And, worst of all, they had Vicki Weaver's death. They weren't going to be able to try Randy Weaver for murder without giving the jury the "good reason" for his wife's death. Their task was making a jury see that Randy Weaver was ultimately responsible for everything that happened on Ruby Ridge. It was not going to be easy.

On the third floor of Boise's glass federal building, Ron Howen's door opened on a simple office with plain, working furniture, ordered files on the desk, and Howen at work—always at work—sometimes straight through the night, banking on his ability to bludgeon his courtroom opponent with sheer drive. He had a decent, serious face and a dusting of gray hair; the thing he looked like most was a prosecutor. He always looked like that, even when he was pressing free weights at the Boise YMCA, a grim, daily workout that gave him a solid linebacker's body and the athletic, purposeful walk of the righteous. All he wanted was to be a prosecutor, and friends marveled at the six-figure offers he turned down from private firms.

Howen was a religious fundamentalist and a family man who hung evidence of his avocations on the walls of his office, including several framed hunting photos. Howen hunted with an old-fashioned longbow, a primal weapon that required incredible strength and discipline, trailing wild game into the mountains of Central Idaho with his son and the llamas they used to pack their equipment.

The other trophies on Ron Howen's walls were encased reminders of his more civilized kills. In one of the frames was an unexploded pipe bomb from the Aryan Nations bombing case. In the other was a silver badge. Up close, it revealed itself as seized evidence from another trial: a shield over a battle-ax and a Roman cross, inscribed with the ancient Gaelic words that translated to "You are my battle-ax and my weapon of war." There were two German words on the shield: Bruders Schweigen.

It all went back to that case, the original Order trial, in which Howen had been co-counsel in a masterful prosecution that tied together the actions of two dozen people in a textbook conspiracy. The strength of that case was that it clearly showed the crimes were committed to further neo-Nazi beliefs, opening up the entire ugly Aryan Nations world to the scrutiny of a jury. Once the prosecutor established there was a conspiracy to commit crimes to further neo-Nazi goals, he could bring in those beliefs as evidence. The formula had worked for Howen in Order II, in the Coeur d'Alene bombing case of Randy Weaver's friend Proctor James Baker, and the prosecution of Elden "Bud" Cutler.

Howen recognized many of the same beliefs in the Weaver case and knew he had a much better case if he could get the Christian Identity religion in front of the jury. Howen thought those beliefs formed the basis for Randy

Weaver's drive to force federal agents into a violent confrontation. He needed to make the jury realize that Randy Weaver and Kevin Harris weren't just responsible for Bill Degan's death but for the entire standoff—responsible even for the deaths of Sam and Vicki Weaver.

That would solve a couple of strategic problems as well. First, of course, it allowed the prosecutors to introduce testimony about the Weavers' beliefs. And second, it opened a door for dealing with Vicki Weaver's death. While testimony about her shooting would take away from the rest of the government's case, it would be more disastrous to allow the defense to introduce her death and make it look as if the government was hiding something. The prosecutors figured it was better strategy to lay the events out forthrightly and show—through the conspiracy charge—that Vicki Weaver died because she and her husband were involved in a criminal enterprise to force a gun battle with the government. As he'd explained to the grand jury: No, Randy and Vicki Weaver weren't members of Order I or Order II, but they were forming their own criminal enterprise to carry on the work of those groups.

And so Howen patterned the indictments after The Order conspiracy cases. As he and Lindquist began piecing together the evidence, they found the Weavers' 1983 interview with the *Waterloo Courier*, right before they left Iowa. In it, Randy talked about establishing a "300-yard kill zone encircling the compound." That seemed to be the beginning of Randy's obsession with violent confrontation, and so the prosecutors decided the conspiracy should begin there, nine full years before the shoot-out.

Conspiracy was the first count charged, because once you proved that, the other charges came together like spokes on a wheel. Through September, Howen worked on an intricate sixteen-page indictment, listing ten separate, circular charges: conspiracy to provoke a confrontation, sawing off a shotgun, failing to appear for court, assaulting and impeding deputy marshals, killing William Degan, intimidating or impeding an FBI helicopter, harboring a fugitive (Randy Weaver), possessing guns and ammunition as a fugitive, violating terms of release, and carrying a firearm in commission of a crime. As they said on old TV shows, Howen threw the book at 'em.

Within the first charge, conspiracy, there were nine objects—or goals—of the conspiracy, things like maintaining a mountain stronghold, collecting guns, and stealing their neighbors' property. There were twenty-eight (by trial there were forty) overt acts, things that might not be illegal but which furthered the conspiracy, actions like moving to Idaho and mailing a letter to Ronald Reagan.

Howen and Lindquist ran their conspiracy theory past Maurice Ellsworth, who approved it. Then they showed it to officials with the ATF, FBI, and the U.S. Marshals Service. The deputy marshals, who had spent eighteen months dealing with Weaver and who had lost a colleague, thought it was a good idea. But U.S. Marshals Director Henry Hudson, himself a former prosecutor, didn't like it. It brought in too much extraneous information and detracted from the important charge—the murder of Bill Degan.

The indictment sent shock waves through FBI headquarters. A conspiracy indictment endangered all the other charges, FBI officials complained. The conspiracy indictment would open the entire case, including the death of Vicki Weaver, to the scrutiny of the defense, and possibly, the public. And, despite earlier reports, top FBI officials knew that Vicki and the other people inside the cabin had never fired at a helicopter. Such a broad case would open up documents and reports that the FBI would much rather remained closed.

Finally, Howen checked with the Justice Department. Two top officials disagreed with the broad indictment but later said they had no supervisory control over the case and so they didn't say anything.

Howen filed the superseding indictment on October 1 and expanded it November 17. As the case moved toward a spring trial, Ron Howen got ready to go into court and prove once more that racist beliefs could fuel a violent criminal enterprise. Only this time, his criminal conspiracy wasn't among a shady group of Aryan Nations members robbing armored cars and murdering Jews. It was a family.

Although they were outraged by the indictment, in a way it was the best news the defense team had gotten since the case began. Nevin had warned the other lawyers of his experiences with Howen and his overprosecution of simple cases like Tommy Fort's gasoline theft. They guessed that—with a federal law officer dead—Howen would reach too far again. When the indictment came out, they saw he had gone further than any of them would have guessed.

Peterson's biggest worry was that prosecutors would simply charge Harris with murder and charge Weaver with failing to appear and selling sawed-off shotguns. Then, he thought, Cooper would testify to watching his buddy die, a pathologist would testify that Harris's gun had killed him, they'd play the tapes of Randy selling guns; that's it, the defense is screwed. Vicki Weaver's death had nothing to do with those charges. The defense attorneys' best chance was a complete retelling of the story, from the government screwups to Vicki's murder, a pattern that just might convince jurors that Randy Weaver and Kevin Harris weren't to blame for what had happened.

As Nevin and Peterson read Howen's indictment, they were amazed at Spence's intuition. All along he'd been saying this case would be about freedom of religion and now, by starting the conspiracy in 1983, the government *was* making it a case about Randy Weaver's beliefs.

To Spence, this wasn't a criminal trial, it was an inquisition. The only way for the government to cover up its barbaric behavior on Ruby Ridge was to demonize Randy Weaver, and the only way to do that was to make him seem like a charter member of The Order. Spence returned to his stone-and-log mansion in Jackson Hole, Wyoming, to prepare for trial. He started with the grand jury testimony Ron Howen had presented. He stared out the window at the Grand Tetons, a jagged ribbon that divides Idaho and Wyoming. One grand jury witness had been James Davis, an FBI agent who painstakingly detailed the crimes committed by The Order, The Order II, and other Aryan Nations members. His testimony covered seventy-four pages of the transcript and had little to do with Randy Weaver. Randy Weaver belonged to none of those groups and had been to the Aryan Nations only a few times. They were trying to bury this little man by associating him with people he wasn't connected with at all. Spence was aghast. Yet, from a strategic point of view, he couldn't have been happier. This was the ground he had come to fight over.

*I*f they expected the far-reaching indictment, the defense team never expected what came next. In November, the U.S. attorney for Idaho, Maurice Ellsworth, applied to the Justice Department's Criminal Division for permission to seek the death penalty. The October indictment had included the possibility of the death penalty for killing a federal officer in performance of his duties, but after studying the case, even Nevin didn't really think the government would attempt it.

Yet Ellsworth wrote that Weaver and Harris's actions were premeditated, "founded in racial bigotry and baseless hatred for authority . . . cold and heinous in its direction at law enforcement. . . . The death of Degan was the direct and planned result of Weaver's selfish and monstrous proclamation that 'The tyrant's blood will flow.' His legal and moral accountability fully supports imposition of the death penalty."

Justice Department officials didn't think Ellsworth was serious. They figured he was just going through the motions, dotting i's and crossing t's. But when Associate Deputy Attorney General David Margolis called Ellsworth and Howen, they assured him they were very serious. They thought Harris and Weaver should be put to death.

In February 1993, the judge in the case ruled the death

penalty didn't apply, but by that time, defense attorneys had gotten the message: they were going hard after Randy Weaver and Kevin Harris.

Another realization came to each of the defense attorneys separately and much earlier, the feeling that they not only had a case, but were staring full-faced at injustice. Each batch of discovery documents first outraged them and—as attorneys—thrilled them: the interview with Deputy Marshal Norris (who said the first shot sounded like a marshal's gun); the early arrest affidavit (in which Weaver and Harris were supposed to have fired from a truck); and the most amazing document of all, the FBI's rules of engagement.

Peterson—who had constructed a computer database to keep track of everything—would jump on the telephone to Spence. "Have you seen this one?"

"Yeah, somewhere, but I don't know where the hell it is," the old gunfighter would growl. "Jeannie, dammit! Where's that document?"

Nevin knew all along he had the most difficult case. He had the shooter, a guy who admitted killing a law officer. Yet Nevin had begun to believe that even the shooting of Degan was defensible, since so much evidence didn't fit the government's side of the story—especially Degan's seven missing shots and the business with the dog. Early on, sitting in Chuck Peterson's office—with Spence on the speaker phone—Nevin had described Kevin's version of the shoot-out ("I mean, what could he do? They were shooting at his friend!") becoming more and more animated until Spence said, "You ought to write that down."

Despite Spence's almost hypnotic insistence that this was a civil rights case, Nevin continued preparing for a case about a shoot-out, a fuzzy two- or three-minute gunfight in the middle of dark woods.

If the other attorneys' opinions were transformed by reading reports and documents, Nevin's epiphany came in October, when he visited the Weaver cabin again.

Spence, Peterson, and Garry Gilman had already been

through the cabin, skulking around and taking photographs of the rock outcroppings, of the Y in the logging road, and of Peterson reaching up to unlock the birthing shed, the way Randy had done when he was shot. Spence said the trip wasn't much help because the FBI had already carted away all the evidence, before they could get a look at it. "It's just a cabin in the woods," said Spence, who'd seen his share of cabins in the woods.

Gilman sensed something else. "The overwhelming feeling I have here is sadness," he told a reporter, while standing on the back porch. "To think that three people are dead over a missed court date seems incredibly wrong and sad."

Nevin and Ellie Matthews flew up to North Idaho in October, after reading Deputy Marshals Cooper and Roderick's account of the shoot-out. Autumn had blown into North Idaho and chill winds shook whistles from the trees. Like everyone else, Nevin was struck by how beautiful and peaceful the ridge top was, and he spent some time just walking around, trying to imagine—like Gilman—how things could have gotten so out of hand. It reminded Nevin of some places he'd been to during his aimless years in Colorado, bikes and tools and junk all over the yard, the clean, full air cut with hints of wood smoke. He began to connect with his client, who was himself just an aimless kid.

The cabin itself was spooky, full of bad vibes even for someone who didn't believe in such stuff. That night, with the October wind howling through the plywood walls, they slept on the cabin floor where Randy and Kevin had lain, where Sara had crawled to get food and take care of the baby. Nevin always liked to visit the scene of a case he was defending. It fueled his imagination and helped him explain what happened to a jury. But this felt different. It was tough staying in this place without being overwhelmed and saddened.

The next morning, they walked down the same trail Sammy and Kevin had taken, through the fern field, down to the Y. They looked for the stump Degan had hidden behind and tried to imagine five or six people with guns in such dense woods and brush. If it was possible to be claustrophobic outside, it would happen there, in that thick forest. Several hundred yards down a steep hill, the government's version of events began to seem less and less likely. For one thing, marshals claimed to have received fire from the cabin after Degan was shot, yet that seemed impossible—unless the Weavers had some sort of weapons that allowed their shots to bend over the lips of ridges and around trees.

They went to the upper observation point where Hunt had photographed the family and down to the lower point where Roderick threw rocks at the dog. That was something Nevin could never figure out. He tried to picture these three guys crawling through the woods, carrying a machine gun with a silencer, suddenly deciding to throw some rocks. He knelt behind the boulder where Roderick had been and threw a couple of stones into the gully between himself and the cabin. It was too far away even to come close to hitting the cabin.

The rock clattered down the hillside, and Nevin stood up. Of course. The silenced machine gun, the rocks, the undercover plan with agents hiding in the woods, even the shoot-out itself. This was a case about a dog.

eighteen

*T*he facts will show that this ultimate confrontation occurred not because of a government perception of Randy Weaver's beliefs, but because of Randy Weaver's actions," Kim Lindquist told the seven women and five men—all of them white—who stared at him from the jury box. They ranged in age from thirty-one to seventy-two, from a ninth-grade graduate to an MBA. Many had served on juries before, several had been crime victims. Most owned guns.

All morning on April 14, 1993, the jury listened as Lindquist, and especially Howen, colorlessly and carefully built their case like legal bricklayers. "On or about" this date the Weavers said this; "To wit," the Weavers purchased that; "In furtherance" of the conspiracy, Randy Weaver did something else. They painted Weaver—in blacks and grays—as a racist, religious zealot commanding an army that also happened to be his family, a man who sought and finally achieved an Apocalyptic holy war with federal agents who did everything they could to avoid confrontation. Finally, on August 21, 1992, the Weavers got their wish, chasing the deputy marshals into the woods and shooting Bill Degan in cold blood.

It was a classic prosecutorial laying out of the facts. And the defense attorneys for Weaver and Kevin Harris couldn't have asked for a better setup. At 1:30 that afternoon, David Nevin stood up to give his client's side, looking for all the world like your favorite college professor, the one all the girls had a crush on and all the boys identified with.

"Kevin dove into the bushes and tried as hard as he could to save his friend's life," Nevin said. "Kevin did the best he could in an almost impossible situation." He told the jury they would hear later about Degan's sad, painful death, and he described it for them himself, to get it out in the open and defuse its power later.

And then, he just told Kevin's story, from the moment he and Sammy Weaver saw the dog running off into the woods—"They think it's probably an animal, deer, bear, elk, mountain lion." Nevin's voice was a revelation to the people who'd never heard him: quiet, honest, and authoritative, a hint of polite Southern lilt, stark contrast to the prosecutors' industrial presentation of the evidence. "It's ten-thirty in the morning on a Friday in late August. It's quite pleasant in the woods." Nevin described Sammy Weaver and Kevin walking through the woods, when suddenly, a camouflaged man steps out of the woods, points a rifle at the dog, and shoots it. And then Nevin's voice gathered strength and authority until it didn't seem possible it was coming from that spindle-thin guy. "The evidence will be there was no legitimate reason to shoot that dog. The dog was just doing what dogs do." Sammy fired at the man, who dove back into the woods, Nevin said, his voice getting louder, his hands more animated.

"The woods just light up. Kevin sees smoke puffs . . . shells fly. Randy sees these guys, and at some point he starts to holler. Sam yells back, 'I'm comin', Dad, I'm comin', Dad.' They cut loose on him." And then Nevin's voice leveled off. "I want you to know, Sam Weaver was shot in the back, running away. Running home."

At the defense table, Randy Weaver stared at his shoes. Kevin Harris began to cry.

Nevin was quiet again. The last thing Kevin Harris wanted was to shoot anyone. But when they started shooting at Sammy, he turned and fired at the marshals. "I don't have any doubt that it's a slug from his thirty-ought-six that kills Mr. Degan." Nevin paused. "I'm sorry about that, but he did that in self-defense." The lawyer described the next twenty-four hours, the family preparing for an attack, the grief over Sammy's death, and the snipers moving into place around the cabin, shooting Randy and Kevin, killing Vicki. "On a jet from Washington, D.C., they have decided they will alter the FBI rules of engagement. Instead of the situation that pertains under Idaho law, they have decided they will shoot to kill whenever they see an adult male with a gun."

Along the way, Nevin planted several seeds for later, like the seven shots Bill Degan fired before he died and the first shot Deputy Marshal Frank Norris heard during the shoot-out, "the distinctive sound of a two-twenty-three"—one of the marshal's guns.

"The evidence will be that they handled this situation terribly," Nevin said. "They botched it. They blew it. I'm here because I didn't want Kevin to be the fall guy for them."

P assing lawyers tried to look nonchalant as they walked by Judge Edward Lodge's courtroom on the sixth floor of Boise's glass federal building. Like little boys strolling past a porn shop, gray men in navy blue suits and tweed jackets peeked into the plain courtroom, and lawyers from all over the city timed their business at the federal court building with the hope of witnessing some major league bombast. By mid-afternoon, David Nevin was finished with his straightforward, reasoned opening, and everyone knew it was time for the big gun. It was Gerry Spence's turn.

Howen and Lindquist anticipated the worst, a

grandstanding mockery of the law. They had fought Gerry Spence's appointment to the *Weaver* case and had tried to get him tossed off the case for talking to the media. Now they stared straight ahead while Gerry Spence pushed his cowboy hat aside, straightened his Western-yoked suede jacket, gathered his notes, and stood.

Outside, on the strip in front of the courthouse, the protesters were gone. A handful had shown up the day before, carrying signs much more professional-looking than the felt-pen-on-beer-box rantings at Ruby Ridge. An angular man in a suit paced with a sign that read: "The Most Despicable Criminals Are Those Who Wear a Badge. FREE Weaver & Harris." A woman marched behind him with another sign: "Hide the Women and Kids. Here Comes the FBI."

But Kent Spence had gathered the protesters, patriots, and white separatists together and asked them to not create a stir. "We really appreciate your support, but we would be better off without it."

Bo Gritz was gone, too. He'd come the day before as well, pacing in a dark leather bomber jacket and chatting with reporters—"Any more questions from the commie-pinko faggots?" Bo had gotten the message from defense attorneys; they didn't want any reminders in the gallery of Randy's beliefs, and so Gritz had personally called Randy's skinhead friends and told them that if they had to come, "be respectful, like you're going to church." A couple of them showed up, sheepish in jeans and T-shirts, their hair grown out into awkward fuzz with no discernible part. The skinheads were nice, misunderstood kids who just weren't needed at this point of the case, Bo said. "There's nothing wrong with a good haircut, but we don't need people thinking this trial is about their beliefs." As for himself, Bo wasn't allowed in the courtroom because he was on the witness list.

Gritz's appearance was a bonus for a scraggly band of mostly older gentlemen who had come to Boise to watch the trial. They carried right-wing newspapers under their arms and wore wrinkled suits that had never been in style, not even twenty-five years before, when that particular strain of polyester was engineered. They sold hastily-pulled-together books and pamphlets about the case (Chapter 14—"Jews Dominate Federal Government") and $6 tapes of Carl Klang's epic "The Ballad of Randy Weaver" ("the army of your enemy/slayed your bride and only son/and nearly killed your close companion/when the shrapnel pierced his lung"). A classical music composer asked for advice on a fifteenth-century funeral piece he was composing ("Which instrument do you think would best represent Striker?"), and at least three radical right poets were inspired to write about the case, including one that rhymed "baby in diapers" with "infidel's own snipers."

Largely ignored by the mainstream media at first, the *Weaver* case reverberated among the far right, among neo-Nazis, tax protesters, Birchers, and the conspiracy-fed old men who rode the elevator to the sixth floor of the federal building in Boise, went through a second metal detector, and waited in line for seats in the back of the courtroom.

Reporters were there, too, regional mostly, but also from Seattle and Los Angeles. They filled a notebook page or two with the prosecutors' technical opening statements and a few more with Nevin's strong opening. They waited eagerly for Gerry Spence to begin.

There were marshals in blue blazers everywhere, a handful in the parking lot, five in the courthouse lobby, several more walking throughout the building, five on the sixth floor, and at least six in the plain federal courtroom. Three uniformed security people stood watch at all four corners of the building and—two years before the bombing in Oklahoma City—the Boise federal courthouse had its first metal detectors, manned by deputy marshals. One day before the tax deadline, Boise citizens walked through a demilitarized zone of federal officers just to pick up a new 1040. Others rode the elevator to the sixth floor and waited outside the packed courtroom, hoping to just hear the famous Wyoming attorney.

*　　*　　*

"We represent a man we're proud to represent." Gerry Spence looked over his shoulder at Randy Weaver, who wore a striped tie and dark suit coat, his skinhead stubble grown into a gray-flecked pompadour, his once-skeletal cheeks filled out from jailhouse food. Spence had to make sure the jury liked Randy Weaver, and to do that, he had to show that *he* liked Randy Weaver, because Spence was damn sure that by the end of the trial, the jury was gonna like *him*. Earlier, he'd sat close to Randy and even put his arm around the smaller man.

Goal number two was to show he was in the same boat as the jury, as a sort of tour guide, leading them all through the complexity and fear to the truth and eventually, to freedom. By the end of the trial, the jury would feel like prisoners themselves, and in that, at least, they could empathize with his client. And Spence would be there to feel their pain. "My proposition here today is that when we walk out of this courtroom together, all of us, that we all walk out, every one of us. The facts and the evidence in this case are going to establish that we all have the right to be free."

And then, the payoff, the quick, dark summation, the refrain they would hear for three months—"This is a case, simply put, that charges Randy and Kevin with crimes they didn't commit in order to cover up crimes the government did commit."

Over the next two hours, he stuttered and misspoke a little, lost his place and overacted as he laid out his case, not quite the smooth delivery of David Nevin. Spence neophytes were confused. Where was the master orator? But those who'd read his books or attended his seminars knew the defense attorney didn't mind stumbling now and again, for the same reason he didn't mind writing about his flaws: his drinking and carousing. Heroes were great for selling books and attracting clients, but in the guts of case building and plot telling, juries and readers alike respond to real human beings.

"Mr. Howen knows the evidence will be they didn't belong to the Aryan Nations," Spence grumbled, looking down at the prosecutor several times and shaking his head. "Yes, they called God Yahweh. They worshiped the same God, the same Jesus, but they were independent thinkers. And they wanted the truth.

"They'd heard [Idaho] was a place where people were independent thinkers. They taught the kids themselves. Taught 'em out of the *McGuffey Reader*. These children were bright and well read."

He was rolling through his rose-colored view of history by the time he got to Striker, the golden Labrador whose barking had started the shoot-out. "He was so dangerous that he could beat you to death with his tail. He and Sammy were close. Striker was his closest friend." Of course, all of the kids had cute, cuddly animals—Rachel's chicken, Rocky, and Sara's mutt, Buddy.

"They lived like ideal people," Spence said. "It was an idyllic life. If the government had left them alone, they'd still be living it today."

But the government "underestimated the moral fiber of this man. Here's what they did," Spence said, speaking for federal agents: "'If we can get a person involved in a crime, then we got 'em and they can either go to jail or the penitentiary for years and years, or they can snitch for us.' That's what they did to Randy Weaver and it's called entrapment because he had never committed a felony in his life. . . . One day, those ATF men came to him and said, 'We've got ya!'"

Spence boomed out each bit of government deceit: the informant, the trick with the broken-down pickup truck, the magistrate telling Randy he could be required to reimburse the government for his attorney's fees if he was found guilty, the probation officer giving him the wrong court date. Was it any wonder, he asked, that the Weavers distrusted the government? And how did the government respond to his fear? By demonizing the Weavers, by telling reporters they were dangerous white supremacists.

"We always demonize our enemies in every war," he said. "Japs, Russians, Germans. For the purposes of the war, we demonize them so we can kill them."

He went over all the events again with that Spence touch. "They threw rocks—these people who will tell you they didn't want to be discovered—big boulders, to see if they could get something stirred up."

The judge had warned Spence about keeping his emotions in check, but the big Wyoming attorney became more animated as he described the shoot-out and, finally, the shot that killed Vicki Weaver.

"It was a perfect shot on her head by a perfectly skilled sniper," he said. He bent over, a make-believe baby in his huge paws, a frown on his suede face. Spence rocked the baby back and forth, his fingers clutched around it. "She bled to death in a matter of seconds. They couldn't get the baby out of her arms." And then his voice caught, and Gerry Spence choked up.

It was just what prosecutors had been expecting. Kim Lindquist objected. "That better not be emotion, Your Honor."

It was a fastball, big and fat, right over the plate, and Spence turned on it quickly, throwing his most disbelieving look toward the prosecutors. "What we have here is something that is sad, and I'm sorry counsel can't feel it."

*C*ynics are well fed in a courtroom. When Gerry Spence finished his opening statements and sat down next to Randy Weaver, the scoring began. Howen was too stiff, too unemotional, even for a prosecutor. A former Marine, Lindquist was effective, but in the battle of number-two hitters, Nevin had shone. Spence's marks were mixed—the expectations were always so high—but everyone appreciated his moral indignation and his emotion near the end.

Of course Spence would argue his emotion was real. In essence, it probably was. Spence often lectured young attorneys to "just tell the truth." Juries, he said, will know if an attorney is being disingenuous. Unspoken was the idea that most important was that an attorney *believe* he was telling the truth, telling and retelling and rethinking his side of the truth until the attorney convinced himself that it was the whole truth. In that way, your clients could always be innocent, and their civil claims could always be just. Gerry Spence tried to teach young attorneys the ability to believe themselves. How else could he describe the Weavers as "ideal people . . . living an idyllic life," when even their own relatives saw them as racist and paranoid, or claim the marshals threw "boulders" to stir up trouble?

Of course, it's an attorney's job to tell one side of the story, and Spence didn't hold the patent on shading the truth. After all, prosecutors had spent part of the morning explaining the Weavers sold their house in Iowa as part of a plan to get into a gunfight nine years later.

It was ironic that in a case that began because of conspiracy theories, that's what good lawyering resembled more than anything else. Both require careful selection of certain facts, rejection of anything that doesn't fit, and the formation of a cohesive explanation of everything that—by its nature—is only part of the truth.

But most juries know their job is to decide whose story contains less bullshit. By the end of that first day of *U.S. v. Randy Weaver and Kevin Harris*, this jury already was sifting through the rhetoric and mulling the basic questions it would use to decide the case: Was Randy Weaver set up? What started the gunfight? Why was Vicki Weaver killed? And the larger questions: Who was responsible for what happened? How do we expect our government to treat those in society with antisocial beliefs?

The jurors filed out of the room and the lawyers—by nature an intelligent and insecure breed of people—felt the shooting pain of doubt: Did I reach anyone at all? By that time, it didn't matter whether Spence's wavering voice was real or not, because, for lawyers on the job, something like emotion is so closely tied to the *effect* of emotion that when the briefcases are closed at the end of the day, only one thing counts: Did it work?

For all the charges of conspiracy, all the philosophy of government power and racial hatred, for all the maneuvering on both sides, the opening statements boiled down to: Who shot first? For the next few weeks, the government would try to prove the Weavers and Kevin Harris chased the deputy marshals through the woods and shot Bill Degan in cold blood. The defense attorneys would try to show that the marshals killed Striker and then turned their guns on Samuel Weaver. Everything hinged on that shoot-out, the witnesses' credibility, the Weavers' mistrust of the government, and most importantly, the ethic, the cohesive view of the world, that Spence and Nevin were trying to get across to the jury.

*L*arry Cooper was some witness. He was poised and authoritative, a good example of what a law enforcement officer should look and sound like. On the first day of testimony, Thursday, April 15, Ron Howen led the deputy U.S. marshal through his impeccable background (Marine Corps, bachelor's degree in criminal justice, police officer, border patrol) to his decision to join the marshals service and the Special Operations Group to his friendship with Bill Degan, who Cooper said "was like a brother to me."

Cooper testified that he had trained Arthur Roderick as well and that he knew all of the deputy marshals on the August trip to Weaver's cabin except Frank Norris. Howen had him list all the gear they brought on that trip and all the weapons. All morning, Cooper described the way they got ready for the mission, how they drove up to the woods and spent several hours watching the cabin. They were getting ready to leave, he testified, when the dog charged down the hill, bringing the Weaver men down with it. He described the pell-mell retreat into the woods, spinning around to shoot the dog, deciding against it, watching Randy run away, and seeing Kevin and Samuel walk toward himself and Degan.

Bill Degan went up on his left knee with his weapon to his shoulder, pointed in the direction of Kevin Harris and Samuel Weaver, and he said, "'Stop, U.S. marshals.'" And Kevin "brought the weapon around to hip level and fired." Cooper said he fired back at Harris, who "dropped like a sack of potatoes." And then he turned his gun on Samuel Weaver. Cooper said he heard two shots—presumably the shots that killed Striker—and heard Samuel yell "You son of a bitch!" and then run away down the trail. Cooper couldn't tell if the boy had a gun. "I decided that Bill didn't fire, that Samuel had not fired at Bill, and that he was a thirteen-year-old kid and I wasn't going to shoot him."

It was a big hole in the government case. Ballistics experts would later say that a bullet from Cooper's gun hit Samuel Weaver in the back and killed him, and yet Cooper still insisted that he saw Samuel alive, running away after he'd fired the last time.

He described Degan calling out to him, said he fired another three-round burst into the woods, and ran to find his friend. Degan was lying on his side, blood in his mouth, gurgling just before he died.

Cooper's lips pursed, and he fought back tears. "I put my left two fingers on his left carotid artery and . . . I get three or four heartbeats and then it stops."

Howen immediately asked the judge if they could break for the noon recess. As they walked out of the courtroom, reporters couldn't believe Howen had blown the chance to humanize the government and use Cooper's emotion. There was little doubt what Spence would've done with a moment like that.

*C*ooper testified all afternoon, detailing Frank Norris's attempts to revive Degan, the yelling and gunfire up the hill, and the exhausting wait for help. The jury seemed captivated as he described the driving rain and the difficulty getting Degan's body down from the ridge. Rather than leave it for the defense, Howen stepped on the other big land mine in Cooper's testimony. Cooper described Jurgenson taking their rifles and counting how many cartridges were missing, and Howen asked if Cooper found anything surprising about that count.

"Yes, sir, when they told me that Degan's magazine was missing several rounds."

"Had you ever been aware of, prior to that time, that he had even fired his weapon?" Howen asked.

"No, sir. I had not."

Somebody had killed Samuel Weaver. Cooper claimed it wasn't him, but the only other person who fired enough times to kill Samuel was Degan, yet the government claimed the first shot in the gunfight was the one that hit him in the chest. So he'd have to squeeze off seven shots, one at a time, with good enough aim to hit someone running away. With a debilitating, fatal chest wound. Prosecutors hoped that such facts—damaging as they could be to their case—would just blend in with the other facts they presented and would be overshadowed by the Weavers' overall responsibility for the shoot-out. Howen asked a few more questions, including one about Degan's funeral, then told the judge he had no further questions.

After seven months of preparation, David Nevin had the first shot at a government witness, and it couldn't have been a more important one. Cooper was the only witness—other than Kevin Harris—to the death of William Degan. In fact, this was probably the most important witness against his client.

But Nevin was stewing about a stunt Howen had pulled earlier. In his opening statement, Nevin had contrasted the marshals' guns—with laser sights—against Kevin Harris's old hunting rifle and its iron sights—a post on the muzzle of the gun and a tiny ring near the breech, two pieces that are lined up to aim the gun.

At one point during his earlier testimony, Cooper changed into the full camouflage that he'd worn the morning of August 21 and held the gun he'd used, the Colt Commando 9-mm suppressed submachine gun—a powerful machine gun with a silencer. Howen asked Cooper several questions about laser sights, a device that placed a red beam right on the target. Then, with uncharacteristic drama, Howen had Cooper point the gun he'd used last August at the prosecutor's chest and turn on the device on top, the device Nevin had called a laser sight.

"Is that not what we commonly call a flashlight?" Howen asked, the plain, white light shining on his gray suit.

"It is, sir."

"It is not a laser sighting device as counsel stated in his opening statement?"

"No, sir, it's not," Cooper said.

"You can buy something like that for about thirty dollars or forty dollars at any Kmart or Fred Meyer store?"

"I believe so."

Nevin was pissed. He felt as though he'd been set up, and he wasn't about to lose credibility with the jury on something like that. When the prosecutor was finished, he leaped up and asked Cooper if he knew the FBI agent sitting at the prosecution table, Gregory Rampton.

"Did you know that Mr. Rampton led me to believe that was a laser sight on—"

Howen interrupted. "I object to counsel testifying at this time."

Nevin came right back in and asked Cooper if he knew "Mr. Howen was going to do a little demonstration with the gun pointed at him?"

"Yes, sir."

"You guys kind of planned that out before you started?"

Cooper said they had.

Nevin would later wonder if his defensive beginning was a bad idea. It was several minutes before the blood drained out of his face and he moved on, right to Frank Norris, the only one of the deputy marshals whom Cooper didn't know before the shoot-out. The attorney tried to paint Norris as an outsider among a group of men who'd worked together often.

Nevin and Cooper sparred over minor points like the law officer's insistence on calling the cabin "a compound."

"It sounds like you're using the term to imply that it's a highly fortified area designed to be defended and strategically protected, that kind of thing. Did you mean to imply that?" Nevin asked

"It certainly was well defended," he answered.

"You know about the luck that Mr. Weaver had in trying to defend that property?"

Nevin also tried to shake the picture of Cooper as an experienced law enforcement officer, impeccable in his observations and cool under pressure.

"Up here on Ruby Ridge was the first time you ever fired a gun in the course of duty, except on the shooting range, right?"

"That is correct, sir."

Nevin asked why the marshals went to a shooting range the day before the shoot-out, and why Cooper needed a 9-mm machine gun with a suppressor. No reason, Cooper answered.

"You just carried a heavier, less effective weapon by accident?"

Nevin bounced back and forth between questions about the dog and questions about the gun, pulling the two strands together carefully, gaining momentum. Wouldn't it be difficult, he asked, to carry on an undercover operation with a dog like Striker around? And wasn't the only useful thing about the 9-mm machine gun the fact that it had a silencer?

Then Nevin asked why they'd thrown rocks at the dog. Just to see what it would do, Cooper answered.

"You wanted to lure that dog up the side of that hill so you could take him out with your silenced gun."

"No, sir."

"You carried that heavy gun that was less capable up there for that purpose, didn't you?"

"No, sir."

It was the best weapon of a defense attorney. The prosecution got to just parade witnesses up on the stand to tell the story, but defense lawyers had a more difficult job, punching holes in the government testimony, then hoping the jury would connect the dots. Nevin tried to show the jury what he'd realized in the woods that day: the marshals were planning to kill Striker all along. That was why they brought the silenced gun and why they threw rocks. They weren't supposed to provoke a gunfight and they did, and that lie called into question everything else they did.

Nevin also pounded at Cooper's first interview with the FBI. In the report of the interview, called a 302, FBI agents quoted Cooper as saying he pointed his gun above the rock he was hiding behind and fired off a three-round burst, possibly the shots that killed Sammy Weaver. Cooper said he never fired blindly into the woods, that he'd never said that to FBI agents, and that his 302 didn't imply that.

During another break, Nevin had Cooper change back into his camouflage clothing, including the net mask. He had Cooper put his pack on, grab his rifle, and stand in front of the model of Ruby Ridge, in front of the jury. Cooper's blue eyes peered out at the jury from two silver-dollar-sized holes in the mask, the rest of his body completely covered in black and camouflage, a floppy canvas jungle hat perched on his head.

"Suppose you were walking through the woods," Nevin said, "and you don't have in mind to assault anybody. . . . And imagine that suddenly you encounter somebody who's dressed like you are. How do you think you'd respond to that?"

"If they weren't pointing a gun at me, I would be curious as to what was going on. I might even decide to leave the area."

But Nevin kept at him. "What if one of them was right here and shot your dog right there next to you, bang, with a gun, just like that?"

Cooper didn't lose his temper. He calmly said, "That didn't happen, Mr. Nevin."

Spence told Nevin he did a great job, but David was stuck with his usual second-guessing. He'd tried to shake Cooper on several points, but Cooper was a strong witness. He stuck by his story, and Nevin worried that he hadn't gotten across to the jury how important

Degan's seven shots were and how ridiculous it was to bring a machine gun with a silencer on such a mission.

He asked a few more questions the next day, and then Spence took over the Cooper cross-examination. He got right to the issue.

"So, the problem we have here is that the government relies on you, the defense relies on Kevin Harris. It is your word against his. That's the situation, isn't it?"

He asked how many times Cooper had gone over his version of events, how many people he'd told it to.

"You've had what, seven, eight months since this event to get your testimony straight?" Spence asked.

"Same as everyone else, sir."

Spence asked about the Y, the fork in the old logging trails where he and Roderick and Norris waited all day for help. "So you will admit you had over twelve hours to sort of get your perceptions about what took place—your stories about what took place—all lined up; isn't that true?"

"Counselor, my perceptions maybe," Cooper answered. "But if you are implying that we sat up there and talked to each other about our stories—"

Spence asked if Cooper told his story to anyone else after Degan's funeral.

"I came back, but I don't recall going into the story of what happened up there."

"The story of what happened," Spence repeated.

"The truth of what happened, Mr. Spence."

"Is it okay if we use the word story to do that?"

"I can't control what you do, Mr. Spence."

"No," the attorney said. "You can't."

Spence took every opportunity to redefine Cooper's testimony and to show the jury the worldview he was creating. Cooper described surveillance of "the compound," but in Spence's words, it became "this little residence with the chickens, the zinnias, the dogs, and the kids." Samuel Weaver became "little Sammy, whose voice hadn't even changed, much to his own embarrassment." Striker became Old Striker and in coming days, That Big

Ol' Yella Lab, and finally, Old Yeller, Who Never in the History of the World Bit Anybody.

Ah, the dog. Spence had plenty of questions about the dog. "Now the dog's big crime was that he was following you, isn't that true?"

Howen objected.

Spence tried again. "Did the dog do anything . . . that was illegal?"

Howen objected.

"Let me put it to you this way. As a member of the Special Operations Group, had you in your training been taught that because . . . you had automatic weapons and you were wearing camouflage gear, you had the right to kill somebody's dog?"

"No, sir," Cooper answered. "That has never been taught to me."

Spence just kept asking about the dog. He asked whether Cooper had a dog, whether he liked dogs, whether dogs generally chase people who run away. "Do you know of anyone in the history of the world that Striker ever bit?"

Cooper admitted he didn't.

"Did anyone ever tell you," Spence asked, "it's against the law to kill a person's dog?"

Spence got to the question that had intrigued the defense attorneys from the beginning. "As a man who has thought about this case a lot, Mr. Cooper, does it make sense to you that Officer Roderick would be shooting the dog *after* Mr. Degan is dead, *after* Mr. Degan is shot?"

Of course, Howen objected.

*O*nly two days of testimony, one witness into the trial, and Spence was already pushing the envelope. He kept trying to ask Cooper if he thought the defense's version of events made more sense than the prosecution's. Howen repeatedly objected, and Lodge sustained most of the objections. But Spence kept asking the same questions, until the judge slammed his hands on his desk and ordered the attorney to stop.

Spence immediately moved for a mistrial, and Howen objected to the motion for mistrial in front of the jury.

"What we've got here is counsel who gets to his feet and accuses me of improprieties," Spence said, pointing at Howen. "When the court sustains him, the court not only sustains the objections, but it sustains in the eyes of the jury his argument that I am improper."

Spence said it would be impossible now to put on a case. "I think I have been irreparably injured, and I ask for a mistrial."

Judge Edward Lodge was a grandfatherly man, white-haired and soft-spoken, a former football star. In Idaho, he had the reputation as a patient and wise judge who stayed on schedule and didn't mess around. He called Spence's motion preposterous and denied the bid for mistrial.

The constant bickering—Spence objecting to Howen objecting to Nevin's objection—had become comical, but Lodge wasn't amused.

"Your Honor," Howen said at one point on the second day. "In my twenty years of prosecuting, I have never had anybody object to the way I number exhibits."

Prosecutors complained that Spence wasn't asking questions. He was testifying, introducing information, and throwing around theories barely disguised as queries. And the questions he did ask, Howen said, were "inflammatory and repetitious."

Spence said the judge's rulings hamstrung him. "We can't have a fair trial if this jury believes I'm some kind of charlatan . . . breaking the rules."

Lodge said he wasn't trying to make Spence look bad; the attorney was doing that himself. "I'm not going to play games with either counsel. This has been a personality problem from day one," Lodge said in the middle of day two. "So start acting like professionals."

As the informant Ken Fadeley took the stand, some of his testimony was already lost to the prosecution. The hail of pretrial motions largely had been focused on the question of Randy Weaver's beliefs. Spence had argued eloquently that the government's plan to dredge up every unsavory deed and belief of the far right was prejudicial. And focusing the case on his religion—even if it was a racist religion—was a violation of Randy's religious freedom. "Our religion," he'd said, "our very belief in God, although probably conventional, is our own business." Howen argued that Weaver's beliefs were the very point of this case, the reason he wanted a shoot-out with federal officers. Judge Lodge tried to negotiate the middle, agreeing that some evidence would be allowed in, but some wouldn't—like the first taped meeting between Randy Weaver, Frank Kumnick, and the fictional biker, Gus Magisono, during which Randy made it clear that he didn't want to join their new criminal gang but also made it clear he had some pretty ugly beliefs.

"Call Kenneth Fadeley."

He looked like the biggest accountant at the firm, the guy leading the company softball team in homers. Bald and thick, with ruddy skin and glasses, Fadeley hadn't seen Randy Weaver since the meeting when Randy accused him of being a snitch. Fadeley had been at home last August when he heard about the shoot-out and Degan's death from an ATF agent who called him. Fadeley had set the telephone down and started crying.

Now, on April 19, the fourth day of the trial, Fadeley settled into the witness chair and began laying out his undercover work, the Aryan Nations World Congresses, the meetings with Kumnick and Weaver.

He described one of the last meetings, during which Randy seemed agitated. "He felt he was being prepared to do something dangerous for the white cause," Fadeley said. "He would give 110 percent to whatever it was he felt he was being prepared to do, and it was dangerous and something he had to do." Fadeley did everything the prosecution wanted, put Randy at the Aryan Nations, placed him in context with a breed of criminals and racists. It was strong, compelling testimony that perfectly set up the tapes.

Jurors put on wireless headphones and listened to recordings of Fadeley's telephone call to the Weaver's house, the meeting at Connies, where Weaver sold him the shotguns, and the final meeting in Fadeley's car. They heard racist small talk and Randy's demand for $300 for a double barrel gun and listened to him say, "I heard you were bad." For hours, they listened to scratchy, fuzzy meetings between the white separatist and the informant—the poor quality of the tapes just adding to the sense of shady, criminal behavior. The language was profane and racist and Fadeley was such a good witness, he apologized to the jury for his part in it.

On one taped telephone conversation, Vicki Weaver answered the telephone. Hearing her voice at the defense table, Randy began crying.

Still, the prosecution's case was as good as it was going to get. They had the strong testimony of Cooper, two credible FBI agents talking about evidence, and now Kenneth Fadeley, exposing Randy's racism and gun dealing. Putting an informant on the stand was usually problematic for the prosecution, because informants often had a criminal record themselves and were getting some sort of deal for their testimony. But Fadeley was different. Whatever reason he had for going undercover—thrills? moral indignation? money?—he was no criminal. He seemed unshakable.

At the end of the fifth day of the trial, Lodge instructed the jurors to ignore a news event that was playing itself out 2,000 miles away, in Waco, Texas. There, a poorly executed ATF raid two months earlier had dissolved into a gun battle that left ten people dead and sparked a standoff between David Koresh's apocalyptic Branch Davidian church and federal officials. Richard Rogers and his Hostage Rescue Team had responded and spent the next fifty days trying to negotiate a settlement. But that afternoon, on April 19, 1992, the FBI sent tanks to open holes in the walls of the Branch Davidian compound for tear gas, and a fire—probably set by the people inside—erupted, killing eighty people. Jurors were to ignore that case, Lodge said, and not to connect anything that was happening in Waco with the case before them.

Little did he know those two cases would be linked forever and would help fuel a movement that would eventually lead to the bombing of a federal building and the deaths of 168 more people in the name of conspiracy.

*C*huck Peterson was scared to death. Kent Spence told him that if the cross-examination of Fadeley went bad, the whole case might be lost. "If they can't believe us about the initial stuff," Kent said, "they'll never believe us when we ask them to disregard a marshal."

Gerry Spence had given Peterson a quick lesson in his trial methodology: creating a story and telling that story with every witness, with every question. "Every time you stand up, you tell the story again." Now, the old master was ready to turn Peterson loose on one of the most important witnesses in the case.

"You know," Spence said, "I've never done this before."

Oh shit, Peterson thought. I'm going to die out there. Initially, Spence was going to cross-examine all the witnesses, but Peterson had worked hard and had prepared for Fadeley, and so Gerry had offered him a crack at the informant. But now, clearly, Spence was having second thoughts. Peterson had done plenty of cross-examinations, but never in place of the best attorney in the world.

"Look," Spence said. "I'm going to be there, and we're going to get through it. Now, you're going to say, 'You lie for a living?' Right? And you're gonna say, 'That's how you make your money?' Right? And 'You like to lie?' Right?"

Peterson hadn't thought of any of those questions. "Wait, I'm here to help you," Peterson said. "You don't have to let me cross."

"No," he said. "I've got to do this."

Peterson started slowly. He called the informant Magisono. "Excuse me, it's Fadeley, isn't it?" It seemed Fadeley had several names. And there were other things that Fadeley said that weren't true. He wasn't a biker, was he?

No, Fadeley answered.

"So that was a lie?" Peterson asked. Then he asked if Fadeley really was a gun dealer.

No, Fadeley answered. That was another part of his cover.

"That would have also been a lie?" Each time Fadeley admitted lying, Peterson turned and scratched another red hash mark on the standing chart. It wasn't long before Peterson had counted thirty-one lies.

Then he turned to Fadeley's undercover work at the Aryan Nations, where he'd talked to guys like Randy and Frank Kumnick and had written down license plate numbers in the parking lot.

"I used to live there in Coeur d'Alene," Peterson said. "There's a little Free Methodist Church down the road not too far there from Hayden Lake. Did you stop there and get all the license plates, too?"

"No, sir."

Peterson asked about the Catholic church, then the Episcopalians, the Baptists, the Quakers, and the Lutherans. "Those people were not considered by you worthy of taking their license plates. They weren't a danger, were they?"

He asked questions about how poor Randy was and how often he said he was just trying to feed his family. Then he asked why only certain meetings were taped. Why, for instance, wasn't the actual meeting taped in which Randy offered to sell sawed-off shotguns? Fadeley said he didn't know before the meeting that Randy was going to suggest that.

But Peterson asked if it was because the informant didn't want any evidence of entrapment. There were several meetings without tapes or notes, Peterson pointed out.

"Didn't ATF tell you, 'Don't record those initial conversations when you're talking to this guy about the money and getting him to do the deal? You don't record those because then there'd be evidence of entrapment that could be used against us.'"

"That's wrong," Fadeley answered.

They talked about the October 11 meeting, in which Randy and Fadeley both pointed at the barrel of the shotgun that Randy was going to saw. Fadeley said that when he pointed, he was only making sure he understood where Weaver had pointed. But Peterson suggested he was showing Randy exactly where to cut the gun off.

And then they started talking about money.

Peterson asked how much Fadeley would get paid for the case.

Fadeley said he didn't know.

"It certainly would have something to do with the number of lies you told throughout the case, wouldn't it?"

"I was told I'd be paid case by case," Fadeley said later. He got expenses, and "after we concluded a case, there may be a monetary settlement possible."

"And you would assume that . . . you would have to get him convicted, right?"

"If he was guilty," Fadeley answered.

"Well, if you don't get a conviction, you don't get any money, right?"

Fadeley considered it. "I would assume so."

"And that's not just your assumption, sir," Peterson practically leaped over the podium. "If Randy Weaver gets acquitted of this gun case, you don't get paid, right?"

"I guess so."

A bonus for a conviction? Suddenly, the perfect witness, the unshakable informant, had a reason to lie. When Peterson turned around, Spence and Nevin were grinning at him.

nineteen

*B*oise, Idaho, is where flat farmland bumps up against rumpled foothills, like a shirt half-ironed. It is the state capital—sophisticated by Idaho standards—a small city of about 125,000 people, who, like the terrain, seem at the juncture of two very different places. They are the wealthiest people in Idaho, the civil folk of the Great Basin region: government workers, corporate heads, and potato farmers, largely Mormon, like their cousins to the south and east in Utah. But they are Idahoans, too, and though Boise is squarely in the southern part of the state, its residents have a touch of the mountainous Panhandle in them, a bit of the West.

For them, the three-month Randy Weaver trial played out like a baseball season—highlights at 5:00 p.m., box scores the next morning in the paper. They began encouraging Nevin and Peterson at the grocery store, and there were Spence sightings all over town. Boise is the law and order of a wild state, and many people still hoped they buried those neo-Nazis. But in Idaho, where antigovernment sentiments were more than a political fad, the details of the

Weaver case turned many people away from the federal government. "Hey," a bank executive said to Spence one day. "You guys are doing a good job."

The city is bisected by the polite Boise River and a walking-and-bicycling green belt, on which bankers jog to work and housewives bike with babies. Downtown, the choices for crossing the river are wide and patriotic—Capitol or Americana streets. Past the Egyptian Theater on Main Street, through a surprisingly funky downtown, the one-ways veer around the Capitol building to a one-hundred-year-old neighborhood, and on the edge of downtown, a seven-story glass rectangle where the rhythm of a long trial had settled, and it wasn't uncommon to see a fuzzy-haired skinhead in a black T-shirt sharing a cigarette with a deputy marshal in a dark blue blazer.

Upstairs, on the sixth floor, prosecutors worked the second week of the trial on Randy Weaver's original gun charge. Herb Byerly, the courtly southern ATF agent, testified about Randy Weaver's violation of the law and tried to clean up the mess Fadeley had accidentally left, saying that he had never promised the informant a bonus if Weaver was convicted, and that Fadeley was going to get his money whether there was a conviction or not. More ATF agents followed—Lance Hart and Barbara Anderson testifying about the broken camper, the ruse they used to arrest Weaver. The government's case was alive again.

The only thing the defense could do was hammer at Byerly about the money and his decision not to tape some meetings, and to show that Hart and Anderson's ruse had made the Weavers even more suspicious.

The failure-to-appear charge was next. Prosecutors called as a witness Stephen Ayers, the young part-time magistrate who had presided over Randy Weaver's court appearance after his arrest. His testimony was damaging to Weaver, especially a tape recording of Randy promising to show up in court and promising to get rid of his guns. But Ayers's mistake came out on Peterson's cross-examination: his suggestion that Randy might lose his land to the government to pay for his court-appointed attorney if he lost his case.

"So you agree that you misstated that law?" Peterson asked.

"In that instance, it appears I did."

Next came the officials who were in charge of Randy Weaver's case after his release. They testified that Randy's court date was changed from February 19 to February 20. His probation officer, Karl Richins, said he accidentally wrote Weaver a letter in which he said his trial was set for March 20.

When a court official showed him the mistake—more than a month after he'd mailed the letter—Richins felt sick to his stomach. "I realized what I'd done, and I was very worried." The image of a bungling bureaucracy began to emerge, in which Randy Weaver was given two court dates, February 19 and March 20, neither of which was right. On February 20, Weaver's real court date, the judge called the case without a defendant.

The jury understood that Weaver could have shown up any of those days and it would have been okay. Clearly, he wasn't going to show up, no matter what day his hearing was. But the mistakes further wounded the prosecution, and to some jurors, Randy's distrust of the government was making more sense.

Next, prosecutors presented Vicki Weaver's letters to the government—the "Queen of Babylon" letters. The U.S. attorney, Maurice Ellsworth, testified about the strange letters, which were full of confrontational language that cut away at Spence's claim that the Weavers were "ideal people."

Spence took his turn at Ellsworth, the conservative-suited bureaucrat who seemed eager to spar with the anti-establishment Spence, in his suede cowboy jacket.

"We've never met, have we?" Spence asked.

"Yes. . . . We met in Judge Williams's chambers at the commencement of this case."

"Oh, yes." Spence looked down his nose where his glasses had slid. "You didn't look so official."

"Apparently," Ellsworth said, "you are a more memorable character than I am."

"Thank you very much."

"That wasn't necessarily a compliment," Ellsworth said.

"Then I withdraw the 'thank-you'."

Spence and Ellsworth went back and forth over the letters, Ellsworth saying how worried he was by the letters, Spence arguing that it was a First Amendment issue. "If I were to call you the Queen of Babylon, it might insult you a little bit, maybe, [after all] you don't look like a queen."

"It would not be my favorite address," Ellsworth answered.

"But you would recognize that I, as a citizen of the United States, have a right to call you the Queen of Babylon if I want to, true?"

"Sure," he answered.

"The Constitution guarantees it, doesn't it?"

"With some limitation." Ellsworth said the letters were so threatening, he'd called the deputy U.S. marshals to assess the danger of whoever had written them.

Spence went over the language in the letters, much of it biblical. "Do you believe . . . we should have a threat assessment of the Bible?" he asked Ellsworth.

"No."

On April 23, the ninth day of the trial, Judge Lodge took a week break to go to a conference, and so the case was postponed until May 3. In just nine days of testimony, Spence had tested Lodge, himself an old cowboy. Lodge tried to keep order as Spence argued incessantly with prosecutors, pushed the limits of courtroom behavior, and requested three mistrials. Toward the end of the second week, he was contrite though, offering at one point to stipulate that Vicki Weaver had written the "Queen of Babylon" letters without the usual firing back and forth of arguments and briefs and clustering around the judge's desk.

"Judge," Spence said, "I am here because I want to earn and get from you some brownie points, because I might need them later. Do you know what I mean?"

"I know what you mean," Lodge said.

"Maybe the record should reflect that I should get some of those points," Lindquist joked, "for giving Mr. Spence the opportunity to ask for those points."

Lindquist and Howen needed them far more than Spence. By that time, the prosecution wasn't just fighting the defense team. It was also battling a cover-up.

*F*rom the beginning, FBI officials realized Vicki Weaver's death could cause them problems. Early reports that the family had shot at a helicopter slowed the demand for an explanation of the shooting, but now, with Howen's broad indictment, the decisions that eventually led to Vicki Weaver's death would be dragged into the open.

There was little doubt that Howen's broad indictment put the FBI in a difficult situation. When prosecutors came looking for evidence, the FBI at first refused to release a handful of documents. Among those objecting to the release was E. Michael Kahoe, the FBI official who was inexplicably put in charge of a review of the shooting, even though he had been involved in the early discussions of the rules of engagement, which stated deadly force could be used against any armed adult. Two years later, Kahoe would be suspended from his job and accused of shredding a document that showed that his boss, Larry Potts, approved the unusual rules. Also objecting to the release was the official under Potts, Danny Coulson, the man who had developed the Hostage Rescue Team. Coulson had been at FBI headquarters when the operations plan—with the rules of engagement—was faxed there. He said he never saw the rules.

When prosecutors asked for that operations plan, Coulson said it had never been approved, so it shouldn't be released to defense attorneys. He said it would disclose HRT secrets. Potts—who, at the least, had approved the *idea* of changing the rules of engagement after Rogers left Washington, D.C.—was named in another memo ordering that the documents were "not to be released." Since the standoff, Potts had been promoted to the number-two position in the FBI.

Strangely, other Weaver documents simply weren't there. A Justice Department probe would later question why there were no notes or records of early discussions about the rules of engagement, a very unlikely scenario in the paper-heavy FBI. By that time, the foot dragging wasn't over the rules themselves (there would be no way to keep those out of the trial) but who at the FBI approved them. Even people in other law enforcement agencies began to whisper about the FBI's actions and the question became Nixonian—who knew about the rules of engagement and when did they know it?

Justice Department officials who tried to help the prosecutors get past the FBI's roadblocks noted the "Bureau's intransigence appears to emanate from Larry Potts's level or above."

*H*owen and Lindquist were stuck trying to defend the FBI's actions in court—without the FBI's cooperation. In September, less than a month after the standoff, Lindquist had flown to Quantico, Virginia, to meet with Richard Rogers and to discuss the actions of the Hostage Rescue Team. When Rogers said he couldn't have the operations plan, Lindquist, the former Marine, said he'd get a court order if necessary. Rogers let him read the plan but not copy it.

In October, prosecutors and defense attorneys in the *Weaver* case agreed to a fairly open discovery policy—the amount of material they would share with opposing lawyers—and FBI officials were angry again. Howen and Lindquist requested reams of reports and other information from the FBI, much of which they were required to share with the defense team. Prosecutors received most of the material, but there were some documents the FBI at first refused to give up:

- an incident report on Horiuchi's shots, including interviews with other HRT members, diagrams, and photographs;
- the flawed review of the shooting incident report, supervised by Kahoe, in which FBI officials concluded the HRT acted lawfully and appropriately and that no administrative action was necessary;
- the operations plan;
- a critique of the U.S. Marshals Service done by the FBI, which concluded, among twelve critical findings, that it was too risky for deputy marshals to take twenty-four trips into the woods around the cabin in the six months prior to the shoot-out.

When asked for that marshals' critique, an FBI agent said that he'd rather see a mistrial than give the report to defense attorneys in discovery. Lindquist tried to explain there would be trouble if they didn't turn it over at that time and it showed up later as part of some Freedom of Information Act request for documents. The agent said the document had come from someone's desk and was not in any official file that would ever be made public. Lindquist was stunned. Was the FBI agent implying they were going to destroy the critique? He "strongly advised against that."

By early 1993, Howen and Lindquist were getting pressure from defense attorneys, who were asking for every shred of paper having anything to do with the case. But the FBI still held back. Howen wrote to the FBI and listed some of the documents they still hadn't received. "In other words," he wrote, "we want access to everything."

Under greater Justice Department pressure, the FBI finally agreed, as long as its officials could black out sensitive information. Prosecutors were never allowed to see the entire FBI files, and it was weeks still before the controversial documents arrived in Boise, the last one on April 12, 1993, one day before the trial started.

Howen was pinched in the middle, buried by requests for documents from the defense—some of which he figured were designed to just keep him busy—and stalled by defensive FBI officials.

There also were problems with the FBI laboratory, which neglected some of Howen's requests for help and moved slowly on others. Blood samples were allowed to spoil, and some of Howen's requests for technical assistance were refused or simply ignored. He asked for an expert to reconstruct the shooting at the Y and was told "there's no such thing," and so he had to find his own expert.

If that weren't enough, Howen was fighting the FBI over who would help prosecutors with their investigation. Upset by the FBI's lack of cooperation, Howen and Lindquist asked the marshals service and the ATF to help with interviews and evidence gathering. FBI officials were outraged. They were the investigative agency. It was improper for the marshals service to investigate the death of one of its own.

When a Boise FBI agent complained about the other agencies' involvement, Lindquist told him, "You guys can't work with anyone." Howen and Lindquist got the entire team together, two people from each agency, for a series of chilly meetings, but the problems only got worse. Howen and Lindquist wanted interviews in Iowa and the local FBI agents volunteered to call agents in Iowa, to conduct the interviews. Howen wanted local agents to go, and so he sent deputy marshals, in part because they didn't have to "produce paper." FBI agents were required to make written 302 reports of every interview. If those interviews could be done without creating documents, it was one less thing the government would have to turn over to the defense. When FBI agents heard the deputy marshals were conducting interviews, they were furious again.

Howen and the local FBI had been at odds for three years, since Howen refused to prosecute a local sheriff after an exhaustive FBI investigation of corruption. One agent called Howen "very pompous and condescending."

The U.S. Marshals Service and the FBI also had a relationship that was strained at times.

On the *Weaver* case, those agencies could barely work together. In the spring, an FBI agent questioned the key prosecution theory that Degan hadn't fired before he was shot. (After all, seven shells from his gun were spread over a twenty-two-foot span at the Y, showing Degan was probably moving for cover when he fired, almost impossible if he'd been shot in the chest.) One of the marshals said he didn't need to hear that and stormed out of the room.

By the spring of 1993—even before the trial—each government agency blamed the others for the Weaver fiasco. FBI agents faulted ATF for investigating the Aryan Nations, which was clearly FBI turf. They blamed the U.S. Marshals Service for the sloppy operations that preceded the gunfight and the U.S. Attorneys Office for its broad indictment. But people in those agencies believed the biggest mistake had been made by the FBI, killing Vicki Weaver. In the middle of all that, Ron Howen was trying to prepare a case.

By the time the trial arrived, some of the principal law officers involved in the *Weaver* case had already contacted private attorneys to protect themselves against prosecution or civil suits. And while two men stood trial for murder, the agencies continued to snipe at one another behind the scenes and to cover their own mistakes. A case that had soured because of impersonal bureaucracy, miscommunication, and competing agencies now stank beyond recognition. In Boise, there was so much mistrust, the FBI agents and deputy marshals kept their hotels secret from one another, afraid the other agents might find out where they were staying.

*D*ave Hunt frowned for two solid days as he went over everything he'd done to try to resolve this case peacefully. After the week break, Hunt's sincere testimony about his efforts to settle the case provided exactly what the government's case needed: someone with common sense and humanity.

It was clear the case had worn on Hunt. He was the only member of the six-man team who hadn't gotten any time off, and during his testimony he looked tired. While the other deputy marshals had returned to their home states, away from the *Weaver* case, Hunt had worked nonstop on it since the shoot-out, gathering evidence and compiling documents in a three-ring binder that quickly became two binders and then three.

Hunt testified about the meetings with Bill Grider and Alan Jeppesen and the letters back and forth, the frustration painted on his face. Just like earlier in the trial, the letters were powerful reminders of the Weavers' beliefs and Hunt's tireless efforts showed that the marshals service had tried to resolve the case peacefully. The jury seemed genuinely sorry for the sad, loping deputy marshal.

When his turn came, Spence handled the witness differently than he had others. He questioned him gently, asking if he'd been troubled since the case ended.

"It's been a difficult case. There's been some problems, yes."

Spence asked if Hunt wanted "from the depths of your heart, to settle this without bloodshed?"

"I had promised the people in North Idaho that there wouldn't be a confrontation, and I wasn't going to allow it."

Spence asked if he had spent nights lying awake.

"That's correct."

Wasn't he torn up, wondering if he should have done something differently?

"That's correct."

And, Spence asked reassuringly, "you did everything to the best of your ability?"

"That is correct. Even more than has been brought out."

"There's only so much you can do, isn't that right, Officer?"

"Yes," Hunt said.

Then Spence got to the point. Hunt needed cooperation from both sides, so that when Ron Howen refused to allow Hunt to negotiate with Weaver, it didn't help matters, did it?

"It required flexibility from both sides." Hunt said he got enough cooperation from the government but not from Randy Weaver.

By the end of Hunt's second day on the stand, Spence's gentleness was turning to sarcasm.

"Well, it must have been quite a thing for you to see this man . . . a man of his principles, holed up on that mountain, wanting to be alone, while the streets of this country were crowded with dope dealers and thugs and murderers. Wasn't that quite an interesting dilemma for you?"

"Objection."

For the next few minutes, that was the answer to most of Spence's questions, but he didn't seem to mind and he just continued asking questions that, coincidentally, showed the jury his view of the case.

Later, Hunt described a meeting in which Bill Grider threatened to kill Randy and another meeting where Grider and his son had shaven their heads.

"You said you saw that Mr. Grider had short hair?" Spence asked.

"Yes."

"And his son had short hair?"

"I believe I said a shaved head or very close cropped."

Spence looked down at the prosecution table next to the podium where he stood. "So we can understand how close, would it be any shorter or closer than Mr. Lindquist's?"

Lindquist—who was bald—nearly smiled. "As long as the record can reflect my hereditary situation."

"Shorter than Lindquist's?" Spence asked again.

"No." Hunt said it was stubble, about a quarter of an inch.

"Now you recognize that Michael Jordan for example and George Foreman and Telly Savalas all have shaved heads?"

"I guess I can recognize that."

"You didn't suggest that they open a risk assessment file on Michael Jordan, did you?"

Lindquist objected.

Spence withdrew that question. "How about Mr. Lindquist?"

Lindquist objected.

"Has Mr. Weaver ever said anything to you?" Spence asked.

From the witness chair, Dave Hunt looked at Randy Weaver sitting at the defense table, and he remembered that day in August, running down the hill toward Degan, with gunfire crashing all around him. "He shot at me," Hunt said.

"Well," the attorney rumbled, leaning back at the podium and reveling in another Spence moment. "We'll see about that later."

As soon as the government had proven its conspiracy charge, prosecutors could begin to put on more evidence of the Weavers' beliefs and their statements about starting an armed confrontation with federal officers. On May 3, nine days into the trial, prosecutors offered to the judge that many of the overt acts of the conspiracy had been proven. Lindquist made his argument in front of Lodge. "The record is now replete with overt acts at the hands of Vicki Weaver that mark her participation in the conspiracy as far as the failure to appear and the efforts of Vicki Weaver and Randy Weaver to oppose the marshals service and law enforcement generally, to bring him to justice."

"As you know," Spence said, "prosecutors never file a case any more without including a count for conspiracy." He argued that the letters were written by Vicki Weaver and not Randy, and that they were protected free speech. One was even a letter to her cousin.

So far, except for Cooper's testimony, Kevin Harris's name had only come up a few times, and with so many letters floating around, Kevin was only named in one exhibit—a phone number in Randy's wallet. Nevin had watched most of the first three weeks of trial, popping up every once in a while to remind the jury that they were seeing no evidence connected to Harris. Now he asked the judge to rule that the prosecution hadn't proven conspiracy against Harris.

The judge ruled that, so far, he saw no conspiracy by anyone.

Hunt's boss, Ron Evans, testified about the drive he and Deputy Marshal Jack Cluff took up Ruby Ridge, where they saw Sara with a pistol and Samuel with a knife and Randy yelled at them to leave.

Spence asked if he remembered Sara's gun, and Evans said he would "never forget it." Then he asked Evans if he remembered saying in his grand jury testimony that he "did not see one with her." Evans also had conflicting testimony about the dog, saying once he could see it nipping at his tires and the next time, that he couldn't see its head. Which of these stories was the truth, Spence asked.

"I believe I was telling the truth in both instances."

As the witnesses paraded through, the rancor increased between Spence and the prosecutors. At one point, Spence asked for a meeting away from the jury to complain about the government's presentation of evidence which he said was "kind of like a sideways lizard on about two legs.

"I want to cry and moan and that sort of thing for a little while, Judge, but it would be better for me to do it outside the presence of the jury," Spence said.

"But probably not as effective," Lodge shot back.

The government's case shifted to a string of people who'd come in contact with the Weavers during their eighteen-month holdout in the cabin, witnesses who testified that Vicki or the children would meet them at the base of the driveway with guns and that the family talked often about Jewish conspiracies and the evil federal government.

"He said if they came to take him, he'd die fighting," said George Torrence, one of Weaver's neighbors.

But on cross-examination by Spence, Torrence said he'd been invited in for cookies and water and that the family seemed very tight-knit, clean, and well fed. They never actually pointed any guns at him, Torrence said. They were polite, and Striker wasn't vicious, just noisy. He said Randy told him he didn't like the German kind of Nazis, but that they had some pretty good ideas.

Next, prosecutors called Randy's friend from the Aryan Nations, Rodney Willey, to testify about the family's resolve to fight the government and the long vigil in which they'd all agreed not to surrender.

But Willey—the big, furry electronics worker—was a boon to the defense. He described Vicki in such warm terms, his nose turned red, and he started crying. She was sweet and lovable, he said, a woman in charge of day-to-day life. "She had the respect of the entire family; they never doubted her authority.

"Her fear was that the government would remove her children from her presence and distribute them in the welfare system and she would never see her family again. She said she couldn't live with that."

Spence drew Willey's testimony out slowly and tenderly. Willey described Sam as a history buff who had memorized the Constitution and the names of all the presidents. He and Striker were like the boy and dog from the Disney movie *Old Yeller,* Willey said. And Sara! She was like a carbon copy of Vicki. "In the absence of Randy and Vicki, she could handle any situation."

But his most compelling description was of Vicki and Randy together. "A love story," Willey called it. "Every time I seen 'em together, they were holding hands. They often embraced each other, and they never argued or fought."

At the defense table, Kevin Harris wiped at his tears.

Howen tried to take back the momentum, asking about some of Sara's poetry, which he implied was racist. And he reminded the jury that there was someone whose name hadn't been mentioned for weeks. "You talked about a number of the family members here," Howen said to Willey. "What can you say about William Degan?" Nothing, Willey said.

Bill Grider was next—one of the strangest and most hostile prosecution witnesses ever called in any trial. When the bailiff asked if he solemnly swore to tell the truth, Grider answered, "My yeas will be yeas and my nays will be nays."

Howen started in on Grider, hoping to get him to confirm that Randy and Vicki were set on having a gun battle with federal officers. But he also hoped that by having Grider testify, the jury could understand what kind of people these separatists were, what kind of man Randy was. Grider glared at the prosecutor and seemed so unwilling to cooperate, he was nearly worthless as a witness. He said he loved Randy and never threatened to kill him. Most of his answers were so cockeyed, the jury didn't know what to make of them.

Trying to establish Grider's ties to the Aryan Nations, Howen asked if he had become associated with any groups in northern Idaho.

"Well, I played softball for Green's Cleaners."

Next were witnesses from Iowa—the old Bible Study members Shannon Brasher and Vaughn Trueman and the newspaper reporter who'd written the story about the "300-yard kill zone." Brasher and Trueman seemed embarrassed to have to testify against their old friend, and they spoke warmly about Randy and Vicki. Yeah, they said, the Weavers believed the world was going to end.

Then came former Idaho friends Terry Kinnison and Sam Wohali. Along with Ruth Rau, their testimony was more damaging—describing Weaver's racist views and wild, confrontational behavior—but defense attorneys attacked their stories, too. The government had set up the Raus' remote phone line for free. Kinnison, they pointed out, tried to get the FBI to investigate Weaver way back in 1985 and had a land dispute with him.

Another government witness was Mike Weland, the weekly newspaper reporter who'd interviewed the Weaver family. He described a spotless house and a friendly family. He said they weren't really threatening, and that he was impressed by their closeness. Also a part-time police officer,

Weland said he talked to the FBI after the standoff began and told them that Vicki Weaver was the strength of the family and that they needed to separate her and Randy to end the standoff. The next day, Vicki was killed.

Even with the trial going so well, Spence complained to Judge Lodge that the jury was being unfairly prejudiced by repetitive testimony about Weaver's beliefs.

On May 18, after twenty days of testimony and constant bickering between the lawyers, the weary judge told Spence he was wrong: "To date, about seventy-five percent of the witnesses called by the government have been favorable to the defense."

*I*t wasn't right to hang all the Weavers' guns on a Peg-Board, Spence argued. "It's a trophy board: 'Look at all the weapons we got from this poor jerk's house, he must be an evil man.'"

Instead, they brought in the guns on a handcart that creaked under the weight and then laid them out on the floor: six pistols, six rifles, and two shotguns. There were 4,500 rounds of ammunition. Two of the 30.06 rifles were loaded with armor-piercing bullets.

But this was an Idaho jury and most of the twelve were comfortable with the guns as they were passed through the jury box. Earlier, during the questioning of thirty-six potential jurors, one of the questions had been, "Do you own guns?" Nearly every hand had gone up.

Next, prosecutors played the surveillance videos of the Weaver family, hoping to show how they responded with guns to strangers and were almost always armed. With 120 hours of video, the judge allowed each side to choose some representative moments to play for the jury—the prosecution's tape showing armed children running out to the rocks and Randy walking around with a rifle, the defense tape showing kids riding their bikes and playing with dogs. When both sides agreed to two minor points—including fast-forwarding the tape past Randy Weaver urinating in the woods—the judge was shocked.

"We're getting off to an awful good start here," he said. "It really worries me, two agreements in a row."

The videotapes were shot from almost a mile away, and the people seemed ridiculously small, living their lives out on that rocky knob, tying up the dog, checking the springhouse, running around with guns. Arthur Roderick narrated the tapes—"I think that's Sara down in the lower right-hand corner; no, I think it's Rachel." They played two hours of tapes, and Kevin Harris couldn't take his eyes off the pictures of the ridge. Randy wouldn't watch at all, until the end, and even then, he turned his head away whenever Vicki or Sam was on-screen.

The attorneys watched the jury to see how they reacted to ten-year-old Rachel carrying rifles under both arms. Some of them seemed bothered. Roderick testified about how often the Weavers were seen on camera with guns, for instance, Sam—thirteen when the video was shot—carried a gun 84 percent of the time.

Spence stood up to cross-examine Roderick. "What percentage of the time did this little boy go out and shoot his BB gun?"

"I don't know," Roderick answered.

"What percentage of the time did Old Yeller wag his tail?"

"I have no idea," the witness said.

The guns and videotapes seemed effective, and Roderick's early testimony was strong. He was handsome and competent, a contrast to the bumbling of government workers who had come before. Even the defense attorneys agreed: this was a guy you wanted to believe. From the beginning, they had seen him as one of the most dangerous witnesses against them. But every day they became more confident. As the trial entered its seventh week, the defense lawyers were even ready for Art Roderick.

An unmistakable glow seeps from a trial lawyer whose case is going his way. The Weaver defense team watched witness after witness stand up and bolster their claim that the Weavers were pretty decent people set up by the government. They couldn't have put on a much better defense than the prosecution's case so far. As Nevin and Peterson grilled government witnesses and drew out testimony that further eroded the case against Weaver and Harris, Gerry Spence leaned forward on his big paws and bellowed, "Good! Good!" loud enough for the jury to hear, as if God himself were doing color commentary on the trial. The lawyers waited eagerly for their own turn at the government's weak case, like kids standing in line to beat the candy out of the piñata.

Most days, the defense team had lunch delivered to the courthouse, where they laughed, joked, and eagerly suggested new strategies, since everything else seemed to work so well. Their personalities complemented each other perfectly: the reasoned, intelligent Nevin, the bulldog Peterson, and the grand old master, Spence, who lavished compliments on his unknown counterparts, once handing Matthews a note that read, "You saved my ass," after his short, to-the-point cross-examination picked up a thread that Spence had missed in hours of wide-ranging questioning. Having worked with lawyers from all over the country, Spence said he'd take that group over any other. His son, an accomplished blues player, whipped out his harmonica and riffed during meetings as the attorneys eagerly asked, "Who's next? Who's next?" They read the witness list and jokingly fought over the chance to point out this witness's inconsistent grand jury testimony or the existence of a certain report by that witness. During recesses, they did their best Ricky Ricardo impersonations, substituting Lucy's name with that of the next witness. "Roderick? You got some e'splainin' to do."

There were pleasant surprises every day. When Tony Perez, Arthur Roderick's boss at the U.S. Marshals Service headquarters, testified about Operation Northern Exposure, Peterson asked him if there was a plan to take care of the dogs.

Perez answered yes. He explained that part of their

discussions of the dangers of the mission involved "taking the dogs out of the equation."

So, Peterson asked again, there was a plan to take out the dogs?

"The plan?" Perez asked. "Yes."

Lindquist tried to make it clear that when Perez said "take out" he meant to take the dogs out of the equation, not to necessarily *kill* Striker. "With regard to the terminology 'take out,' did you use that terminology with regard to the children?"

"Yes," Perez said, "I did. It's even in my notes."

"Did you mean take the children out, meaning to kill the children?"

"Absolutely not," Perez said. But the damage was done. There was a plan to do *something* with Striker. And it wasn't likely they were going to find the dog a new home.

As the defense team worked on its theories of the case, there emerged a sort of continuum of what they each believed. On one end was the government acting malevolently, intentionally setting Randy Weaver up from the beginning because of his beliefs, then intentionally provoking the gunfight, then intentionally killing Vicki Weaver because they knew she was the strength of the family. On the other end of the continuum was the government screwing up. It also started with an ATF setup, but then said the marshals accidentally got into a gunfight after Roderick shot the dog and the FBI sniper took a poorly advised shot at Kevin Harris and never saw Vicki Weaver. In between were other possibilities, like the idea that Horiuchi saw Vicki Weaver but didn't kill her on purpose.

Vicki's death especially divided the defense attorneys, some of whom thought it was an accident. Spence said it was intentional and argued that the jury had to believe that, but Peterson thought they would be outraged enough by the rules of engagement and by an FBI sniper who fired wildly toward a cabin filled with children. Spence listened to suggestions from all the lawyers, and even though he was clearly the boss, it was a fairly democratic system.

Working with Spence sparked creativity in the other defense attorneys. It was like watching someone juggle, Nevin said. When he was done, you didn't completely know how to do it, but you sure wanted to try. Many of Spence's methods were similar to ideas the younger defense attorneys already used. Still, they studied how he drew attention away from his client to himself. He came into court in his suede Western coats (the open pockets stuffed full of pens and notes to himself), bolo ties, cowboy boots, and a ten-gallon Stetson hat that he ceremoniously set on the defense table every day. With his plain but powerful speech, he fairly cried out to the jury: "Look at me! I'm real, just like you are!" When he lost his belt, Spence cinched his pants up for days with cheap suspenders and then told the court when he'd finally bought a belt. Juries didn't want to see some smooth lawyer, he reasoned; they wanted to see themselves.

The other defense attorneys adopted some of Spence's strict trial regimen, exercising every day and getting plenty of sleep. They even tried some of his patented bean stew, the main staple of his trial diet. The beans gave Spence a resounding flatulence, which he shared with the deputy marshals who sat behind the defense table—Spence spinning in his chair, leaning back with that half-grin, "Sorry." Even that was part of his persona. Real people fart.

Other lawyers might have great rapport with the judge or the other attorneys or the witnesses, but Spence was a jury man, always had been. Most times, his questions weren't as much to the witness as they were to the jury: "Didn't you hear that little boy say the last words that he spoke in his life, in a little high child's voice—" He usually got 75 percent of the descriptive question in before he was interrupted by the objecting prosecutors.

Spence worked behind the scenes to make sure he was getting himself across to the jury in the best possible light. "Judge," he offered one day. "May I have a second? . . . Judge, I feel pretty bad about a feeling I have that the jury thinks I'm dragging my feet or somehow delaying this trial. And Judge, the jury gets that impression. You jump on me—I'm sure I have it coming—but I was just feeling bad."

Lodge agreed that he wanted the trial to speed up. "I think the government is going to have to be more selective on their witnesses, which are wasting a lot of time. There is a tiny bit of relevance, but the defense, in my opinion, is getting ten times more out of these witnesses than the government."

"Judge, I agree with that," Spence said, rocking back on the heels of his cowboy boots. "I want to tell a story. When I was in the sixth grade, my teacher would punish other people and punish me even though I—"

Lodge cut him off and said they needed to keep moving. That's interesting, he said, "but it has nothing to do with the case."

The judge was right. Despite the defense team's glee, the most important part of the case was still to come—the shoot-out, the standoff, and Vicki Weaver's death. Even with everything that had gone wrong, prosecutors still had time to convince the jury that Kevin Harris and Randy Weaver were cold, premeditated murderers.

But while the defense team was celebrating, Lindquist and Howen were grimly overworked—still wrestling with the uncooperative FBI and with defense lawyers whose rhetoric was twisting their case beyond recognition. As the trial progressed, the FBI agents began to notice that Howen seemed to just stare off into space at times. Even Lindquist began to worry about him. He was working all the time, and for the first time Lindquist could remember, Ron Howen was making mistakes.

With the jury out of the room, David Nevin addressed Judge Lodge. "Late yesterday afternoon, after the court concluded, the prosecuting attorney Mr. Howen came and provided me something that I think is—to call it important would be an understatement. It is pivotal. I think in many ways, the whole case turns on this information."

Howen had failed to tell the defense attorneys about a witness he interviewed weeks earlier, a member of the Idaho State Police Critical Response Team named David Neal, one of the men who climbed Ruby Ridge and brought down the deputy marshals and Degan's body. As they reached the marshals that night, Neal said, the first thing Roderick told him was, "I shot the dog."

Howen said later he didn't think Roderick meant he shot the dog *first*. But he took notes of the Neal interview and decided to call him as a witness anyway, to get it all out in the open. But later, he decided not to call Neal, and he forgot to tell the defense about him until now, a month into the trial and the day before Roderick was supposed to testify about the gunfight.

Nevin said he called Neal the night before, as soon as Howen told him about the interview. "It was Captain Neal's distinct impression that Mr. Roderick told him that the first shooting which occurred, the first gunfire which occurred at the Y, was Mr. Roderick shooting the yellow dog Striker."

Not surprisingly, Spence was even more outraged by the breach in discovery. "If Mr. Roderick fired the first shot and killed the dog, the fact that the boy returned the fire and then was absolutely massacred by Cooper and Degan, shot in the back as he ran home, establishes what [Randy Weaver] stated from the very beginning."

The defense attorneys asked for a day off from testimony to go interview Neal further. At the same meeting, they said that Howen had also just given them an FBI report of an interview with Cooper in which he claimed to have seen Kevin Harris running down the road after firing at him. During his testimony a month before, Cooper had said it was Sam Weaver running down the road. Spence said they could have challenged his testimony with the FBI report. He said the writing was different in that report and another one that had been provided to the defense, and Spence suggested Cooper had written one of the reports himself.

"So what I'm trying to say is, we're being ambushed,"

Spence said. He called it "an obvious effort of the prosecution to cover up what we have claimed from the beginning was the murder of two people, and that cover-up is beginning to fall apart."

Howen said an FBI agent had just found the notes in his drawer a couple of days before and that he had just received them. "I wish, Judge, that I had known about these notes. I did not know." He offered to turn over his own notes from the Neal interview—which he wasn't required to do—and to call Neal and help set up an interview for the defense.

Lindquist stepped in. "If Mr. Howen were any less of a man of integrity than he is, counsel would have never heard about the oral statement of Mr. Neal, yet they now know about it."

Judge Lodge was angry at Spence for accusing a cover-up without any proof, but he expected that out of Spence. What he couldn't understand was a good prosecutor like Howen failing to turn over material that was clearly covered by discovery rules.

"One thing we have always been proud of in Idaho is the camaraderie between the lawyers and the ethics of the lawyers," Lodge said. "The court is very disturbed by what has happened here.

"The court has felt during this trial that there has been a lot of pressure on counsel." Lodge said that every morning there were new filings on his desk, as if the two sides were working all night. "The court wants both sides to take stock of what has happened here and make doubly sure that this does not happen the rest of the trial."

And with that, they broke for the Memorial Day weekend. But neither side got any rest. The prosecution had to try to resuscitate its case. And among other things, the defense had to resolve the Sara question.

twenty

"W ell, how do I look?"

She looked beautiful, of course, an olive green dress cinched in the back, her long black hair pulled up, except for two ringlets that were spun with baby's breath and hung down either side of her face. She looked like Vicki. Julie and Keith Brown felt a tug as if this were their own daughter getting ready for the junior prom.

Sara Weaver was going to the prom. Julie still couldn't believe it. That was how far these girls had come, how hard they'd worked to adjust. The difference had struck Julie earlier that day, as she watched her niece at the florist, looking for a boutonniere to pin on her date. Suddenly, she spotted him across the flower shop, looking at corsages, and she ducked, laughing and hiding in the greenhouse until he was gone.

Now, thirty minutes before he came to pick her up, Sara was still fidgety in that charming, teenage way. She preened in the mirror until her girlfriend showed up and decided Sara's dress was all wrong, raced home, and came back with another one—short, black, and sheer. Sara fired back up the carpeted stairs to her bedroom, changed, and came back into the living room. "Well?"

Julie and Keith sat back on the couch with the thoughts of any anxious parents before a date (Will she have fun? Is she ready? Is he nice?). Yeah, that dress looked even better. A few minutes later, her date showed up, and Sara Weaver left for the prom.

It hadn't always gone so well. Sara and Rachel arrived in Iowa angry and bitter. For the first month, they lived in Fort Dodge, on the farm with Grandpa David and Grandma Jeane. But the Jordison house was so small, and the girls needed so much attention, it just didn't work out. Randy's family helped out, but Sara and Rachel were miserable in the flat heartland, restless so far from their father. Sara stayed up late most nights, talking to supporters on the telephone and wishing she could do *something*, anything to help her father. Rachel looked for mountains to hike and trout streams to fish, but her first trip to the Des Moines River only produced a sucker, an unclean and unholy animal if ever there was one. In October, Sara dragged her younger sisters to Colorado to live near the mountains with Randy's nephew. All through their tough transition, Julie and Keith watched the girls from a distance and knew they belonged in Des Moines, with them. But Sara, especially, was so full of anger, they weren't going to force the issue. Even Randy wanted the girls to live with Keith and Julie. But Sara and Rachel knew what awaited them at the Browns' house in Des Moines. School.

They weren't in Colorado a week before Sara called Keith, who was at the recording studio he managed. They were uncomfortable with Randy's nephew and his wife, and they wanted to come home, Sara confided. "Can you come get us?"

The girls were quiet on the way back to Iowa, and Keith told them it was time to decide where they were going to live.

Rachel, who had just turned eleven, looked up at her uncle. "I kind of want to live with you and Aunt Julie."

"Well, if you live with us, you have to go to school," Keith said.

Rachel's eyes flared. "But if you go to school, you'll get AIDS."

This is not going to be easy, Keith thought.

At the Browns' house, set back from a suburban street just outside Des Moines, Julie hugged the girls and said they could stay, but they had to follow a strict routine.

"That's what I want," Sara said.

Julie and Keith talked about it privately and wondered if it would be fair on their own two girls—Emily, who was fourteen, and Kelsey, who was four. Keith was hesitant at first and pointed out there was a chance Randy would spend the rest of his life in prison and they would have the girls forever.

Almost immediately, the girls convinced him it was the right thing. Perhaps he expected angry, raving Aryan children. Instead, they were just girls, looking for someplace to belong. Keith could see how hard the girls were working to adjust to Iowa and how little their upbringing had prepared them for civilization.

The girls were in Des Moines only a few weeks when Keith took them to a nearby mall and watched Rachel ride the escalator up and down three times. She'd never seen one before. Another time, he went to an army/navy surplus store, and Sara walked along the aisles, impressing the store owner by naming all the guns.

After resenting their arrival at first, Keith now wished he could make them feel more welcome. So, he joked, the family would no longer be the Browns or Weavers. They were "the Beavers." It stuck.

The girls started school in November, horrified at first about what might happen to them, and ready to run at the first sign of brainwashing.

"I'm not going to learn anything there," Sara insisted. The first few afternoons, she came home frustrated, saying she had nothing in common with the people she was meeting. She still wouldn't see a psychologist or a counselor, and so Keith and Julie encouraged her to deal with it her own way—through poetry. She wrote simple, rhyming verse, stark poems about the mountains and about missing her brother and mom.

School immediately improved for Sara. She got *A*'s on her first report card and was student-of-the-month in February. She made quick friends—a slew of nonconformist, mostly liberal students who took her to poetry readings at Java Joe's, a downtown brick coffeehouse with eighteen-foot ceilings, burlap coffee bags hanging from the walls, and a single stool on a small stage where one night Sara stood up and read her poetry.

She was amazed at the first movie she saw—*The Last of the Mohicans*. It was incredible! She got a job at the Cineplex in the nearby mall and went to movies whenever she had the chance. Posters of actors Brad Pitt and Brandon Lee hung in her bedroom, where she listened to tapes of Aerosmith and Nirvana. Her favorite movie was *What's Eating Gilbert Grape?*, about a quirky young man who falls in love with a girl passing through his small town. The boy wants to go off with her, but he can't leave because he is saddled with a mentally retarded brother and his horribly obese mother. Sara even got to meet the author of *Gilbert Grape* at a writing conference. She began to think that's what she wanted to be—a writer.

Rachel was getting *B*'s at her elementary school and making plenty of new friends, but she seemed easily betrayed, coming home from school sometimes and proclaiming yesterday's best friend "stupid." Raised without much of anything, she hoarded candy and toys in her room and looked hurt every time she had to share something. She did talk to a school counselor, eating lunch in the counselor's office every day and refusing to go to the cafeteria. She wouldn't say why she hated the cafeteria so much.

That spring, Julie figured it out. She and Rachel were on the campus of Drake University one day when they wandered over to the cafeteria. Julie could see that Rachel was too terrified to eat there. It was the noise. Rachel had never heard anything like the echo of assembly-line eating, the clang of trays and silverware in a school cafeteria. To a girl who'd eaten every meal at the kitchen table or on a rock overlooking the forest, it was deafening.

Elisheba was a darling, active baby, with blondish hair and big eyes that tried to take in the world. Vicki had held her during every meal, and so she was a clinger, hanging on to Julie's hip. She took to calling Julie "Mama" almost from the beginning and cried whenever her aunt tried to leave the room.

"No," Julie said tenderly. "It's Aunt Julie."

By spring, the Browns were amazed by all the girls' progress and tired from their own effort. There were plenty of fights—over eating, cleaning their rooms, and the inevitable arguments over white separatism. Sara made it clear from the beginning that they wouldn't change their core beliefs, and Julie and Keith said that was fine. Conspiracy newspapers and newsletters came to the house, and Sara pored over the Bible at night and talked on the telephone with skinheads, especially David "Spider" Cooper, who'd gotten to know Sara after she came down from the cabin.

As she and Keith debated how to handle Sara's friendship with David, Julie thought often of Vicki, and she vowed not to drive Sara off the same edge where Vicki had gone. Gingerly, she and Keith tried to show Sara their lifestyle but were careful not to infringe on her beliefs. Whenever it was possible, they showed the girls that some of what Vicki had taught them simply wasn't true. Through the girls, they found out just how far Vicki's beliefs had gone—alongside the dark fear of hospitals, schools, and churches was an all-encompassing theory of the world that included not just Yahweh and the Queen of Babylon, Jews, niggers and Aryans, but also space aliens and angels.

The Weaver story died down in the mainstream media, but right-wing newspapers kept writing about it, and money came in from all over the country. Crumpled twenty-five-dollar money orders arrived in the mail alongside scribbled checks for ten bucks. Bo Gritz alone raised thousands of dollars for the girls. He and Jack

McLamb stayed in touch—Bo calling once just to try talking Keith and Julie out of putting the girls in public school. Every time the shy McLamb saw the girls, he gave them a hug and pressed a fifty-dollar bill into their hands, even though the family knew he wasn't well-off himself. So much money came in to the trust fund and Randy's defense fund (a total of $55,000 over two years), the Browns hired an attorney to manage the girls' portion, only allowing them to spend some of it on trips to see their father and other expenses related to the case. Sara wrote a thank-you note ("May our creator bless you") for each donation, until that became overwhelming and then she just wrote "Thank you" on the checks themselves, figuring they'd be sent back to the people who donated.

Once, a check for $800 arrived in the mail from the Ku Klux Klan. At Julie's urging, the old leftist Keith Brown—his hair still reaching his collar—was learning to be tolerant of intolerant beliefs. But even handling a donation from the KKK was more than he could stomach.

"We're sending it back!" he told Sara, who didn't argue.

"If you send it back," Julie reasoned, "they'll just use it to burn a cross or something." They thought about sending it to the United Negro College Fund but didn't know how the group would feel about getting a check from the KKK. The Browns didn't spend any of the donated money on themselves, but in this case, they made an exception. They were moving into a bigger house, so they decided to use the money to hire a moving truck. Keith took some solace in knowing that the KKK had paid to move an old sixties liberal and his family across town.

There were Asian families in the Brown's new neighborhood and people of all sorts of ethnic and racial backgrounds who performed at Keith's studio. One of Rachel's close friends was an Asian girl from the neighborhood. The girls seemed tolerant of everyone, and race was rarely an issue.

Anger and fear, however, were never far from the surface.

That spring, Sara couldn't take her eyes off the

television as she watched accounts of the ATF raid and the FBI siege on the Branch Davidian compound in Waco, Texas. They were doing it again! Her heart was in her throat as she watched the horrible video of the compound burning down, and she raged about the murderous government. For the first time in months, Keith and Julie saw the same angry young woman who'd first come to Iowa.

"How do I know you're even really my uncle?" Rachel asked one day, during an argument over whether or not Rachel had to mind Keith. It really seemed to worry Rachel, and she asked Julie the same question while she had her arms full of Elisheba and was trying to cook dinner after a long day at work.

Randy helped convince her. He called often from jail and was always supportive of Keith and Julie. He'd tell the girls to mind if they were being troublesome and he deferred to the Browns on decisions like school.

The girls were becoming more and more relaxed, less and less political. The radical-right newspapers usually went into the garbage unopened now, and Sara was more eager to talk about gardening or movies than she was religion or race.

When a sidewalk was poured next to the Browns' new house, Keith and Julie sent the kids out to scratch their names in the cement. Of course, Sara'd never done that before, and she didn't trace small initials in some corner of the sidewalk. Her aunt and uncle smiled when they saw that she'd written in huge, block letters: "The Beavers."

She was the perfect witness. That was Gerry Spence's side of the argument as he debated Chuck Peterson over the Sara question: Should they put her on the stand? In a way, the question mirrored the larger question of whether they should put on a defense at all.

Peterson had interviewed Sara Weaver way back in October, and her story had been so compelling and had rung so true, it helped erase any doubts he had about

Randy. But putting Sara on the stand was just too dangerous, especially when the case was already going their way. Why take a chance?

But she had such a great story to tell, Spence argued. And if anyone knew great stories, it was Gerry Spence. Who could look that cute young girl in the eye and think she was telling anything but God's own truth?

Over the Memorial Day weekend—while David Nevin interviewed the Idaho state policeman who'd been dropped in their laps—Peterson and his wife flew to Spence's house in Jackson Hole, Wyoming, a town alone in beauty and Western trendiness: the London Fog and Ralph Lauren outlets not far from Spence's garrison-style office—with its garish eagle sculpture screaming from the roof.

The girls flew to Wyoming, too, just a couple weeks after the prom. They'd been to Spence's sprawling mansion—like a lodge with its exposed timbers and stone fireplaces—and were uncomfortable being in what was essentially a huge cabin. During their first visit, they were still offended by images, and every wall seemed to hold a ceremonial mask or one of Spence's vivid paintings. The creepiest one was a self-portrait of Spence without a shirt on, spread out Christ-like, his flesh being torn by birds. Every television in the house was tuned to CNN, and the girls weren't allowed to change the channel. Finally, they were led into a high-ceilinged room with huge rocks built into the walls and a big television where they were allowed to watch a movie. Spence called it the media room. But the media room was kind of spooky itself, like a cave. Rachel entertained herself by trying to catch a mouse that was squeaking in the basement.

One morning that weekend, Spence knocked on the door to one of the guest rooms and told Peterson to get up, they had work to do. It was 5:00 a.m. For the first three hours, they kicked around ideas for closing arguments, and then they began debating whether to put Sara on the stand. They were winning, Spence argued. She would be the knockout. They could get a very compelling story in front of the jury without having their client testify. And if the prosecutors got tough with her, it would look as if they were picking on a kid. She was attractive and smart, and most of all, she was telling the truth. The truth. Spence liked that. Big deal, Peterson said. They had nothing to gain and everything to lose.

In the afternoon, they got Nevin on the phone and asked his opinion. The attorneys had all become so good at arguing with one another and playing devil's advocate, it was hard to know if everyone believed the side they were spinning. But Nevin finally agreed with Spence. They should put her on.

"You guys are nuts," Peterson said. "Bring her up here and let me cross-examine her for ten minutes."

They brought Sara up to the second-floor office, and Spence told her he wanted to hear her testimony. Sara laid out the whole terrible story—from hearing the dog bark to seeing her mother dead—just like Spence said she would. It was an unnerving story to be sure, and she wore honorably and sadly the pain her family had been put through.

Then Peterson took over, playing the part of prosecutor. "You wore a swastika as part of your everyday existence, didn't you?"

"Yes," she answered.

"You think a swastika is a pretty good thing to wear, don't you?"

"Yes."

"Did you steal a pipe from your neighbors?"

"Yes."

"Did you think it was okay to march in front of the Raus' house and call their kids niggers?"

"Yes."

After a few minutes of riddling her with harsh questions, Peterson turned to Spence. "I'm not even Gerry Spence, and I can do this," Peterson said. "Don't you think Howen or Lindquist can do this?"

In the mock witness chair, Sara wiped at her tears. Julie Brown took her out of the room, and the lawyers were quiet after she was gone. They didn't have to say anything.

They could never put her on the stand. She was just too damn honest.

"I heard a shot to my left," Arthur Roderick testified. "I knew it was a heavy caliber weapon. . . . The dog just stopped. He froze to look back where the sound was coming from."

"What happened next?" Lindquist asked.

"I shot the dog."

"Why did you shoot the dog?" the prosecutor asked.

"A couple of reasons." Roderick said the dog had looked back at him, like it might attack. "But mainly, that dog had just led Kevin Harris and Samuel Weaver a half-mile . . . to right where we were at. I was afraid if we did get into the woods and we continued down that trail, that dog would keep alerting and lead them right to us."

If anyone could pull off that story, David Nevin figured, it was Arthur Roderick. During his second stint in the witness chair, Roderick maintained the mystique of a law officer, square-jawed, handsome, intelligent, the kind of guy who inspired calm: "I'm glad he's out there protecting me."

He answered defense attorneys with such disdain, it fired from his dark eyes and dripped from his Boston accent. "That's what I said, *Counselah*."

But by May 24, Nevin was in full battle mode, reacting to whatever the government threw up there. He's lying, Nevin thought. That's all. He's just lying. Nevin's job was to prick holes in Roderick's testimony, and no matter how small the holes were, Spence would try to crawl right through them. And so Nevin asked Roderick if he realized the first shot he'd heard was the beginning of a firefight.

"Yes."

"And you paused at that time to shoot the dog?" Nevin asked, enough disbelief to show the jury something he'd contemplated during his trip to Randy Weaver's cabin: Who would pause to shoot a retreating dog in the middle of a gunfight?

There comes a time when every attorney has piled the bricks as high as they'll go and just has to ask, "Is this a house?" So, after subtly tracing around the question, Nevin got right to it—again and again. If he could show that Roderick shot the dog first, then the rest of Kevin's story fell into place—Sammy yelling "You shot my dog, you son of a bitch," and opening fire on Roderick; Degan firing to protect Roderick, and Kevin firing to protect Sammy. Self-defense.

"In fact, sir," Nevin said, "you shot the dog first, didn't you?"

"No. You have heard my testimony." Roderick was steady and calm, condescending. "No, I did not shoot the dog first."

Nevin asked about the Idaho state policeman's interview with Roderick, in which Roderick told him he shot the dog.

"I just started telling him what happened, which road was here, which road was there, and I said, 'Watch out, there's a dead dog up the trail there,' and then he asked me a question, 'How did Degan get shot?' And I told him."

"Didn't you tell Captain Neal that you shot the dog first?"

"No," Roderick said. "I didn't say that I shot the dog first. . . . I don't know how many times I've got to answer this question."

Of course, television lawyers are the only ones who ever break a witness with a direct question like that, but as Nevin sat down, he hoped he'd given the jury enough doubts about Roderick's story. Because Roderick himself wasn't about to crack.

Spence took a shot, but it quickly dissolved into an Abbott and Costello routine between a Boston cop and a country lawyer.

"You told the ladies and gentlemen of the jury that you came sneaking out there in the middle of the night—"

"I did not say anything about sneaking," Roderick said.

Glasses hung from a masking-taped strap around Spence's neck, and he set them on the bridge of his nose and looked down at the witness. "You did sneak, didn't you?"

"What do you call sneaking? You're putting words in my mouth."

"Sneak," Spence said, shuffling for a dictionary, "to go stealthily or furtively."

"Are you asking me what I said or what you said? I did not say I was sneaking. You said I'm sneaking."

"Now don't get excited."

"I'm not getting excited. I'm just explaining to you what your question was."

"You asked me, didn't you?" Spence tried. "Or have you forgotten? You asked me what I meant by sneak."

"Correct, but I did not say that."

"I didn't ask you that," Spence said.

"Yes, you did."

Spence chased the witness for a couple of hours like that, hurdling Lindquist's constant objections.

"I'm ready to go home," Spence said finally.

They had one more shot at Roderick the next day. He and Cooper had drawn diagrams of their positions at the Y, supposedly independent of each other. But when Nevin copied one of the drawings onto a seventy-five-cent transparency, then lined the two drawings up, they were nearly identical. "It must have been magic" that they drew the exact same thing independently, Spence said to Roderick on the twenty-fifth day of testimony.

Art Roderick was tired of watching his story picked over by these lawyers, who made him lay out his story over and over in different ways and then tried to find inconsistencies between the different versions. "The truth is the truth," he said finally.

Spence paused long enough for jurors to think about every discrepancy in the government's case so far. "Yes," he said finally, and then he paused some more, in case anyone had missed it. In his deepest, most solemn voice, he finished. "It is."

Frank Norris was the last deputy marshal to testify and the defense's best chance to prove Roderick and Cooper were wrong. Throughout the trial, defense attorneys had been building toward Norris's testimony, trying to paint him as an outsider among a group of old friends.

Howen led Norris through the direct testimony, again tripping his own land mine rather than leave it for the defense. What was the first shot he heard that day, Howen asked.

Norris knew it was coming. All the other deputy marshals had testified that it was a heavy-caliber shot, but they all sounded alike to Norris, and he had told the same story since the first day. He told it again now. "I thought it sounded like a two-twenty-three."

Ellie Matthews took just a few minutes for his cross-examination, establishing that Norris told FBI agents it was the "*distinctive sound* of a two-twenty-three," the guns the marshals and Samuel Weaver were carrying, and not the 30.06 that Kevin Harris used to kill Degan.

Spence was next. "So what you heard would be consistent with Mr. Roderick shooting the dog, followed by two shots, also of a two-twenty-three, consistent with Sammy, when he hollered, 'You son of a bitch! You killed my dog!' and fired two shots at Mr. Roderick. That's all consistent with what you heard, isn't it?"

Howen objected. But Spence knew the jury had gotten it. He asked only a couple dozen more questions, managing to squeeze the words "distinctive sound of a two-twenty-three" into five of them. When he was done, Spence said, "Thank you, Mr. Norris." And he meant it.

C ould it get worse? Each time it seemed Ron Howen made some progress, the muscular prosecutor was forced to apologize for evidence problems or for a witness he'd forgotten. On May 25, he had especially sour news. A bullet the government wanted to enter into evidence had apparently been picked up by an FBI agent below Ruby Ridge, then placed back in about the same place later, and photographed.

Spence and Nevin had been ridiculing the bullet—they called it the "magic bullet"—from the beginning. It was in almost-pristine condition, unlike the other bullets recovered from the scene of the shoot-out, which had slammed into trees or rocks, bending and twisting the bullets. It was an important piece of evidence because it matched Sara's .223 rifle, and the government might have tried to show that it proved other family members fired at the Y. But at least two searches of the crime scene with a metal detector had failed to turn up the bullet, and it wasn't found until the last day of the standoff. Defense attorneys argued from the beginning that it was planted. Now, they said, there was proof.

"I have been raising all kinds of unprecedented Cain in this case about the magic bullet," Spence railed. "Today we are told that the photos [of the bullet] we were given were reconstructed photos!"

More evidence came in doubt that afternoon, when an FBI agent testified that he'd mapped the scene without using triangulation—the standard method of marking evidence by measuring the distance from two fixed points. Because of the thick brush, the FBI agent had measured the distance from only one tree or one rock and could only really fix evidence from one direction.

Spence asked for precise distances between shell casings at the Y. The FBI agent said he could only give estimates.

"Nobody knows today where anything was?" Spence asked.

"Generally," the agent said, "that's the way it was."

Other FBI agents testified that evidence was lost or misplaced. They were all minor errors that might have gone unnoticed in any other case, but this wasn't any other case. In some ways, it had become The People vs. the U.S. Government. And the most critical testimony in that case was still coming. The first part had asked: Was Randy a gun dealer and neo-Nazi; the second: Who shot first; and now the trial moved into its final question: Why was Vicki Weaver killed.

*   *   *

*D*uke Smith knew. The associate director for operations in the U.S. Marshals Service, Smith had been on the jet with the FBI's Richard Rogers when the rules of engagement were devised and was in the helicopter that Lon Horiuchi said he was protecting when he shot Randy Weaver and Kevin Harris and killed Vicki.

Matthews blistered Smith as much as his laid-back personality would allow. "Isn't it true, Mr. Smith, that the so-called rules of engagement were primarily established because the United States marshal was dead?"

"No, sir."

"There was no ongoing firefight, was there, when you arrived in Idaho?"

"That seems to be the case," Smith said.

The rules established that lethal force should be used against any adult with a weapon. Matthews pointed out that the marshals' own statistics showed the Weavers almost never left the cabin without guns. "Weren't you aware that your actions virtually guaranteed snipers shooting somebody on that hill?"

Smith said the rules were just guidelines.

Spence took his turn at the piñata. "You made the rules of engagement before you even talked to the men that were there. Isn't that true?"

"That's true."

Spence asked his questions incredulously, with a how-could-you-be-so-stupid tone that burned guys like Smith, who weren't used to having their authority questioned. "After a year and a half of no confrontations, no injuries, nobody pointing a gun at anybody," Spence asked, "didn't you want to know at the time you were making your rules of engagement what had happened to precipitate a sudden explosion like this?"

Smith's biggest blow to the prosecution was his description of his ride in the helicopter, which Horiuchi said he heard just before he began firing. Smith traced on a model their low path, which took the helicopter south and west of the cabin, most likely out of rifle range and with only a quick glimpse of the cabin itself. Horiuchi had testified the helicopter was behind him, *north* of the cabin. If Smith was right, then Horiuchi was protecting a helicopter that wasn't there.

The defense attorneys asked if the helicopter had been assaulted or hindered in any way (the exact wording of the charge against Weaver and Harris). Smith had to admit that no, in his opinion, it hadn't been. In fact, he admitted, it wasn't even there.

"Did the helicopter you were riding in ever come behind those snipers over here?" Matthews asked.

"I don't believe so."

*R*ichard Rogers finished the first half of his testimony, and David Nevin leaned toward a couple of reporters and said, "I don't know about you guys, but if there's a fire, I'm following him out of here."

The commander of the FBI's Hostage Rescue Team, Rogers was also the author of the rules of engagement. He was tough and sculpted in a perfectly fitted blue suit, an Irish-looking guy aging better than most of his contemporaries, with thinning hair but a gaze that looked like it could bore holes in younger agents. He answered every question with confidence, a twenty-two-year FBI agent who knew he was in the right.

Lindquist asked him if the modified rules of engagement were a mandate for snipers to shoot any armed adult.

"Absolutely not," Rogers said. "It's giving them the tools to be able to protect themselves."

Spence took the defense's first crack at Rogers. "The people that get on these [hostage rescue] teams are people who know they are going to be trained to shoot other human beings, aren't they?"

"Well, that's certainly part of it."

"And they know when they volunteer that they may have to kill a citizen?" Spence asked.

"Or a terrorist."

"Now listen," Spence said in the challenging voice he saved for people of authority. "We're going to be here a long time. Just answer my question, and we'll get along." Spence explored his theory that the snipers had been ordered to kill Vicki Weaver. "Would you expect [your snipers] to hit somebody in the head every time at two hundred yards with a ten-power scope?"

"If they're not moving," Rogers answered. But he insisted her death was an accident and that they didn't know about it until almost a week later.

"Would there be any reason why, if Mr. Weaver was standing next to the front door of his home and screaming at the top of his lungs, that nobody could hear him cry out that his wife had been killed?"

"Well, sir, clearly it didn't happen."

Rogers was up again the next day and Spence dove right in, asking about federal agents' fears of booby traps on Ruby Ridge. "If there were booby traps," Spence said, "the children and the dog and the chickens and all the rest running around the yard must have a very good knowledge of where they are, so they [didn't] step on them."

The rules of engagement were next. "So you knew that under these rules of engagement, a true ambush had been set up by the federal government, isn't that true?"

"Of course not," Rogers answered.

"Did you use that rule of engagement at Waco?"

Finally, even the unflappable Rogers was fired up by Spence. He'd led the HRT's response to the standoff with the Branch Davidians also, and at Spence's mention of that case, Rogers flushed red. "Judge, you know, I resent the implication that man has made concerning Waco." He spun back to Spence. "Are you aware of the fact that no shots were fired at Waco?"

Ellie Matthews asked if Rogers ever considered the laws of the state of Idaho.

"In what regard, sir?"

"In establishing your rules of engagement," Matthews said.

"No, sir. I don't operate under state law. I operate under federal law, which supersedes state law." The answer went over like a bomb in a courtroom and a jury box filled with Idaho citizens.

"**M**r. Lon Horiuchi."

The muscle came in first, four buffed, flat-topped FBI agents in boxy suits, eyes shifting immediately to the conspiracy theorists in the back rows of the courtroom. There were already a half-dozen deputy marshals inside, but with the trouble between the two agencies, no one was surprised the FBI provided its own security. The burly agents had been pacing the halls of the federal building since Rogers first testified, but they were even more visible with Horiuchi's appearance. A neo-Nazi group had published a "Wanted: Dead or Alive" poster featuring Horiuchi, and the FBI was keeping him in a safe house in Boise during his testimony.

Spence objected to the FBI agents, saying the jury might interpret them as a sign that Randy and Kevin were dangerous. "They sit with Uzis under their arms. . . . Their presence is very, very conspicuous. They look mean. They don't smile. They stand around the hallway so you can't walk through the hallway."

Kevin and Randy glared openly at Horiuchi as he sat down in the witness chair. Behind Horiuchi, agents set up the heavy, two-plank door from the Weaver cabin, with its little windows on top, covered by curtains, the bullet hole visible from anywhere in the courtroom. The crude door looked strange so far from Ruby Ridge, here in the middle of a courtroom in the middle of a city. Kevin and Randy fidgeted and bounced their knees as if they might leap over the defense table at any time. The FBI sniper never seemed to look at them.

Every inch a soldier, Lon Horiuchi said "sir" every time he answered a question. He testified evenly, without a hint of emotion, that he heard a helicopter and thought Kevin Harris was moving to a position to shoot at the helicopter.

He said he was trying to kill Kevin Harris when he wounded Randy Weaver. He was also trying to kill Kevin Harris when he killed Vicki Weaver. Horiuchi pointed to a hole in the door's curtain, which corresponded with the hole in the window as proof that the curtains were pulled and that Horiuchi couldn't see Vicki Weaver when he fired.

"Did you intend to shoot her?" Lindquist asked.

"No, sir."

A lot of guns had passed through the courtroom over the previous thirty days, but none like this one. Lon Horiuchi's sniper rifle was camouflage-painted and bigger than anything they'd seen, with an extra-thick barrel and a huge scope on top, a bipod screwed to the bottom of the gun.

"This is the gun you shot Mrs. Weaver with and Mr. Weaver with and Mr. Harris with," Gerry Spence asked, "isn't it?"

"Yes, sir, it is."

Spence was almost as professional as Horiuchi as he asked the sniper about every detail of the rifle, the orders he was given, and the kind of bullets he used. There was no playful banter with this witness, fewer loaded questions.

On Horiuchi's second day on the stand, Spence asked him about the rules of engagement, specifically how they related to the dogs, which weren't to be shot unless they threatened FBI agents. The Weavers, meanwhile, could be shot just by coming outside with weapons.

"From your standpoint, the dog had better rights than the human?"

Lindquist objected.

There were no doubts about Gerry Spence's sincerity during his seething cross-examination, a string of knifing questions that left the row of reporters with dry mouths. Horiuchi sat upright, staring straight ahead, answering like a machine. Usually a blur of whispered comings and goings, the courtroom was still.

"You intended to kill both [Kevin and Randy], didn't you?"

"Sir, if they came out all at one time, we were intending

to take them all out at one time, versus waiting for one individual to come out and take him piecemeal. Our normal procedures are whenever you have more than one subject, you try to take them out one at a time."

Spence laid out his own theory. The snipers knew that the armed personnel carriers would bring the Weavers out into the open. They had plenty of documentation from the marshals about what happened when the Weavers heard a vehicle on their road. The snipers got in position and waited for the family to come out, so they could gun them down. "You testified yesterday the reason you didn't shoot them [at first] is because you hadn't expected them to come into sight, that they surprised you?" Spence asked.

"Yes, sir."

Spence cocked his head. "I take it that had you known they were going to be in that position, had you been ready, you would have shot at that point."

"Probably not, sir."

"Well, I'm confused. You tell me the reason you didn't shoot is because they surprised you."

The prosecution objected, but Spence was moving too quickly to argue the point.

"You saw somebody you identified as Kevin Harris?"

"Yes, sir."

"You see him in the courtroom?"

"Yes, sir, I do."

"That's the man you were going to kill, isn't it?"

"Yes, sir."

"You wanted to kill him, didn't you?"

"Yes, sir."

Spence had the 6-foot-2-inch Harris and the 5-foot-8-inch Weaver stand up and he asked Horiuchi how he could mistake them. He asked if they were told that Vicki Weaver was more zealous than her husband. And then he asked about the second shot. "Just before you shot, you knew the door was open, didn't you?"

"At that time, yes sir."

"Didn't you know that there was a possibility of someone being behind the door?" Spence asked.

"There may have been, yes, sir."

"You shot twice and both times you made mistakes; is that correct?"

Lindquist objected and again, Spence moved on without much argument. He considered asking Horiuchi if he was sorry for what had happened, figuring he'd give some answer like, "Sir, I am not trained to feel sorry." Then again, he might just say yes.

"You heard a woman screaming after your last shot?" Spence asked.

"Yes, sir, I did."

"That screaming went on for thirty seconds?"

"About thirty seconds, yes, sir."

"I want us to just take thirty seconds, now pretend in our mind's eye that we can hear the screaming—" Spence was quiet and whether they wanted to or not, everyone in the courtroom watched the plodding second hand on the wall clock straight across from the jury, one, two. . . . With the door behind the sniper, a jagged hole in its window, the jury was as close as it would get to the horror of Rachel and Sara Weaver, who screamed and screamed until they were out of breath while their mother lay dead on the kitchen floor.

*I*n the next year, FBI officials reviewing the case would decide no one needed to be fired or prosecuted, in part because Lon Horiuchi wasn't actually following the modified rules of engagement. Since he firmly believed that a helicopter was in danger, FBI brass said, Horiuchi was in fact following the bureau's standard rules—shooting to protect someone else.

Nevin proposed to the sniper that there was no threat to the helicopter and Horiuchi said yes, there was. "But you were waiting to kill them irrespective of a threat, weren't you?" Nevin asked.

"Based on the rules, sir, we could."

The defense attorneys challenged virtually every part of Horiuchi's story, Nevin hauling the door out, standing behind it where Kevin would have to have been to be struck in the arm and chest by the gunshot. Earlier, Horiuchi said he'd seen Kevin flinch. Nevin wondered how Horiuchi could have seen anyone flinch behind the door if the curtains were indeed drawn.

"Mr. Horiuchi, I'm going to put my arm right next to the door," Nevin said. "Would you do me a favor? Would you say bang?"

"Bang."

Nevin shook. "Did I flinch?"

"I can't tell, sir."

"You can't see me, can you?"

"No, sir."

Lon Horiuchi finished testifying the afternoon of June 4, on the thirty-second day of the trial, almost two months after it had begun. On the way, he paused and looked over the shoulder of a courtroom sketch artist who had drawn his steady, mechanical face. Later, Horiuchi asked if he could get a copy.

Ron Howen walked back to his office and slumped at his desk. A package was waiting for him. There was a cover letter from an FBI agent saying the package was in response to an April 13 subpoena for any records generated by the shooting review team. The package contained some documents that already had been turned over, but also notes of interviews with FBI agents that had *never* been given to the defense.

Most troubling were two crude drawings by Horiuchi. Scribbled on a hotel notepad, they showed a stick figure approaching a door, a dot on his chest, crosshairs on the door. In one of the windows of the door were two semicircles— two heads. Howen set the drawings down on his desk. Horiuchi had just finished testifying, and now the prosecutor was going to go into court and hand over notes and drawings that showed two heads where the sniper reported seeing none. Of all evidence problems and delays, Howen knew this was the worst. It was the low point of his career.

He checked the date on the cover letter. It had been stamped in the FBI mail room May 21, 1993. Two weeks earlier. The FBI had mailed this vital package fourth class.

twenty-one

D avid Nevin had gone over the scenario so many times in his mind that he could practically hear it. "Defense calls Kevin Harris." The jury leans forward in their chairs as Kevin walks past the judge and sits down in the witness chair, looks at the jury with his perfectly guileless face, and waits. No lawyer games. Nevin just asks him, "Kevin, what happened?"

No, he couldn't do it. Howen would eat him alive.

But Kevin was such a good kid! They would *have* to see that. Nevin had looked that jury over, face by face, and he knew that if they could just hear Kevin's story in Kevin's own words—instead of through some self-conscious attorney or through the piecemeal impeachment of government witnesses—they would realize he was telling the truth. He'd given the same story ever since the FBI interviewed him in his hospital bed, never once deviating or denying that he'd shot Bill Degan. Kevin would sit down in that chair and say what he'd said from the beginning: "I didn't want to shoot that man."

The decision had been looming there throughout the

trial. During every crisis over this witness's testimony or that piece of evidence, every split-second reaction to the government's case, the larger questions were there, haunting him: Should he put on a case? Should he let Kevin testify? Kevin didn't take much of an interest in the strategy of his defense, but he made it clear he wanted to testify.

The knowledge that somebody's life rests in your hands can paralyze an attorney, and Nevin had done enough death penalty cases to know what that fear felt like at its dreariest. But life in prison was often a victory in death penalty cases. With Kevin, freedom was the only acceptable answer. In some ways, that put more pressure on Nevin. This kid's life was at stake. Shouldn't he be allowed to testify?

So he ran it past Spence, who turned down the corners of his wind-weathered mouth and shook his head. "You can't put Kevin on."

That convinced him. He had to put Kevin on.

Maybe it was just his own personality; maybe David Nevin was too self-tortured to completely trust *anyone*. After all, Spence had been nothing but generous to him. (Perhaps too generous? Stop it, he told himself.) But from Spence's first phone call—"David? I hear you're a great attorney."—Nevin had played a game with himself, allowing somewhere in the back of his mind for the absolute worst and then half expecting it. He worried about being so completely overshadowed by Spence that the jury forgot Kevin Harris even had a lawyer. That hadn't happened, and as the trial wound down, he knew he'd held his own. He worried that Spence would be impossible to work with, like some legal rock star, trashing motel rooms, demanding bourbon and women after every cross-examination, dismissing the rest of the team like some studio band. That hadn't happened either. Spence had been gracious and complimentary, a great attorney to work with.

So maybe Spence wasn't just outfoxing the prosecutor but was outfoxing Nevin, too. Maybe Kevin's testimony would somehow hurt Randy's case. Worn down from months of preparation and weeks of grueling trial, Nevin imagined Spence as some Machiavellian figure who had it all wired and who knew the jury needed to punish someone and that someone would be Kevin. That was the worst thing Nevin could imagine: Randy walking out that door and the jury nailing the quiet kid who wouldn't even have been there if it hadn't been for Weaver.

He knew that Spence and Peterson were leaning toward resting their case without calling any witnesses. It made sense for them. They'd presented a pattern of government bumbling and misconduct against Randy. They had a paid informant, screwups on Randy's court date, and a parade of witnesses who said Randy and Vicki were great people. And, when it came right down to it, their guy didn't shoot anybody.

It was harder for Nevin. He still had the shooter. He thought he'd convinced the jury that the deputy marshals' version of the shoot-out made no sense, but at the same time, who knew? Maybe they would get back in deliberations and say it didn't matter who shot first. They wanted to punish someone. Kevin.

Nevin wasn't some inexperienced hack. He was one of the top defense attorneys in Idaho. But Spence was something else—the best in the world. Who knew what went on in a mind like that?

Spence came over for dinner a few times during the trial and hunched down on the couch with Nevin's two sons, playing video games. He sidled up to Nevin at one point and said he wanted to confide something. "The big boy needs some attention." Spence smiled. "The little one's such a star." And the thing was, he was right!

But the more generous Spence became, the more Nevin played devil's advocate with himself. For weeks, all six attorneys had debated whether or not to put on a case. It came up at meetings, over lunch, while jogging, and the lawyers routinely jumped from one side of the argument to the other. Nevin decided to convince Gerry that Kevin needed to testify—in part to see how Spence reacted. He explained that Kevin's story would be unimpeachable because it had never changed.

"You're right," Spence said finally. "You should put him on."

What? Nevin was tortured. Perhaps it was all in his mind, but why the sudden switch, why did Spence now say it was a good idea to put Kevin on? What the hell did that mean? Maybe Spence only advised him not to put Kevin on so he then would change his mind and put him on. That would get Randy's story out there without having to put Randy on the stand. It was so perfect!

"I can't put him on," Nevin said. They'd done a good enough job cutting away at the government case that he didn't need to do it.

"Right," Spence said. "You can't put him on."

There is a point at which a lawyer must trust his client. Tougher, Nevin realized, was the point at which he trusted another lawyer.

He remembered watching Spence screw up a cross-examination once. He settled in behind his big Stetson, a funny curl on his mouth, and Nevin said to himself, "What the hell was that? That's the best attorney in the world?" He couldn't imagine what that pressure must be like. Spence's power wasn't in his trial strategy or his craftiness; he wasn't smarter than the other lawyers. He worked hard at those things, but his real strength was those moments of clarity when he spoke so intuitively and powerfully to a jury that he cut through all the posturing and legal language. The questions of guilt and process fell away, and his words sounded suspiciously like truth.

Consciously, Nevin had always liked Spence. Now, by the end of the trial, he even convinced his subconscious. Still, that didn't help with his decision.

His taciturn co-counsel Ellie Matthews was no help either. He listened to Nevin's back-and-forth ravings as if the young attorney were speaking a lost Chinese dialect. "Just make a decision," Matthews said.

*   *   *

*J*udge Lodge was furious about the Horiuchi drawings and interview notes. He fined the government for one day of the defense team's costs, even sixty dollars an hour for the Spences and Garry Gilman, who were working for free. "It does appear that it is somewhat of a pattern on the part of people, agencies outside of the District of Idaho," Lodge said. "It seems to be totally inexcusable . . . to send something like this fourth- class mail when a trial of this nature is going on, with the cost of time and human tragedy that is involved." Lindquist wondered if someone at the FBI was doing this on purpose.

They dragged Horiuchi back from Washington, D.C., put him on the stand, and asked him about the two heads he'd drawn in the Weaver's window. Horiuchi said they represented Randy and Sara Weaver, who were running ahead of Harris into the cabin. The sniper said he added the heads because the FBI agent who interviewed him asked where he estimated Randy and Sara were when he fired. He insisted the curtains were drawn and that he couldn't see them or Vicki Weaver.

Spence pointed out that in his drawing, Horiuchi's crosshairs weren't on Kevin Harris, but were on the window of the door. "Let's assume . . . the curtain at the bottom of the window was open approximately seven or eight inches," Spence asked. You could have seen the movement of those two heads across there, couldn't you?"

"If the curtain was open, yes, sir."

"You could have seen Vicki Weaver standing there, couldn't you?"

"Yes, sir, if the curtain was open."

"You did see her, didn't you?"

"No, sir, I did not."

Nevin asked why, if Horiuchi's drawing was to estimate where Randy and Sara were, he drew only the tops of their heads in the window. If he couldn't see through the window and he couldn't see through the door and he was trying to illustrate where they might be, wouldn't he draw their whole bodies behind the door? Unless of course he did see those two heads in the doorway.

"No, sir."

All that was left was the science. Ballistics and forensics experts were often the spine of a prosecutor's case, the place where the other evidence could be connected to the government's theory. But since the FBI laboratory had virtually ignored the case, prosecutors were forced to go out and find their own experts to examine the evidence. And if there was ever a case that needed experts, it was this one.

Seven weeks into the trial, there was still no consensus about what had happened during the shoot-out with the marshals. For nearly a year, the news media had described it as an Old West shoot-out, but this was no quick draw on opposite ends of a corral, good guys versus bad guys. It was a confusing blur of shots, overlapping positions, and wild firing.

The defense team's story remained the same: Roderick shoots Striker. Sammy gets mad, fires at Roderick, and misses. Degan shoots Sammy in the arm, and Sammy turns to run home. Kevin shoots Degan to protect Sammy. And Cooper shoots Sammy in the back.

Government witnesses told a less concrete story: Degan rises on one knee, identifies himself, and tells Sammy and Kevin to freeze. Kevin wheels and kills Degan. Nearby, Roderick shoots the dog. Cooper sees Degan go down, fires at Kevin, and mistakenly thinks he hits him. Degan fires seven shots after he's hit, one of which hits Sammy in the arm. Then, somehow, Sammy is shot and killed as he runs away. Ballistics evidence clearly pointed to Cooper, but he claimed to have seen Sammy running away after he finished shooting. Since the bullet passed through Sammy and was never found, it was impossible to say for certain.

Dr. Martin Fackler, a surgeon and wound ballistics expert, bolstered the government's version of the gunfight, testifying that Degan *could* have fired his gun seven times even after being hit in the chest by the bullet that eventually killed him. He said Degan probably could continue fighting for three full minutes. "Even if the blood had stopped to the heart," Fackler said, "he still could have emptied a whole magazine of ammunition." Since Degan was a trained marksman, and only one of his shots had hit Samuel Weaver (in the arm) from fairly close range, Fackler concluded that Degan was probably injured before he fired the seven shots. After two months of claiming Harris shot first, the prosecution was finally explaining how it could be possible.

It was a ridiculous theory, defense attorneys said, assuming that because Degan had missed, he'd already been shot. Peterson asked why Fackler hadn't interviewed Cooper or talked to the pathologist in the case or even visited the scene of the shoot-out. He implied Fackler was a hired gun, an expert called in to simply agree with the prosecution's version of events.

During one of the breaks, Kent Spence had been turning Degan's backpack over in his hands when he discovered what looked like another bullet hole that government experts had missed, a shot that appeared to have come from behind Degan and went clean through the backpack without hitting him. The defense attorneys suggested that it was friendly fire and that it called into question the shot that killed Degan, too. Maybe, they surmised, Kevin only *thought* he killed Degan. The jury didn't buy that, but the new bullet hole added to the picture of government bungling and couldn't hurt in the reasonable doubt department.

Fackler also testified that the shot that killed Degan severed his axial nerve, which would paralyze the upper left side of his body. Peterson began a series of demonstrations, with Fackler assuming Degan's position on the floor and Peterson strapping on Degan's backpack and walking around with a pointer coming out of the bullet hole. They talked about directionality, cavitation, and bullet trajectory, a technical blur of information that led to one point: How could Degan squeeze off seven shots, one at a time, traveling over a distance of twenty-two feet, after being shot in the chest and having his upper left arm deadened? After all, his body had been found with his left arm still in the gun strap, defense attorneys reminded, so he couldn't have held the gun up with just his right hand.

Peterson asked if it was possible Degan fired one shot *before* he was hit. Two? Three? Four? Seven? Each time Fackler said it was possible.

Fackler's testimony contradicted Larry Cooper's version of events. Cooper insisted he saw Sammy running away after he had fired his last shot. Yet Fackler testified that ballistics evidence pointed to Cooper's bullet as being the one that tore through Sammy's back and killed him.

Defense attorneys especially attacked Fackler's contention that Degan must have been wounded before he fired a shot, since he missed Samuel. After all, how did Fackler know who Degan was firing at? If Degan was shot by Kevin Harris, why didn't he shoot at Harris instead of turning his gun on a fourteen-year-old boy? Couldn't he have just been firing into the woods to cover his colleagues?

"Have you been told that Mr. Cooper claims he fired point-blank . . . at Kevin Harris?" Nevin asked.

"No," Fackler admitted. "I have not been told that."

When Cooper fired at Harris, he was closer than Degan was to Samuel Weaver. And yet Cooper missed with three shots. "Would that change your opinion as to whether or not that was an easy shot by Mr. Degan?"

Lindquist objected. But the point was made. In northern Idaho that weekend, Cooper missed at least three times, and Horiuchi missed twice, and the government claimed the two people it did kill were by accident. Defense attorneys proposed that missing with six shots wasn't anything special for a government agent on Ruby Ridge.

After two days of complex, medical testimony, Fackler's opinion seemed to boil down to this: It could have happened this way; it could have happened that way.

The next witness, a private ballistics expert named

Lucien Haag, told the same story. Again, his testimony in some respects contradicted Cooper's, the government's only witness to the shoot-out. Haag concluded that according to the ballistics evidence, the government's version of the gunfight was only the most likely of several possible scenarios.

And that was it. After thirty-six days of testimony by fifty-four witnesses, Ron Howen stood and said the government's case was finished.

*T*he attorneys gathered in a scrum around Judge Lodge's desk. "Your Honor," Spence said, "we're not gonna put a case on. We want to go ahead and rest right now, and then we can argue our motions later."

Lodge didn't seem surprised. In the back of the huddle of lawyers, Nevin cleared his throat. Perhaps he'd known all along that he wouldn't put Kevin on the stand. Now his mind games about Spence seemed pretty silly, and he realized that if they won this case, it would be because they all hung together, refusing to allow the case to be defined by the prosecution, redefining it themselves, creating one ethic, and presenting it to the jury; they were a team in the sense that attorneys rarely get to experience. The government hadn't proved a damn thing. Spence had rested cases before, most notably, during his defense of Imelda Marcos. Nevin had never done it before, but now he was ready. He reasoned that if he did put on a case, the jury would just choose between the two cases—who had done a better job, him or the prosecutors? But if he simply rested, the jury would be forced to consider whether or not the government had proved the charges beyond a reasonable doubt. Nevin believed strongly that was the most important principle in our legal system, and so he felt confident when he said, "We're not gonna put on a case either."

Lodge looked incredulously at Nevin. "You're not putting on a case *either?*"

Nevin's mouth went dry. Oh my God, he thought, what have I done? Even the judge thinks I should put on a case.

Back at the defense table, Nevin tried to smile at Kevin Harris. The judge asked for the defense to begin its case, and Nevin stood. "In view of the evidence that has been presented, Mr. Harris waives his right to present evidence." It was dead quiet in that courtroom. And then there was this buzz, as people in the gallery behind him whispered. Out of the corner of his eye, Nevin saw a juror's mouth fall open and saw others stare at him, confused. In the gallery, Kevin Harris's mother and girlfriend began crying. That's it, Nevin thought, I've killed this poor kid.

Spence allowed just the right pause before standing up. In his deep voice, the words sounded like an official proclamation. "In view of the evidence that has been presented and the evidence that has not been presented, the defendant Mr. Weaver also waives his right to present any defense and rests at this time."

Kim Lindquist had taken on as much of the responsibility for the case as he could, and still Howen was working himself to the point of exhaustion. He'd labored for ten months nonstop on preparation and on the trial itself. He was pulled apart by what Lindquist saw as the defense team's unrealistic discovery requests and the FBI's reluctance to turn over documents. Howen wasn't eating well and had stopped working out at the gym. Earlier in the case, he was getting by on just a few hours of sleep each night, and now it seemed as if he never slept. It was no secret around the U.S. Attorneys Office that all the time he'd devoted to the Weaver case had caused problems with his family life, too. There was concern about the effect of the stress on Howen. Lindquist hated seeing such a good, strong prosecutor undone by the pressure in this case.

At midnight on June 11, Howen called his best friend and talked about the grueling trial. He said he'd probably be working until 5:00 a.m. that morning, preparing to fight the defense team's argument that all the charges should be dismissed.

In court that morning, defense attorneys argued that the government hadn't proven its case at all.

"No one claims to have seen Kevin Harris shooting except for Mr. Cooper," Nevin argued. "And everything that Mr. Cooper says about the circumstances that he claims occurred there have been shown by the government's own witnesses to be wrong."

After a ten-month apprenticeship, Peterson railed against the conspiracy charge with Spence-like overstatement. "This is perhaps the newest low in the history of American jurisprudence . . . sinking to the level now of prosecuting families."

Howen argued that the charges had been proven, beginning with the first count, the conspiracy indictment he'd so carefully crafted. As he spoke, his left hand shook. "We presented evidence that there was a prediction and a prophecy going back to 1983 in Iowa that there was going to be this violent confrontation with federal law enforcement agents who were deemed by the defendant and his wife as being Satanic or Luciferian."

Howen went through each of the charges, getting to number seven: that Kevin Harris had harbored a fugitive, Randy Weaver.

"And the actions Mr. Harris took in terms of the harboring charge involved him making security patrols with a firearm, transporting food and supplies and mail up to the residence. He took up . . . residence with them . . . and . . ."

Howen paused, shuffled through his notes and looked over at Lindquist, who smiled reassuringly, raising his eyebrows. Howen looked back at his notes and swallowed.

"I'm sorry, Judge," he said. "I can't continue." He sat down, his hands pressed together between his legs. Deputy marshals and local FBI agents patted him on the back, and Lodge called a recess.

When the lawyers came back, Howen was gone. The judge asked Lindquist if he wanted to finish Howen's argument, but he said no. Then Lodge dismissed one count against Weaver: being a felon in possession of a firearm and one count against both Weaver and Harris: threatening to shoot at a helicopter.

Ten months earlier, Lon Horiuchi had first testified before the grand jury that he accidentally killed Vicki Weaver while trying to protect an FBI helicopter.

June 15, 1993. Closing arguments. Rachel Weaver rested her head on her aunt Julie's shoulder, lifted her hand to her chest, and flashed a small wave to her father, who waved back. Sara and her dad made eye contact, and he gave her a wide smile and a slow, easy nod. The girls looked away from Randy to the big, handmade door taken from their cabin, which leaned against a wall where it had been placed as evidence just a few days before.

The courtroom was packed with friends, relatives, reporters, lawyers, and a growing number of people who now saw the *Weaver* case as representative of federal law enforcement abuse. Federal law officers were there also, including Deputy Marshal Dave Hunt, who joined Lindquist at the prosecution table where Ron Howen was conspicuously absent, another person worn down by the *Weaver* case. Behind them, the crowd spilled out into the hallway and into an adjacent courtroom, where sound was piped in for those who didn't get into the main courtroom. "I think half the lawyers in Boise are here," one attorney said.

Most had come to hear Spence. He stood up at the defense table and scanned the courtroom, going through the ritual that he repeated every time he argued for the last time in a case. Basically, it consisted of this: Gerry Spence got scared out of his wits.

He'd fought it when he was younger and tried to hide it from the jury. But they knew. And he'd become a better lawyer when he admitted that fear to himself and, finally, to the jury. Now, it was part of his method. During a long trial, he tried to force the jury to make a decision—not about the evidence so much but—about him. And so if they found against his client, then he had to face the truth that *he'd* been rejected and that an ideal that he had believed and represented to the jury had been called a lie. There was no match for the fear of failure that haunted a lawyer who never lost.

At the defense table, the other attorneys shook hands and whispered confidently to Kevin and Randy. Peterson straightened Kent Spence's tie, and Nevin watched the jury to see if they still hated him for resting Kevin Harris's case.

"Due to unforeseen circumstances," Kim Lindquist said, "Mr. Howen is not with us." Lindquist spread his notes over the podium and ended the way the case had begun, with a reasoned explanation of charges that had lost much of their power since the first week of the trial. But Lindquist brought fire to his closing argument and reminded the jury that a deputy U.S. marshal was dead. To Gerry Spence's horror, Lindquist, the tight-lipped ex-Marine, stood up there and told a story.

"This trial has a theme," Lindquist said. "This is the story of . . . two people, who had the purpose or the resolve to defy the law and then resist the enforcement of that law with violence. Every aspect of this trial is reflected in that theme."

Lindquist blamed it all on the Weavers' Old Testament, white separatist religion. "They believed they were dealing with Satan himself. It became the center of their lives. It became a self-fulfilling prophecy to legitimize their beliefs. If there was no persecution, the core of their religion would have been false, a fallacy. Ridiculous."

Lindquist wrote hatred in red letters on the evidence pad. He talked about the old Bible study group and the newspaper story with the "300-yard kill zone." "That's Iowa," Lindquist said, "and the conspiracy has begun." He held up the shotguns, spoke about the ATF investigation and the attempt to make Weaver an informant.

"Randy Weaver told them to pound sand. He'd rather wait until they came to get him—Hear my words—he'd rather wait until they came to get him, at which time he'd resist with violence. Because that was his resolve." He spoke about the failure to appear, the Queen of Babylon letters and the U.S. Marshals Service's attempts to negotiate a settlement.

"If any of these marshals had gone up there . . . and said, 'We are U.S. marshals and we're here to arrest you,' you know what would've happened. . . . The suggestion has been made they were there to ambush the Weavers. [Why would] they go up there without bulletproof vests . . . and leave the sniper rifle behind?"

He talked about the shoot-out and said, "Kevin Harris murdered Bill Degan while Bill Degan was in the performance of his official duties."

The defense attorneys, he said, used "histrionics and false expression of emotion." Vicki's death was a tragic accident, he said. "I've heard speculation; I've heard sarcasm; and I've heard lots of cynicism. But statements from lawyers are not evidence."

And then Lindquist returned to his theme, the Weavers' resolve to have a confrontation with the government. "Why is Vicki Weaver dead? Because of that resolve. Why is Bill Degan dead? Why is Sammy Weaver dead? Because of that resolve."

*D*avid Nevin told the jury he'd seen the surprise in their faces when he rested his case. He tried to explain to them that he didn't need to put on a defense if the government didn't prove its case. "The government's case against Kevin Harris is false," he said. "The government set out to prove Kevin Harris wheeled and fired and killed Bill Degan for no reason.

"They failed for one very good reason. It's not true."

Nevin said the conspiracy charge was "a preposterous contention."

"If Mr. Weaver had wanted to have a shoot-out with the U.S. government, he could have come down any day and done it." For a couple of hours, Nevin clicked off the inconsistencies in the government's story, the witnesses who changed their minds and the physical evidence that didn't mesh with the marshals' versions of events. He sketched for them Kevin's story and the marshals' stories and asked which one made more sense. "I guess you can boil it all down to this: Cooper says he didn't shoot Sammy, and he did. He says he did shoot Kevin Harris, and he didn't."

Nevin said he also knew they had the choice of convicting Kevin of manslaughter or second-degree murder, and he urged them to go all the way, to rule that he'd acted in self-defense and that he wasn't guilty. Nevin closed with a quotation from George Washington: "Government is not reason, it's force. Like fire, it's a dangerous servant of a fearful master."

Throughout the trial, some jurors had been more impressed with Nevin's measured, thoughtful arguments than with any of Spence's performances. But after Nevin finished, every eye went to the old country lawyer, who was busy conjuring the fear that he used like adrenaline.

Nevin had given the jury the reasons to acquit, and now Spence wanted to give them the desire.

Lindquist's closing had been about five times better than Spence figured it would be. He'd actually given the jury a way to convict in this case: Ignore the contradictory evidence and find these people guilty because you don't like them. Spence wondered all over again if he still had it, if he could reach this jury and make them see what he saw, see what he *believed*.

A friend could tell Spence was nervous, and she called him over to her. "Come over here and let me tell you a joke."

"Don't tell me a joke," Gerry Spence said, "tell me how to be real."

*H*e shook Randy's hand, shook the hands of the defense team, and—as he always did—let the jury in on his fear as if it were some secret. "I've been at this for over forty years, and I never begin any case the way I feel right now. . . . I think to myself, can I do what I need to do now? . . . I need to be the best lawyer I can be in the next two hours and thirty-five minutes."

And then he talked about the jury. Two jurors were in their seventies, three in their sixties, and only one was younger than forty-three. Spence said he hoped he didn't offend them, but he guessed their average age was forty. "You may be the most important jury that's come along in a decade," he said. "This is a watershed case . . . a case kids in law school are going to read about."

During the trial, Judge Lodge had ruled that Spence had to remain in one place, something the lawyer called "a spastic embrace" with the podium. Now Spence was allowed to move around the courtroom and he ran like an unleashed dog. He walked over and knelt next to Randy, looking into his eyes. "Randy, I'll tell you what you're guilty of. You're guilty of being one stubborn mother. You are guilty of being afraid." He looked up at the jury. "And aren't we all guilty of being afraid?"

Spence threw everything he could think of into his argument. This was no time to be subtle. He accused the government of a cover-up and spun out new theories—Vicki was murdered because they thought she was a witness at the roadblock. Degan's own men shot him. He told a story about a talking swan and another about a boy who crushes a bird. Federal agents were "the Waco boys" and "the new Gestapo in America." He introduced his wife Imaging and had her stand, introduced Weaver's daughters and had them stand twice ("I want him to walk out and be free with his little children."), clapped his hands in front of Lindquist's face and yelled, "Wake up!" at the prosecutor.

Like a man stumbling around in a dark room, Spence's argument was all over the place until—as always—he found the switch and turned on the lights.

His voice boomed through the courtroom. "Marshals aren't supposed to shoot little boys in the back!" Sara shuddered. "A little boy whose voice hadn't even changed!" Rachel sat up straight and squeezed her aunt's hand. "This is a man who has been the victim of a smear and had his wife and son killed. And I don't want him hurt anymore.

"This *is* a murder case," Spence said. "But the people who committed the murder have not been charged, and the people who committed the murder are not here in court.

"Randy Weaver was not a criminal," Spence said. "He had no propensity to commit crimes. This is a man who never even had a traffic accident, never even had a traffic ticket. Never been charged with a crime of any kind and honorably served his country.

"I want to talk to you about . . . punishment. Randy Weaver would willingly go to the penitentiary for the rest of his life if he could have his boy back. Randy Weaver would go to the penitentiary for the rest of his life . . . if he could have Vicki back. Hasn't he been punished enough? Doesn't this terror and this horror have to end sometime? Shouldn't it end with you, and shouldn't it end without having to compromise? Shouldn't this jury have the courage to stand up and say, 'No, they overexercised their power.' I ask you to do that."

twenty-two

Cyril Hatfield knew exactly why he was on this jury. He was here to make sure Kevin Harris and Randy Weaver didn't go free.

Gerry Spence had noticed Hatfield right away. Attorneys have a sense for that sort of thing, he said, which jurors they're reaching and which ones they aren't. After forty years as a lawyer, Spence could guess who his allies were, and he just knew he could count on the red-faced, seventy-two-year-old financial planner who smiled constantly at him and seemed to mouth his approval of Spence's most stirring speeches.

"Aw, bullshit," Cyril muttered while he smiled at Gerry Spence. Or: "The son of a bitch is grandstanding." Spence couldn't have been more wrong about Cyril.

The jury had been admonished throughout the trial not to talk about the case, and so they showed up for deliberations fresh, without a clear idea what any of the others were thinking—except of course Cyril, who'd mumbled his dislike of Spence loud enough for some of them to hear. And so they weren't sure what to do when

Cyril walked into the jury room the day after closing arguments and began campaigning to be foreman. Some jurors wondered if he was too set on convicting the men to serve as the mediator for the rest of the jury.

Not that the rest of the jury was necessarily in favor of acquittal. In fact, Dorothy Mitchell, a forty-five-year-old teacher, had no idea what to do. She walked back into the jury room that first day and started crying. It was impossible! They weren't ready to decide this case. After a decade in education, she figured that some jurors simply hadn't been able to process the mass of evidence: the fifty-four witnesses, the hours of videotape, the gruesome autopsy photos, the drawings, guns, and letters, bullets, batteries, and belt buckles.

She had been devastated when the defense rested its case. Like many of the other jurors, Dorothy Mitchell had been waiting for Spence or Nevin to put this case in some sort of context. Her jaw had dropped when Nevin rested. He was such a decent guy, she just wished he could explain to them what happened. She began to think about the arguments he and Spence made during opening and closing arguments and in every cross-examination— particularly Spence, whose sense of the absurd allowed him to deflate the far-reaching government indictment by asking uptight marshals why they didn't just pat their knee and call old Striker over for a scratch behind the ears.

The prosecution had given them no framework to settle the case, Mitchell feared, no plausible story to serve as a touchstone while sifting through the evidence. Instead, the jury was given some vague conspiracy and a litany of technical, seemingly unrelated acts within it. The evidence was mixed-up and confused, and there were too many questions that hadn't even been asked, much less answered. Some jurors wished they could just walk over to Randy Weaver and ask him: "Why didn't you come down? What were you thinking, letting Rachel tote those rifles around? Do you really believe this stuff about lost tribes of Israel?"

The judge sequestered them on June 15, a dozen people who had already been together two months and who celebrated one another's birthdays, chatted about the weather, and made penny bets about how long the frequent trial delays would last. The mound of pennies had grown as a wall chart behind the jury-room door was updated with the running scores, and breezy April gave way to the dry heat of summer. Everything changed with sequestration and the realization that they could talk about the case now. It was as if you were suddenly forced to spend every waking moment with the people in your office until you solved an impossible riddle.

A judge eager for a speedy verdict must have chosen the decor for the jury room in the federal courthouse at Boise. It was spectacularly uninviting—turquoise carpet and institutional green walls. The room heated up like a furnace, and on Saturdays, with the courthouse closed, there was no air-conditioning. In the center of the room, the jury gathered around two long wooden tables, strewn with coffee cups, snack trash, and twelve sets of complex, sixty-nine-page jury instructions. There was a coffee machine, a coatrack with magazines, and a ceramic water fountain. There were no windows.

Cyril was elected foreman on June 16, the first day of deliberations. Dorothy Mitchell, who was a veteran of one of the toughest contract battles in the history of Idaho schools, pushed hard to get Hatfield elected foreman, hoping his duties as moderator would keep him busy and blunt his drive for conviction.

Before them were eight charges:

Against both Weaver and Harris: Conspiracy to provoke a violent confrontation, which carried a maximum sentence of five years in prison; assaulting and resisting federal officers, ten years; and first-degree murder, life imprisonment.

Against Weaver: Making illegal firearms, five years; failure to appear in court, five years; committing crimes while on pre-trial release, ten years; and using a firearm to commit a violent crime, five years.

Kevin was also charged with harboring a fugitive, which carried a sentence of five years.

Each count also contained the possibility of a fine of up to $250,000.

Cyril pushed quickly for a vote. But a couple of jurors stopped him. Dorothy Mitchell knew they weren't ready for a test yet. They had to put all the confusing information into some sort of context. And if the prosecution wasn't going to teach this material, then perhaps she would have to.

*F*irst came the illegal shotgun charge. Arguments began almost immediately, and three camps emerged. Cyril had his law-and-order team: retired judge's secretary Ruth Sigloh, sixty-two; a housewife named Eunice Helterbran, sixty-seven; and Karen Flynn, a pregnant, thirty-one-year-old accountant with an MBA. They saw the government's mistakes, but they weren't ready to forgive Weaver and Harris just because Gerry Spence said so. Still, Hatfield was worried. They weren't the strongest team. Sigloh, sitting at the edge of the jury box near some scary antigovernment people, had suffered a stress attack during the trial and was on medication. Helterbran had an arrhythmia and was on medication to still the pounding of her heart, and Flynn was about six months pregnant. Hatfield had high blood pressure made worse by this awful sequestration. He hadn't felt so cooped up since he was a Marine on Guadalcanal, waiting in the dark during bombing blackouts.

That first week, there were only two jurors who stayed out of the arguments, Mary Flenor, a fifty-two-year-old supermarket inventory auditor, and Leonhard Fischer, a sixty-seven-year-old semiretired businessman.

Six jurors leaned toward not-guilty verdicts. They were led by the Dorothys: Mitchell, the forty-five-year-old teacher, and the quieter Hoffman, a sixty-year-old college tutor. Janet Schmierer, forty-seven, assembled disk drives at Hewlett-Packard, and Frank Rost, sixty-eight, was a retired farmer and a carpenter. Gerry Anderton, seventy, was a retired heavy equipment shop manager whose libertarian beliefs fit squarely with Weaver's distrust of government.

Finally, there was forty-three-year-old John Harris Weaver—Jack—whose youngest brother had been murdered years before. The burly pressman had to be drawn out, but he eventually let it be known he, too, believed the government had not proven its case. While other jurors were surprised the defense rested, Weaver had been more shocked when the prosecution ended its case. He'd been expecting Howen and Lindquist to bring forth their star witness, someone who'd make the whole case clear and credible. This case wasn't about a band of conspirators, he thought. It was about a screwed-up family.

Those jurors slowly coalesced into a team eager to teach the government it couldn't shove people around just because they believed differently. The acquittal coalition began redefining the arguments as debates on government wrongdoing.

"The government is not on trial here," Hatfield said, trying to keep jurors focused on the narrow questions in the jury instructions.

They started with the gun charge. Discussions formed, dissolved, and re-formed around the two tables, the men hoisting the guns over and over and sneaking out to smoke, the women trying to order the debate. On an easel chart, the accountant, Karen Flynn, called on members of the jury to help her list the pros and cons, illegal gun charge versus entrapment. "Your side and my side," she called it, looking at Gerry Anderton.

Anderton jumped on her choice of words. "Listen here, sweet pea. It's not your side and my side, it's our side," he lectured.

Flynn was taken aback and started crying.

They got nowhere on the gun charge, and by the third day, Friday, June 18, the discussions of the failure-to-appear charge—Dave Hunt's dogged pursuit of a peaceful solution versus Randy's very real distrust of government—

had begun to drag as well. One juror was feeling ill and was asleep by early afternoon. The other jurors took a break to read their notes and review the evidence. They worked Saturday but ended the short first week without a formal vote. A couple of informal straw polls made it clear this was going to take some time.

Dorothy Mitchell began making charts of the facts they could agree on, relying heavily on long recitations from the notes kept by Mary Flenor, the inventory auditor. They studied physical evidence and argued over interpretation. Whenever someone tried to speed things up, the methodical Schmierer insisted on replaying all the evidence. By the end of the first week of deliberations, everyone was angry.

Hoffman motioned Hatfield aside during a break and said she'd heard a rumor that Cyril told an alternate juror that he was going to send Weaver and Harris to prison. She accused him of being prejudiced.

"I want you to be fair," Hoffman said. "If you're not, I have a letter I'm going to send to the judge."

"Do you really think I said that?" Hatfield asked.

"I have no reason to believe you didn't," she said.

Cyril was losing control of the jury. The undecided jurors were swinging toward acquittal, and the conviction minority felt it was being interrupted and intimidated. Dorothy Mitchell led the small class of adults through the case, jabbing emphatically at the air with her pencil, unnerving jurors who feared she would throw it at them.

Despite Mitchell's junior-high-teacher work ethic and Cyril's attempts to steer the discussion, they got off the point, arguing for hours one day over whether or not the Weavers would make good neighbors.

"Randy Weaver *hated* the government," Helterbran said.

"That's no crime, I hate the government," said Gerry Anderton.

"Yes, that's obvious," she said disapprovingly.

At least he wasn't like her, Anderton thundered, seeing government as "God Almighty and the bureaucrats that are doing all these things as the disciples!"

Mitchell and Schmierer settled the jury on a painfully slow process. Baffled by the prosecution's impressionistic tale, they started at the beginning and retold the evidence to each other in chronological order, pulling facts from memory, notes, and evidence, and assembling the story of the case in some logical form. It took more than eight days.

The retelling proved invaluable for some, but a few of the younger jurors were frustrated by the pace. Their retired counterparts seemed to have nothing better to do than wallow in this case, extending their fifteen minutes of fame, eating free meals, and staying in trendy Boise, away from the blistering sage flats of their rural hometowns. They were forbidden from talking about the case while anyone was in the bathroom and—with the endless pots of coffee and elderly jurors—some days seemed wasted on the toilet.

Hatfield, meanwhile, was annoyed with some of the acquittal forces, especially the two Dorothys and Janet Schmierer, who wouldn't let them vote and were like Spence's echoes in the jury room. They seemed to be convincing the moderates, Cyril realized. Still, he was sure he had the votes to prevent acquittal on the serious charges: murder, selling an illegal shotgun—the big stuff.

A t the end of deliberations each day, jurors rode an elevator to the courthouse basement where they were loaded into two Chevy Suburbans and seated against windows that were covered with heavy brown paper.

A driver they'd nicknamed "Crash" after he dinged the other jury vehicle raced them back to the Red Lion Hotel. As they blazed through the parking lot, jurors peeking through slits between the paper caught glimpses of television cameras, neo-Nazi skinheads, and the odd-lot constitutionalists on hand to observe the trial.

Like herd animals, they fell into eerily similar and fitful sleep patterns: nodding off right away and then waking up thinking about the case at 2:00 a.m. A few began to feel they were being spied on by the federal agencies curious about the outcome of the case, but they mentioned it only to friends on the jury, fearing they'd be misunderstood.

Meals were late, causing real discomfort for old jurors with blood-sugar and blood-pressure problems. People got grouchy. Sports buffs got to watch the NBA final but were ordered away to dinner at the end of the exciting game. The *Idaho Statesman* newspaper was provided for jurors to read, but marshals had clipped out humorist Dave Barry's column—a parody of a John Grisham legal thriller. Jurors laughed when they noticed their keepers had by accident left in an article about the *Weaver* case. In time, the daily paper had so many articles cut out of it, they called it the "Doily Statesman."

The first few days of sequestration, the U.S. marshals guarding the jury couldn't seem to make up their minds: first the jurors were permitted to open the drapes. Then they had to pull the shades. Then they were allowed to open the shades but were asked to pull the sheer drapes. Jack Weaver's copy of Tom Clancy's *Patriot Games* was taken away because it featured heroic action by the Hostage Rescue Team. Dorothy Hoffman's anthology of American literature was confiscated, too. It included "An Occurrence at Owl Creek Bridge," the story of the hanging of a Civil War spy.

Music cassette players were banned because messages could be smuggled in on custom tapes, but CD players were permitted. Most TV was out-of-bounds, and movies were mostly out of the question, except lifeless choices brought in by the marshals, like *High Road to China,* a cheesy Tom Selleck adventure flick.

Reporters sniffed around for any information about the jury, what counts they were debating, their health, even what they ate. The interest made the pregnant accountant, Karen Flynn, laugh. "We ought to send out for a twelve-pack of Moosehead beer and a gross of condoms, really yank their chain," she said.

During the long days of deliberation, jurors were taken to the courthouse roof for fresh air breaks. Seven stories up, it looked down on the army fort built in Boise in the late 1800s, now preserved with lovely lawns and tall trees. Immediately behind were swaths of dry brown desert and foothills and peaks that hemmed in the Boise Basin.

Jack Weaver wrote in his diary that he felt like a prisoner of war, he had so little contact with the outside world. Adding to their discomfort was the hazy reality of the case, which never seemed to leave their minds.

By the end of the second week, two walls of the room were festooned with fact and time-line charts, the table was piled with candy wrappers, coffee flotsam, and other garbage, and the courtroom had been taken over, too, as a place to take breaks and to act out or pace off different parts of the case. With so many jurors still struggling to visualize the scene, Hatfield sent a note to the judge asking him to take the jury 450 miles north—to the mountains of North Idaho—to walk the trails and examine the scenes of the gunfights.

Judge Lodge said no to the travel request, and the jury went back to work, taking its first vote on the tenth day of deliberations. It came out 8–4 against conviction on the failure-to-appear charge. They backed away, tackled the firearms charge again, tabled it, and then discussed the murder charge against Kevin Harris. Dorothy Mitchell suggested they act out the two versions of events. Jack Weaver got down on his knees and showed how difficult it would be for Degan to fire seven times after being shot. Others stood in for the marshals or the dog. Finally, they voted, agreeing that Harris wasn't guilty of first-degree murder, second-degree murder, or voluntary manslaughter. But jurors stuck on involuntary manslaughter. Still, they were making progress.

The foreman, however, was running out of steam. Cyril had been on Valium earlier in the trial, which eased his stress. But now he was out of the drug and—without knowing it— he was suffering from post-traumatic-stress disorder, reliving his terror from World War II in the incarceration of being sequestered. By Monday, June 28, he was jittery and anxious, fighting a nervous breakdown. During the morning fresh air break on the roof of the courthouse, he told the jurors that the lunch break would be longer than usual. Back downstairs, Cyril called Eunice Helterbran aside. "Stick to it. Don't let them do it," he said to her. Then he told the others he was going to see about moving up an appointment with his doctor. He never came back.

They had to start over, Judge Lodge told the jury, as attorneys and spectators shook their heads. Kevin Harris's mother began crying. Two weeks into deliberations, Lodge replaced the exhausted and sick Hatfield with an alternate, Anita Brewer. But, he said, she didn't have the benefit of their earlier deliberations. "Therefore, you must go back and have a complete and full discussion of all evidence."

After a two-week absence, Ron Howen reappeared that day, June 29, in time to hear that the case had gotten to someone else. A fiercely private man, Howen refused to talk about where he'd been, but friends said he'd suffered from exhaustion. He listened unemotionally as Lodge sent the jury back to start over.

Gerry Spence tried to be level about the loss of Hatfield, but it ate at him. "He's been someone who's been involved with this case with all his heart and soul," Spence told reporters from under his felt Stetson.

Earlier in the deliberations, the lawyers had been encouraged when a Spokane newspaper reporter tracked down an alternate juror who had asked to be dismissed from the case. "I felt like a little kid that finds out there is no Santa Claus," said the dismissed juror, Gena Hagerman. She said she thought that the only way the shoot-out could have started was with Roderick killing the dog first and that she was outraged by the FBI's actions. "I found out what the FBI and the marshal service is all about. I found out they were capable of doing things I thought were not possible."

Maybe they'd reached all the jurors, the defense team
had gushed. But now, with Hatfield gone and the jury
settled into its third week of deliberations, they began to
fear a hung jury. Or worse.

David Nevin wanted to hang himself. He was used to
second-guessing—hell, it was practically a hobby—but
this was ridiculous. He wasn't sleeping. He wasn't eating.
He and Chuck Peterson jogged in the mornings, and
Spence came over for dinner and they all tried to reassure
one another, but at times, Nevin was inconsolable. I
should've put him on, he thought, I should have put Kevin
on the stand.

Almost every day, the lawyers checked in at the federal
courthouse, walking past a violently bored pack of
television and newspaper reporters, who—at one point—
challenged some loitering skinheads to a game of football.

The reporters cornered Spence any time he walked
past, looking for some insight or humor to save what had
become a repetitive story. ("The jury deliberated for the
eleventh day in the trial of . . .") Spence had seen a lot of
juries in forty years. Back in Wyoming, when he was
young, he had been permitted by a judge to lie on the
floor of a men's room above the jury room, where he and
the opposing lawyer could listen to jury deliberations.
He said that lying on the floor of that men's room in a
suit was one of his better legal lessons, an understanding
of juries that he carried to that day. The Weaver jury, he
said, "looked brighter, less frayed, more together than
expected."

T hey were losing it. The jury had already spent more
time sequestered than Randy Weaver and Kevin
Harris had spent under siege on Ruby Ridge. And
now, the judge was telling them to start over. Some of the
jurors were beaten.

Anita Brewer was a quick study. Still, it took a week to
bring her up to speed.

The eight women and four men elected Jack Weaver

foreman. He was an unlikely choice, a quiet guy who spent the breaks during the early part of the trial withdrawn and reading a book. Jurors had to tease him into conversation. But earlier in deliberations, when Eunice Helterbran was accusing the others of being antigovernment, Jack Weaver had delivered a stirring speech.

"While I respect her love of country," he said, "we should not let patriotic fervor stand in the way of the truth. For her to say that I or anyone else is antigovernment is unfair. My brother was murdered. If anyone is dead-set against setting dangerous criminals free, it's me. The Founding Fathers of our nation wrote the Constitution and Bill of Rights with the idea that the citizens would be sensible enough to recognize the excesses of too much government power and gave us the tools so we could change or even abolish our form of government."

When he was elected, Jack Weaver joked that the only people he'd ever been in charge of were his kids. That might have been just the experience needed for this jury. He faced two stubborn coalitions that had moved from honest disagreement into angry dislike after weeks of debate, emotional outbreaks, and recriminations. Janet Schmierer had so angered both sides with her assertiveness, they shouted her down when she tried to speak and all she could do was slip notes to one of the Dorothys.

Most jurors were disgusted by the white separatist beliefs of Weaver and Harris, and they all agreed the government had made mistakes. But the central disagreement remained the same. One side blamed the government, while a smaller group thought Weaver's failure to appear in court caused the entire ordeal.

Sometimes, Jack Weaver wished they could ignore the precise jury instructions and use common sense to deal with Randy Weaver. "If I could have convicted him of gross stupidity, I would have," Jack Weaver wrote in the diary he kept during the trial.

After Cyril left, Eunice Helterbran was the only holdout on the murder charge, still hoping to convict Kevin Harris of involuntary manslaughter. Seemingly shy and easily intimidated, the housewife wouldn't budge. "I don't want to send the message that it's okay to shoot U.S. marshals," she said. Others still wanted to convict Weaver of some of the more serious crimes, too.

But Anita Brewer, a medical technician, changed the dynamic of the jury with her emphasis on scientific methods. In her job, any lab samples offered to a doctor had to be handled extra carefully. In this case, the government was trying to send two men to prison using evidence that had been lost, mishandled, and collected in a way that would not stand up to the rigors of a high school science class. Brewer's reasoned approach contrasted sharply with the Spence rhetoric of some other jurors and made acquittal more attractive to the remaining members of the conviction coalition.

Jack Weaver insisted on verdicts. He would not let the jury hang and force another group of citizens to go through this hell. Finally, on July 2, after fifteen days of deliberations, they began voting. The first one was fairly easy; the outside world would make a laughingstock of any jury that ruled Randy Weaver was planning to appear in court. Jack Weaver asked jurors to ignore their emotions and concentrate on testimony and other evidence. The only question was: Is there reasonable doubt?

They voted 12–0 to convict Randy Weaver of failure to appear. They also found Randy guilty of committing a crime while on pre-trial release.

Next they voted on the murder charge against Kevin Harris. Frank Norris's testimony ("the distinctive sound of a .223") had convinced Karen Flynn that they had to acquit. Now, even Eunice Helterbran admitted that if Norris heard one of the marshals' guns first, then the government couldn't prove that Harris had fired first. While some jurors still wished they could punish Harris for Bill Degan's death, there was certainly reasonable doubt

The vote was 12–0 for acquittal on murder and manslaughter.

The next day, the jury was ready to vote again and reached verdicts on three more counts: not guilty of assaulting and resisting the other marshals and not guilty of using guns in the commission of felonies. Then the big one: Weaver not guilty of murder. Again, the marshals' story just didn't fit with the evidence.

On Monday of the fourth week, they wrestled with count seven: did Kevin harbor a fugitive? Not wanting the outside world to know where they were in the proceedings, the jury sent a vague note to Judge Lodge.

His answer was equally vague. "Did that answer your question?" an impatient bailiff asked in the hallway between the courtroom and the jury room.

"No," Weaver said. The jury finished the day without voting. That night, they watched a home video of Mary Flenor's family picking ripe tomatoes from her garden and imagined going home.

The next day, July 7, they reached a not-guilty verdict on the harboring charge. Foreman Weaver pressed quickly on, to count one, conspiracy. He was stunned to find there was broad agreement. A conspiracy between a family that started in January 1983, lasted until 1994, and was furthered by moving to Idaho? It was ridiculous on its face.

Jack Weaver had expected two to three days of deliberations on the conspiracy charge alone. But within an hour, he could see the rest of the jury felt as he did. They had charged these guys with everything but speeding. In football, he thought, the prosecution would have gotten fifteen yards for piling on.

But with that many acquittals on the flip charts, the proconviction caucus thought Randy Weaver deserved more punishment. Even some of the moderates said Randy's stubbornness had caused this mess. They debated the only remaining charge, that he sold sawed-off shotguns. The acquittal coalition insisted Randy had been entrapped, but jurors couldn't agree on what the ATF agent had said to his informant about reeling Randy in.

Intimidated by Lodge, they wrote an apologetic not[e] asking for a replay of Chuck Peterson's mocking an[d] skeptical cross-examination of Herb Byerly, the ATF agen[t] who'd set up the sting. Peterson beamed as he prepared t[o] listen to the testimony the jury had asked to hear.

Lindquist stared out the courtroom window at th[e] scorched foothills, while Howen watched the jury for reaction. They had filed into a side courtroom, notebook[s] crammed with instructions and notes that they'd puzzle[d] over for close to a month now.

A court clerk read from the transcript Peterson'[s] bristling cross-examination and Byerly's answer. "'[I] instructed Mr. Fadeley about entrapment . . . that he wa[s] not to go out there and entice someone, providing ther[m] with undue rewards for violating the law.'" Peterson aske[d] Byerly why he didn't tape-record every one of Fadeley'[s] meetings with Weaver before the gun sale. "You wanted t[o] make sure . . . there was no record of what your informan[t] was saying to this man," Peterson needled. Byerl[y] disagreed. Then Peterson asked Byerly about th[e] government's practice of paying informants a bounty fo[r] cases in which their testimony helps win conviction.

Outside the courtroom, Chuck Peterson had troubl[e] hiding his smile as Gerry Spence slapped him on the back[.] At the least, the jury was *talking* about entrapment. Peterso[n] always suspected that Spence had given him the weapon[s] charge and the failure-to-appear because he'd figured the[y] were losers. Then, if they lost those cases, Spence coul[d] always say he was still undefeated. It was the best way t[o] stay unbeaten: avoid the bad cases and assign away the ba[d] charges. Now, less than a year after he'd considered quittin[g] the law, just a few months since he'd been scared to death t[o] cross-examine a witness in front of Spence, Peterso[n] allowed himself to believe he might actually win one.

W[h]y wouldn't the ATF tape all those meetings[?] Anita Brewer was insistent that professiona[l] agents of the federal government's top polic[e]

gency could document their cases better. Unless
Peterson was right, and they didn't tape those meetings
on purpose. The Dorothys could see the entrapment issue
eating at the other jurors. They went back to their hotel
on July 7 still disagreeing, but there were some cracks
developing in the three jurors who still wanted one more
conviction.

Back at the hotel that night, they took the winnings
from the penny pools—about three dollars—added a buck
each and bought fifteen tickets for the multistate Powerball
lottery, agreeing to share their winnings. They also had a
cake delivered and threw a baby shower for Karen Flynn,
who was getting ready to deliver the baby boy she carried
through the long trial and deliberation.

Jack Weaver slept well. An exhausted Dorothy Hoffman
cried herself to sleep. The bad diet, lack of exercise, and
stress were wreaking havoc with her system, and she was
hoping she could make it through deliberation over the last
count.

At 7:45 Thursday morning, July 8, Jack Weaver asked for
any comments and instantly regretted it. Frank Rost hauled
out a sheaf of yellow legal pages and Weaver knew he'd
made some more middle-of-the-night notes and was going
to give a speech. He'd done it before with little impact.

Usually, Rost made his point, wandered a little, and then
would pipe down, but this time he went straight for the
jugular, calling Flynn, Helterbran, and Sigloh closed-
minded and unwilling to let go of their first impressions of
the case. The proconviction jurors yelled back with equal
vigor until Weaver jumped in between the hollering jurors
and shouted "Sit down and shut up!" to Rost, a man older
than his father.

Two jurors escorted Rost to the nearby grand jury room
while Weaver called a twenty-minute break. Checking in
the men's gathering room, he found Rost with his feet up
on a chair and one hand over his face, feeling miserable.
Jack Weaver told him they could get a verdict if he could
get things settled down for a vote. At 8:25 a.m., he called
for a cease-fire. Then he polled the jury on their last count.

"To acquit?"

Six hands went up immediately, then came two more, three more and the last one. Finally, they had all twelve. They voted again on every charge, just to be sure. Kevin—not guilty on all five charges. Randy—not guilty on the five most serious charges, guilty of failing to appear and committing crimes while on pre-trial release. The jurors joined hands. The teacher, Dorothy Mitchell, burst into tears. A few others cried as well. It was 9:45 a.m. on the last day of the longest jury deliberations in the history of Idaho federal courts. They were done.

*T*he jurors were nervous as they filed into the courtroom one more time. It was packed with reporters, lawyers, and other observers. Dorothy Mitchell smiled at Kevin Harris and then looked at the prosecution table. Her heart went out to Dave Hunt, the deputy marshal who'd tried so hard to solve this case peacefully. He sat there next to Ron Howen and Kim Lindquist, big black bags under his eyes, an expression of overwhelming exhaustion on his face.

Fourteen deputy marshals guarded the doors in the courtroom. Randy Weaver and Gerry Spence looked at each other one more time, and the Wyoming lawyer nodded confidently. The judge asked the jury if they'd reached a verdict.

"We have, Your Honor," Jack Weaver said. He gave the two-page jury form to the judge, who read it for the longest minute in Boise, then handed it back to be read aloud.

At the prosecution table, Dave Hunt felt as though he'd been kicked in the head. "Not guilty. Not guilty. No guilty." Time slowed down, and he felt like he was losing his bearings. The murder charge. "Not guilty." A whoop went up in the courtroom, and Hunt didn't hear anything else. You can't get away with murder, he thought. Not in America. Hunt's head swung about, and everything seemed to move around him in soundless slow-motion—Kevin Harris and Randy Weaver hugging, the defense attorney smiling. It had all come to this. Dave Hunt thought about everything he'd tried to do and couldn't believe it had ended this way, with Degan dead and Kevin and Randy going free.

Nevin felt redeemed. Tears streamed down Kevin Harris's face as he hugged Randy, who had smiled and nodded as each verdict was read and who mouthed "Thank you" to the jury. Kevin's mom waved a cigarette at her son; it had been ten months since he'd been able to have one. As Nevin congratulated Kevin, Randy, and the other defense lawyers, a deputy marshal leaned toward him and said they would need to take Kevin back to jail for processing. Nevin, who'd come to be known as the cordial, polite lawyer on the defense team, said, "The hell you will. He's going out the front door with me."

Then Nevin turned to Lodge. "Your Honor, I'd assume Mr. Harris will be discharged at this time."

Lodge looked back at the verdicts. "Yeah, I'd guess so."

Dave Hunt didn't hear any of that. He sat bewildered and entranced until Kim Lindquist pulled at his arm and said, "Dave. Dave. Come on." By then the courtroom was emptying, and Hunt stood and walked silently out.

For the first time in ten months, Kevin Harris left the courtroom without handcuffs or guards. He looked for Nevin. "Help, where are you?" He leaned on his mother's shoulder and sobbed. Outside, he was surrounded on the courthouse steps by reporters and well-wishers. "I just want to thank the jury for everything," he said. "I had total faith in Yahweh the Creator."

Gerry Spence watched with a wide grin. Randy had gone back to jail until September or so, but he would be out soon. Spence talked to reporters and signed copies of his new book, which had just been released with impeccable timing. The book was a collection of essays on governmental and corporate power called *From Freedom to Slavery*. The first chapter was a letter to a Jewish friend, defending Spence's decision to represent a white separatist and talking about government's misconduct in the case. Spence scribbled his name on the title page and handed the book to a constitutionalist who'd driven all the way from Portland in a beat-up Pinto station wagon.

Spence autographed another book as he spoke to reporters. He said the case wasn't over. "There is a dead mother who died with a baby in her arms. There is a little boy, four feet, eleven inches tall, who died with a bullet in his back. Who is going to be responsible for these deaths?"

twenty-three

On October 18, 1993, Randy Weaver was sentenced to eighteen months in jail—fourteen of which he'd already served—and a $10,000 fine. With good behavior, he'd be out by Christmas. Earlier, the judge had dropped one of the charges Randy was convicted of—committing crimes while on pre-trial release—because he'd been found not guilty of those crimes (murder, assault, and the rest). After four years of investigation, at a cost of several million dollars, the U.S. government managed to convict Randy Weaver of failing to appear in court.

At his sentencing, two jurors—the Dorothys—testified that since the trial ended they'd visited Randy and met his family and they thought he should go home. Keith Brown also testified, saying Sara and Rachel were doing incredibly well, but insisting they needed their father back. Spence gave another stirring speech and asked Judge Lodge to limit Randy's sentence to the fourteen months he'd already served.

"This is a man who has learned a great deal," Spence

said. "The only evidence before Your Honor—and I think the truth—is that this family needs its daddy."

But Ron Howen was back from his bout with exhaustion, and he was ready to fight some more. He asked the juror Dorothy Hoffman if she'd ever sought out Bill Degan's family, and he recommended that the judge sentence Randy to three years in prison. He said that no matter what happened in court, the standoff and the three deaths were ultimately Randy's fault. Howen read a 1990 letter from Weaver's attorney Everett Hofmeister, a letter that wasn't allowed into testimony during the trial. "'The present course you are following is suicidal,'" Hofmeister wrote two years before the shoot-out. "'In a few short years, you will look back on what happened as a needless disaster.'

"You just want to weep when you read a letter like that," Howen said.

"Since this trial has concluded, every morning I've gotten up and looked at myself in the mirror," the prosecutor continued. "And I've asked, 'What did I fail to do in order to convince that jury the defendant was guilty? . . . Well, it is my responsibility, and I must bear it.'"

Howen said he had begun to realize that he and Weaver were somewhat alike. He was raised in a German Mennonite home with separatist beliefs, Howen said, but he was also bound by the Bible's admonition to submit to authority. And he said he was taught to love his neighbor. Howen's voice cracked as he said he was even trying to love Randy Weaver. "It has been the hardest thing for me to do in my life."

After the sentencing, Ron Howen was transferred out of the criminal division of the U.S. Attorneys Office. Kim Lindquist left Boise altogether and took a job in Bogotá, Colombia, working drug interdiction cases at the U.S. embassy.

In other federal agencies, many people blamed Howen for losing the *Weaver* case, because of his insistence on filing a broad indictment. A sweeping Justice Department investigation found the indictment was flawed. Howen was "overzealous" and used "faulty judgment." His statements before the grand jury bordered on testimony and the decision to seek the death penalty was overreaching, the report stated.

Ron Howen refused to talk publicly about the case. He bought some farmland and set to work as one of the hardest-working attorneys in the civil division. Ten days after Randy Weaver was sentenced, Judge Lodge shifted some of the blame away from Howen though, finding the FBI in contempt of court and fining it $1,920 for its delays and obstructions.

"The actions of the government, acting through the FBI, evidence a callous disregard for the rights of the defendants and the interests of justice," Lodge wrote. "Its behavior served to obstruct the administration of justice. . . ."

D ave Hunt was tired. After the trial, he decided to quit the U.S. Marshals Service. "I don't want to do it anymore," he told his wife. It wasn't anything he could really explain.

The closest he could get was describing his feelings when he came home from Vietnam. So many guys died over there, and so many more sacrificed years of their lives and limbs off their bodies. Yet when Hunt left the Marine Corps and enrolled at Indiana University, people either ignored Vietnam vets or reviled them. He couldn't understand that in 1971, and he couldn't understand this twenty-two years later. "Feds Lose Big," one newspaper headline screamed after the trial. Hunt knew mistakes were made in the case, but he hated the way it had become a symbol for federal law enforcement run amuck. It was an awful feeling to think you were doing the people's bidding and find out the people hated you. He wondered: What exactly did Billy Degan die for?

The other deputy marshals resumed their careers. They talked occasionally with Hunt about the case and said they were most upset that so few people believed Kevin Harris fired the first shot. Art Roderick sometimes wished the bullet that had whistled by his stomach had hit him so that people would see just how dangerous the Weavers really were. He and the other marshals agreed that everyone had forgotten the other victim in the case, Degan. His family remained quiet about their loss.

In the end, Hunt didn't quit the U.S. Marshals Service. He just kept getting up and going to work. But he didn't go after any more fugitives. Instead, he did some accounting for an organized crime and drug task force and took on administrative chores. Other deputy marshals tried to cheer him up: "You won your case. It's okay." Weaver *had* been found guilty of failure to appear, the warrant Hunt spent all that time trying to serve peaceably. But Hunt didn't think he'd ever shake his empty feelings, and every time an Idaho resident complained on the radio or in a letter to the editor—"Why didn't they just go up there and ask Weaver to come down?"—Hunt relived it all over again. Newspaper and magazine stories seemed to portray Randy Weaver as some sort of folk hero, someone who stood up to the oppressive government. Hunt pictured Randy hiding behind his kids—sacrificing his family because he didn't want to face the charges, right or wrong, filed against him. It made Hunt sick.

The case stayed in the news forever, and just when Hunt thought people had forgotten it, some new government report came out or a new lawsuit was filed or some guy wanted on a minor charge announced with bravado he wasn't going to be taken alive. All over the United States people who despised the U.S. government drew inspiration from Randy Weaver. An accountant refused to testify before a grand jury in a drug case; a developer refused to file paperwork for a project; a tax protester refused to come out of his house. They all talked about the inspiration of Randy Weaver.

Throughout the West, "freemen" and militia members vowed to battle traffic tickets and zoning ordinances with violence. In Montana, some local officials felt terrorized and threatened. The sheriff of one county talked about a

ugitive living on a nearby ranch with semiautomatic
veapons who had dared the sheriff to come get him. The
utgunned sheriff asked for help, but federal officials
lidn't want to get involved in such cases because they
eared being drawn into a violent confrontation over some
ninor issue. Such reluctance had a name among federal
gencies. They called it Weaver Fever.

Of course, Bo Gritz didn't win the presidency. He
didn't even win Boundary County. Gritz received
just 46 of the 3,850 votes cast in Boundary, 1.2
ercent, not even half as many votes as Randy Weaver
ot when he ran for sheriff. Some folks blamed his poor
howing on the fact that he was quoted as saying Weaver
vas a "punk" in an interview published after the
tandoff.

But in the growing radical right-wing survivalist
novement, Bo Gritz had become a respected leader. He
poke at "Preparedness Expos," and held paramilitary
eminars called SPIKE—Specially Prepared Individuals
or Key Events—all over the country.

After the Weaver-Harris trial, Gritz moved from his
ome in Nevada to the hills near Kamiah, Idaho, a few
ours south of Naples, where he bought two thousand
cres with some partners, subdivided the land into thirty-
cre parcels, and sold them for $3,000 an acre. He called it
Almost Heaven, a covenant community for people sick of
overnment and ready to prepare for Apocalypse. For
everal months, magazine and newspaper reporters doing
he requisite story on the emerging survivalist right wing
cheduled a stop at Almost Heaven so Gritz could call
hem "faggots." Not that Bo needed the publicity. There
vas no shortage of people willing to pick up a rifle and
humb their nose at the government.

After the Oklahoma City federal building was
ombed in the spring of 1995, Gritz said it was a
nasterpiece, "a Rembrandt," and therefore was more
han likely the work of the government trying to

discredit white Christian patriots. Gritz said it was th
beginning of a reign of terror against such people, th
beginning of the end.

*I*ts front porch sagging, Sara's gardens grown over wit
weeds, Randy Weaver finally sold his cabin and twent
acres in the spring of 1995 to a lifelong friend of Kevi
Harris. Except for law officers, the defense attorneys, and
few Weaver friends, it had been vacant since Weaver and hi
daughters walked out the door—frightened and solemn—i
August 1992.

The Weavers' friends Jackie and Tony Brown took car
of the cabin for a while after Randy was arrested. Durin
that time, they noticed every now and again a fresh set o
car tracks on the old logging road or some other evidenc
that someone had been nosing around. They worried tha
people would break off parts of the cabin as souvenirs.

People stopped in the Naples General Store or th
North Woods Tavern fairly regularly and asked how t
get up to the cabin. Sometimes they were given directions
other times not. A sign warned: "No Trespassers. N
Sightseers."

But many people couldn't resist. They wanted to stan
on the rock outcropping and see just how defensible th
knob was. They wanted to see the signs and cans used fo
target practice. They walked down to the Y in the loggin
road and tried to gauge the distances and the positions
tried to find proof that the marshals had been lyin
Sometimes they hiked across the gully to where Lo
Horiuchi had aimed at the house. They stared across tha
stretch of ground and looked for proof that he'd done it o
purpose, although if they had come that far, they likel
didn't need much evidence to believe that Vicki Weave
had been murdered. Some of them stood on the porch an
imagined it was *them,* that it was *their* wife holding *the*
baby.

For most Americans, the Weaver case faded in the wak
of fresher news. But like the siege and catastrophic fire :

Waco, Ruby Ridge's impact was actually spreading as time went on, an untreated wound festering among the angriest and most frightened Americans. When the Treasury Department released its report into the ATF's involvement in the Weaver case (finding no wrongdoing), it was a slap to those people. Each investigation that followed only increased cries of a cover-up.

Exactly two years after the final raid on the Branch Davidian compound, on the morning of April 19, 1995, the federal courthouse in Oklahoma City was carved in half by a fertilizer bomb that killed 168 people. When the primary suspect—Timothy McVeigh—was identified Americans were shocked by his motivation. He had been a soldier, a quiet American who was seduced by the same radical-right thought, the same old conspiracies. The bombing itself seemed patterned after *The Turner Diaries,* the novel that inspired The Order more than a decade before.

There was little surprise when news accounts indicated that Timothy McVeigh had visited the site of the gun battle and fire at Waco. In the summer of 1995, one of McVeigh's coworkers told the *Washington Post* that he had visited Ruby Ridge. There, according to the *Post,* McVeigh had conducted his own inspection of the scene of the gunfight, the cabin, and the sniper position, and he no doubt decided that federal agents had intentionally killed Vicki and Samuel Weaver.

*I*n August 1994, Chuck Peterson, Garry Gilman, David Nevin, and Ellison Matthews filed civil lawsuits on behalf of the Weavers and Kevin Harris, charging the government with wrongful death for killing Samuel and Vicki Weaver and with violating the rights and property of Kevin Harris and Randy Weaver. By the time they'd finished compiling all the damages they sought, the total claim was more than $300 million.

In the meantime, Nevin and Peterson, who'd become close friends during the trial, returned to what they jokingly called their ham 'n' egg law practices. It was tough to go from getting an acquittal on murder and conspiracy charges with Gerry Spence to arguing a case in traffic court. But, in truth, things would never be the same for the Boise attorneys. For one thing, business improved a little. Spence sent prospective clients to both men, and the publicity from the case brought other people to their firms.

In the Boise Yellow Pages, Chuck Peterson's ad read: "Co-counsel with Gerry Spence, *U.S. v. Weaver.*" He was even hired by the National Rifle Association to help prepare for congressional hearings on the Branch Davidian standoff in Waco.

Occasionally, Peterson felt twinges of something else, the feeling that his biggest case might be behind him. It was a new kind of pressure, like a lighter version of what Spence must feel whenever he came into some backwater like Boise and was expected to make magic. Suddenly, Peterson was the well-known attorney, the guy who won *U.S. v. Weaver*. He would never be allowed to lose gracefully again. In Great Falls, Montana, Peterson gave a case the Spence treatment and was promptly growled back into place by a judge who said he didn't allow such theatrics in his courtroom.

David Nevin didn't feel all that changed by the case. Already one of the best criminal lawyers in Idaho, Nevin attracted a few better-paying clients, like a Boise financier who was arrested for assault. Racked with the second-guessing temperament of a perfectionist, Nevin had nonetheless always believed in himself, that he was a great attorney. His experiences with Spence only reaffirmed his own theories of defense law, and he never doubted that bigger cases would come his way. After the Oklahoma City bombing, Nevin was approached about being the lead attorney for the primary suspect, Timothy McVeigh. Nevin considered it but didn't want to move to Oklahoma and leave his wife and two sons behind. He said he wasn't interested.

He stayed in touch with Kevin Harris, who moved to a

small town in Washington State, got a job as a welder, and did his best to stay out of the spotlight.

A few months after the verdict, Nevin was invited to speak to law students at the University of Idaho, where he had gotten his degree. These law students were far more conservative than the recovering hippies he'd gone to school with, and Nevin wondered if they would grasp the themes of religious freedom and government misconduct that he'd come to find so inspiring in the case.

Nevin told the law students about his difficult decision to rest without calling any witnesses. Nevin told them about the verdicts and how a deputy marshal had leaned over and told him that Kevin would have to go back to jail before being released. Nevin told the law students how he'd convinced the judge to set Kevin free right then.

"Someday," he said, "I hope that all of you get the opportunity to walk out the front door with somebody who's been charged with murder but isn't guilty." The law students leaped up and began clapping. David Nevin started crying.

A man in Florida—one of Randy Weaver's many supporters—paid Randy's entire $10,000 fine. Weaver received letters of support from as far away as Korea, Japan, and Europe, and more than a few letters from interested single women.

On December 17, 1993, Randy got out of jail. He spent a couple of days in Boise and then he flew home to Iowa to be with his daughters. Delayed by a snowstorm, his plane landed two hours late on the night of Sunday, December 19. The hallway was dark when he came walking toward his daughters, and they squealed when they saw him. He hugged the girls and they all stood around the Des Moines Airport, crying and smiling. Rachel, who was twelve now, patted her dad's jailhouse potbelly, and he rubbed her head. Sara wiped her eyes with one hand and kept the other around her dad's waist. After a few moments, Elisheba went to her daddy.

They drove back to Keith and Julie Brown's house, where Randy told stories about being in jail and sneaked out to the garage to have a cigarette. He stayed up until 3:00 a.m., talking about jail, the government, and his heroes, Jesse James and Robin Hood. He said it was ironic that people called *him* a hero. "I just wanted to be left alone." But, he added, "I believe in sticking up for your rights. If people learn that, well, that's okay."

He was awake the next morning by 6:00, scanning the news on television and watching his daughters get ready for their last day of school before Christmas break. Then he borrowed a Cocoa Beach T-shirt from Keith and went to see his probation officer, who said Randy wasn't allowed to have any firearms or to leave southern Iowa. After school, Randy packed up his daughters and drove to a house he'd rented in Grand Junction, a town of 880 near his family's home in Jefferson.

They celebrated Christmas—although they called it X-mas because they still believed it was a pagan holiday—and Randy set about raising his daughters.

He said he never wanted to move back to Idaho and would never go to his old cabin again. After his probation ended, he hoped to move away from Iowa, to a ranch in Montana or the Ozark mountains. He was pretty sure his daughters would want to go with him. He was glad they were doing so well in school, but he never really thought school was that important, he admitted. "The girls will meet men, get married, and become wonderful homemakers like their mother. . . . I'm a chauvinist, I guess. But that's their calling and that's what's best for them."

Randy lived off donations from his supporters and Social Security benefits from Vicki's death while he waited for money from his lawsuit against the government. Randy found a girlfriend and tried harder to keep his beliefs to himself. Still, his attorneys thought he talked too much, and some people in Grand Junction accused him of spouting hatred in a bar one night. Randy consented to a few interviews, also. A photo that ran in the *New York Times* and *Time* magazine showed him in a leather coat, his thumbs hooked in his back pockets, a steely, James Dean look in his eyes. An Iowa television station called him "the man who fought off hundreds of federal agents"; *Time* called him "the Rebel of Ruby Ridge."

He seemed bitter some days, aimless others. He railed against the opinion that he'd put his children in danger, saying that everything he did was to protect his family. And he talked about getting *his* story out sometime. After his release from jail, Weaver said he didn't fear the government anymore, but he hoped that, someday, the Justice Department would come clean about what it had done. "It has to, so Vicki and Samuel didn't die in vain," Randy said. "I want them to bring the truth out. They know the truth, and I want it to be made public. . . . Everybody's watching this case, and they've got to do what's right."

In August 1995, nearly three years after the standoff, the Justice Department settled the claim filed by the Weaver family, paying them $3.1 million to compensate for the loss of Vicki and Samuel Weaver. Each of the Weaver daughters was paid $1 million and Randy Weaver received $100,000. In settling the case, the government refused to acknowledge any wrongdoing.

During the trial, Des Moines television stations occasionally showed old videotape of Sara and Rachel. One day toward the end of the trial, Julie Brown walked into the living room and found Sara curled up in the fetal position, sobbing and refusing to go to school. She said everyone would see her on television and would talk about her. It crushed Julie to see Sara like that, so she and Keith called the TV stations, brought them the girls' report cards to show them just who they were hurting. They asked the stations to please stop showing videos of Sara and Rachel. They just want to be normal kids, Julie Brown said.

A few months later, Gerry Spence called the Brown and said Tom Brokaw wanted to do a piece on the *Weaver* case for his newsmagazine show *Now*. Spence said it was a great chance to show what the government had done and he said the girls needed to be interviewed for the show. Keith Brown explained how traumatic that would be for them, especially for Sara.

Spence pleaded, "Keith, you've gotta put the babies on."

But the Browns were still the girls' legal guardians, and Keith said no. So Spence went on without the girls, strolling along the logging road on Ruby Ridge with Tom Brokaw, talking about government and people's rights. He pointed to an imaginary rabbit, and while Brokaw looked for the animal, Spence said Randy Weaver was sort of like that rabbit. As long as you left it alone, it was a peaceful animal. But if you stuck your finger in its home, it would fight back.

For the next two years, Spence continued to talk about the case and to decry the Justice and Treasury Departments' investigations. He was a regular on the Larry King television program and on others and became even more famous. He wrote another book—his sixth—called *How to Argue and Win Every Time,* a sort of self-help book featuring tips and stories from the master. It was a bestseller.

Spence had represented Randy Weaver free of charge. But after the trial he signed Randy and Sara to a contract that gave him exclusive rights to their story. He talked for a while about writing a book about the case and about negotiating movie offers with big producers, but a couple of years later, nothing had come of it. After the Oklahoma City bombing, Spence expanded a chapter about the *Weaver* case in the paperback version of a book he'd written two years earlier, *From Freedom to Slavery.* (In the book, fourteen-year-old Samuel Weaver was a "child without an adult hair on his skinny white body." Striker was, of course, "Old Yeller.")

After O. J. Simpson was arrested and charged with killing his ex-wife, Spence talked less about the *Weaver* case. He seemed to be on every television talk show.

doing play-by-play about the Simpson trial. "Larry," he'd say, "it's like . . ." and then he'd toss out some endearing analogy about geese as an explanation for the complicated rules of discovery.

In the summer of 1995, Spence got his own cable talk show, broadcast from his Jackson Hole log mansion, in which he spun O.J. trial strategy with his country-lawyer friends. On his TV show, Spence seemed nervous and uncomfortable. But he was also disarming, and you got the feeling he didn't mind being a little rough, that once he warmed up to the great big jury out there, once he figured out a way to be *real*, he had the potential to be damn entertaining.

T he washing machine—the one David Jordison had painstakingly rebuilt and was waiting to bring to Vicki—was still in his garage three years later. In a way, the Jordisons themselves were like that, too, constantly reliving the visit they never got to have with Vicki. They were like clocks frozen on the hour and minute of an earthquake.

For almost seventy years, David Jordison had understood the world. Now it made no sense at all. David, Jeane, and Lanny had all come to believe that Vicki was killed on purpose, that the FBI instructed its agent to kill her, and that he aimed at her face and pulled the trigger.

Julie knew that her parents would never heal as long as they believed the FBI had killed Vicki on purpose. Julie herself went back and forth. Once she began realizing some of the awful things the government did in the case, intentionally shooting an unarmed mother didn't seem so implausible.

But she thought her parents could live with what happened if someone could just convince them Sam's and Vicki's deaths were essentially accidents, a series of misjudgments and blunders. For a while, Julie hoped someone in the FBI or the Justice Department would just call her parents and explain how it happened and what they were doing to fix it. It would be easier for them all to heal if they felt like the truth was really being told. For a while, she thought Gene Glenn—the FBI agent in charge of the standoff—could help her parents understand. But they never heard from him after the standoff, and they'd all come to believe that he knew all along that Sammy had been shot in the back and that Vicki had been shot in the face. They figured he was only acting when he choked back tears and told them she had been killed.

Julie hoped an upcoming Justice Department investigation into the case would bring the truth out. It had to at least acknowledge that the FBI had made horrible mistakes, that it gave its agents the power simply to kill people without provocation. Even if they hadn't killed Vicki on purpose, they'd drafted rules that gave the snipers permission to kill anyone they wanted. The Justice Department investigation had to at least punish the people responsible. That seemed like the only thing that could help the family move on.

"Nobody, thank God, was following the rules of engagement," FBI director Louis Freeh said in January 1994. He said the rules were "poorly drafted, confusing, and can be read to direct agents to act contrary to the law and FBI policy." Fortunately, he said, the sniper Lon Horiuchi wasn't following those rules when he shot at the Weaver family. He was protecting a helicopter, which fell under normal FBI rules of engagement, Freeh insisted.

But Freeh said no FBI agents committed crimes or engaged in any intentional misconduct. He doled out minor punishment to fourteen current and former agents, half for poor evidence gathering and failure to cooperate with the U.S. Attorneys Office.

Danny Coulson, who had been in charge at FBI headquarters when the rules of engagement were faxed back to Washington D.C., was given a letter of censure.

E. Michael Kahoe, who had been involved in researching the rules of engagement and later oversaw the flawed FBI review that determined nothing had been done wrong, was reassigned to an FBI office in Florida. He was censured and suspended for fifteen days. Another agent involved in the review was suspended for five days.

Richard Rogers, the head of the Hostage Rescue Team during the Weaver and Waco standoffs, voluntarily accepted a reassignment. He was censured and suspended from duty for ten days. In a carefully worded release, the FBI said that "his drafting and recommending of rules of engagement that arguably directed agents to act contrary to FBI shooting policy and law, though not causally related to the shooting death of Vicki Weaver, demonstrated performance below the level expected of a person in his position of responsibility."

Larry Potts, who had been in charge of the FBI's Criminal Investigative Division and had approved the general idea of the rules before Rogers ever left Washington, was given a letter of censure, the same punishment Freeh gave himself once for losing a cellular phone. Potts had been promoted to acting deputy director, and at the same time he censured his old friend, Freeh recommended that he be named permanent deputy director, the number-two position in the FBI.

The most serious punishment was given to Gene Glenn, the Special Agent-in-Charge of the FBI's Salt Lake City office and the on-scene commander at Ruby Ridge. He was censured, suspended for fifteen days, removed from his position and reassigned to Washington, D.C. Freeh ruled that Glenn was ultimately responsible for the rules of engagement and the failure to cooperate with prosecutors in the case.

Lon Horiuchi received no punishment. Freeh said the sniper acted "in defense of other law enforcement officers . . . to protect the lives of those agents."

In Iowa, Vicki Weaver's family was stunned. Letters of censure? Fifteen-day suspensions? How could the FBI say that the rules of engagement had nothing to do with Vicki's death? Randy, Sara, and Kevin were running for cover toward the cabin when Horiuchi fired and killed Vicki Weaver. How was that threatening a helicopter? Besides, the helicopter charge had been dropped. At the least, Horiuchi had fired at a cabin with children inside. At the most, he'd intentionally murdered an unarmed woman.

Gerry Spence called Freeh's actions a hand slap and a cover-up. A few months later, Larry Potts was permanently promoted to the number-two position in the FBI.

For more than a year, U.S. Attorney General Janet Reno promised to release a 542-page Justice Department report on the case, the result of an exhaustive probe by five lawyers and nineteen FBI investigators. Among other findings, the report said Kevin Harris and Randy and Sara Weaver were running for cover when Horiuchi fired his second shot and weren't endangering the helicopter. That shot "violated the Constitution," according to the report. It recommended that the case be turned over to the Justice Department and evaluated for possible prosecution. Yet the report wasn't released, and the Justice Department decided not to file criminal charges against any of the FBI participants. Later, the Justice Department rejected the finding that the second shot was unconstitutional.

Randall Day, the Boundary County prosecutor, was conducting his own criminal investigation, looking at possibly charging Weaver, Harris, and the government. Day asked the Justice Department not to release the report until he was done with his probe.

By the spring of 1995, the report was still under wraps. But it had been leaked to a couple of newspapers. One of them, *Legal Times,* posted most of the report on the Internet. Justice Department officials said the report was, in essence, a meaningless, out-of-date document.

Gene Glenn refused to be the scapegoat for Ruby Ridge. In the spring of 1995, he wrote a letter to Justice Department officials, charging that the FBI's investigation was unfair and was designed to protect top officials like Larry Potts. He insisted that on August 22,

1992, Potts approved the rules of engagement during a telephone conversation.

In May, another Justice Department investigation was begun, this time into the alleged FBI cover-up over who actually approved the shoot-on-sight orders. After reportedly failing a lie detector test, E. Michael Kahoe was suspended for destroying or altering documents that may have shown Potts approved the rules of engagement. With congressional hearings about to begin on the Waco siege and more hearings promised on Ruby Ridge, Louis Freeh announced that Larry Potts was stepping down as the FBI's number-two official and would be reassigned to a position training young agents. Freeh stopped short of saying Potts had done anything wrong, blaming his demotion on the *publicity* over Potts's role in the *Weaver* case.

In August 1995, Potts, Danny Coulson and two other FBI officials were suspended while a criminal investigation was launched into the allegations that they tried to cover up their roles in approving the rules of engagement.

For three years, the FBI had maintained that fine balance, punishing its agents for mistakes, but insisting those mistakes had—in essence—nothing to do with the death of Vicki Weaver. They continued to insist that Lon Horiuchi hadn't been following the modified rules of engagement when he fired his two shots and therefore, any misconduct stemming from the drafting of the rules was harmless.

However, the unreleased Justice Department report concluded the shot that killed Vicki Weaver was unconstitutional. "We cannot fault Horiuchi alone for these actions," the report concluded. We are persuaded that his judgment to shoot . . . *was influenced* by the special Rules of Engagement, which he had no role in creating but which he was instructed to follow."

In Boundary County, prosecutor Randall Day confided to one magazine writer that he wished the case "was closed right now." He turned down an offer from Gerry Spence to help prosecute the case and painstakingly continued investigating all three deaths himself. Meanwhile, federal prosecutors continued their own investigation and contemplated charging FBI agents with Vicki Weaver's death.

In the fall of 1995, the Senate Judiciary Subcommittee conducted hearings into the debacle at Ruby Ridge, finally bringing together officials from the various agencies to explain their roles. Senators reacted with disbelief as the case was laid out in front of them and then issued a report that plowed little new ground. The subcommittee acknowledged that there were some contentious points that might never be solved, yet it echoed the earlier report, that the government action was full of errors and possible misdeeds and that the second shot appeared unconstitutional. Despite harsh criticism of the FBI, it was the ATF that drew the ire of subcommittee chairman Arlen Spector, who proposed abolishing the agency because of its role at Waco and Ruby Ridge.

Randy Weaver invested some of his government settlement in a car lot. Appearing before the panel in a denim shirt, blue jeans, and tennis shoes, he gave dramatic testimony before the government he'd once vowed to fight. With Spence sitting mostly silent behind him, Weaver repeated his claim that federal agents set him up, purposefully killed his son and wife, and then covered up their crimes. Framed by the cracked window through which her mother was shot, Sara Weaver tearfully said that the curtains on the cabin door were open and that Lon Horiuchi must've seen Vicki Weaver just before he fired.

The ATF informant, Ken Fadeley, testified from behind a screen, with his voice altered—both for his protection and, he said, so he could continue his undercover work. Several senators expressed disbelief over his claim that Weaver had initiated the gun deal. The ATF also was roundly criticized for targeting Weaver and for falsely reporting to the U.S. Marshals Service that Weaver had a criminal record and was a suspect in a bank robbery. Still, ATF agents continued to deny wrongdoing.

During their testimony, deputy marshals Cooper and Roderick said they believed Randy Weaver had accidentally shot and killed his own son, a theory never made public before. There was no evidence to support that

contention—since the government's own experts said
Cooper's gun killed Samuel Weaver. In the end, deputy
marshals said the Weavers deserved the blame for what had
happened. Later, the Marshalls assigned to the Weaver case
were given the service's highest honor.

FBI agents were even less forthcoming. Some agents, like
Horiuchi, invoked their Fifth Amendment right against self-
incrimination and chose not to testify. And the agents who did
testify, like Potts, denied any wrongdoing, pointing fingers in
every direction but their own. FBI director Louis Freeh
testified that, if he was on the hill that day, he wouldn't have
taken the shot that Horiuchi took. He also said the rules of
engagement would never be altered again. But in the end,
subcommittee members expressed outrage that no FBI official
stepped forward to claim responsibility for revising the rules.

In Boundary County, the Weaver cabin fell into disrepair
and then was remodeled. Weaver and his daughters returned
to film a television news program and then quickly left again.
Long grass covered the meadow where the initial shooting
had taken place and new families moved into the creases
between the Selkirk mountains. Someone even reopened the
Deep Creek Inn, although things near Naples, Idaho were
never quite the same. There, and throughout the West, the
Senate hearings received mixed reviews. Many people were
glad to see the Ruby Ridge case finally aired, but there was a
sense among others that, at some level, it would never be
resolved.

Besides, everything had become so distorted on Ruby
Ridge, so bent by perception and misunderstanding, there
were no guarantees that if the truth ever did come out, anyone
would recognize it.

epilogue

Sara Weaver walked onstage at the Des Moines Art Center during the 1994 Annual Poetry Festival. She stepped up to the microphone and introduced herself before reading one of her poems, entitled "Remembering."

"My name is Sara Weaver," she said. "Two years ago, my mom and brother were murdered. Today is my brother's birthday. He would have been sixteen. I wrote this poem for my mother . . . I guess the reason I titled it "Remembering" is because I seem to be doing a lot of that lately."

After Randy Weaver was released from jail, Rachel and Elisheba lived with him in a modest house in Grand Junction. But Sara stayed in Des Moines to finish her senior year of high school. She visited her dad on weekends, but during the week she wrote poetry, worked at the movie theater, got good grades, and was a happy, normal high school student. Still angry and sometimes bitter toward the government, she talked less and less about it.

Even though their dad wouldn't go back to the cabin, Sara and Rachel did go back to Idaho, to get some of the things ney wanted to keep from that place—like their mom's rugs nd quilts. The first time she went back, Sara said, she just sat n the big rock outcropping near the door and sobbed.

In the spring of 1994, Sara graduated from high school. Her last quarter, she got *B*'s in algebra and art, and *A*'s in dvanced composition, physical education, and overnment.

After the standoff, federal agents had taken her dog, Buddy, to the pound, but some friends rescued him and ared for him until Sara could get her own house in 1994. The Weaver girls had about $24,000 left from the donations ent by supporters, and Sara used some of the money to buy tidy one-story house near her father's place for $8,000. She put a park bench in the front yard, hung flowers, and lanted a huge garden in the back. She got into recycling, ontemplated going to college, and thought about writing a ook. Like her father, Sara hoped to move to Montana omeday. She missed the mountains.

With the Weaver girls gone, Keith Brown sold the tation wagon and bought a convertible Camaro. He and ulie resumed their lives, but they worried that Sara, achel, and Elisheba would have no one to provide an lternative to their separatist beliefs. Still, whenever they aw the girls, they seemed happy and well adjusted. Rachel ot *A*'s at the grade school in Grand Junction, and she alled one day to have Keith and Julie mail her certificate rom a drug awareness program she'd completed.

Sara began dating David Cooper, one of the skinheads who had protested during the standoff, a friendly onstruction worker who grew his hair out and moved to owa. At first, Julie and Keith disagreed over what it meant. Keith said he'd seen her progress past the racism and aranoia they'd grown up with, and now she was falling ght back into it. "I'm not going to condone that," he said.

"You don't have to," Julie said. With all she'd been hrough, Sara needed time and space to develop her own eliefs, Julie said. And they needed to respect whatever ara did and to realize that she wasn't defined by her eligion.

Keith and Julie found themselves arguing over tolerance, of all things—whether they could be tolerant of someone who wasn't, whether they could love someone without agreeing with them. Julie had already lost her sister because of the walls that are put up between people, and she wasn't going to lose her niece that way.

Julie decided that people like Randy and Vicki need to be tied to the world somehow, connected by family or friends who keep them safely away from the edge.

After the bombing of the federal building in Oklahoma City, it seemed to Julie Brown the whole country was living her dilemma: how to keep people on the fringe tethered to society. Releasing the truth about Ruby Ridge would be a start, she thought.

Julie saw Sara occasionally. David Cooper made her happy, but she seemed torn between two places, the world she'd grown up in and the world she lived in now. In the end Julie stopped worrying about Sara's beliefs and just hoped she would find someplace in between where she could be happy.

She drew encouragement from a copy of Sara's graduation photo that hung in their house. "I couldn't have done it without you," Sara wrote alongside the picture. "Love Always and Forever, Sara Beaver."